Neutron Imaging and Applications

Neutron Scattering Applications and Techniques

Series Editors:

Ian S. Anderson
Neutron Sciences Directorate
Oak Ridge National Laboratory
Building 8600, MS 6477
Oak Ridge, TN 37831
USA
andersonian@ornl.gov

Alan J. Hurd
Lujan Neutron Scattering Center at LANSCE
Los Alamos National Laboratory
PO Box 1663, MS H805
Los Alamos, NM 87545
USA
ajhurd@lanl.gov

Robert L. McGreevy
ISIS
Science and Technology Facilities Council
Rutherford Appleton Laboratory
Harwell Science and Innovation Campus
Chilton, Didcot OX11 0 QX
UK
r.l.mcgreevy@rl.ac.uk

Neutron Applications in Earth, Energy and Environmental Sciences
Liyuan Liang, Romano Rinaldi, and Helmut Schober, eds.
ISBN 978-0-387-09415-1, 2009

Neutron Imaging and Applications: A Reference for the Imaging Community
Ian S. Anderson, Robert L. McGreevy, and Hassina Z. Bilheux, eds.
ISBN 978-0-387-78692-6, 2009

Ian S. Anderson · Robert L. McGreevy
Hassina Z. Bilheux
Editors

Neutron Imaging and Applications

A Reference for the Imaging Community

Springer

Editors
Ian S. Anderson
Oak Ridge National Laboratory
Oak Ridge, TN, USA
andersonian@ornl.gov

Robert L. McGreevy
Rutherford Appleton Laboratory,
Oxfordshire, UK
r.l.mcgreevy@rl.ac.uk

Hassina Z. Bilheux
Oak Ridge National Laboratory
Oak Ridge, TN, USA
bilheuxhn@ornl.gov

ISBN 978-0-387-78692-6
ISBN 978-0-387-78693-3 (eBook)
DOI 10.1007/978-0-387-78693-3

Library of Congress Control Number: 2008936128

Cover illustration: Cover graphics courtesy of Dr John Katsaras, National Research Council, Chalk River Laboratories, and Dr Taner Yildirim, NIST Center for Neutron Research.

Printed on acid-free paper

springer.com

Preface

The first successful experiments in Neutron Radiography were carried out (to our knowledge) in 1935, just a few years after the discovery of the neutron, by H. Kallmann and E. Kuhn using a small neutron generator [1]. Not surprisingly, the field has developed and diversified over the last 70 years so that neutron imaging, in the broad sense, is now routinely used in a wide range of applications. The aim of this book (one of a series on the applications of neutron scattering [2]) is to introduce to the reader, whether novice or experienced researcher, the basic techniques used to image objects using neutron beams and to give a flavor of the vast range of applications where these imaging capabilities provide unique insight (Fig. 1).

Fig. 1 Left to right: Pressure gauge with metal backplate; fire hydrant and test tubes filled with H_2O and D_2O imaged with gamma-rays (*top*) and neutrons (*bottom*) [3]

Traditionally, "neutron imaging" is used to describe the direct production of images by transmitting a beam of neutrons through an object onto a detector (e.g., film), i.e., exactly the same as is done with X-rays to image broken bones. An extension of this two-dimensional method (radiography) is to take many images of the same object in different orientations, and then to use the set of images to reconstruct a three-dimensional image (tomography). Although this may still seem to be "direct" imaging, in fact the image is "re-constructed" by software in a computer and sophisticated mathematical processes can be used to enhance particular features or generate virtual slices of the imaged object. We refer to such approaches as "constructed" or indirect imaging methods.

There are many other ways in which virtual pictures or images of an object can be rendered from more indirect measurements. Ultrasound imaging of an unborn baby involves the reconstruction of an image from scattered sound waves. As the size of the object being imaged decreases then simple direct imaging becomes effectively impossible, as the wavelength of the radiation being used becomes of comparable size. "Indirect" imaging methods such as electron diffraction can produce essentially the same "images" of crystal structures as those produced by an apparently "direct" method such as transmission electron microscopy, and indeed it is common to use the combination of these two techniques to enhance the images.

For example Fig. 2 shows a TEM/STEM image of Si_3N_4. The sample progresses from crystalline at the bottom to amorphous structure at the top. In addition there is a layer of lanthanum atoms on top of the crystalline part of the sample. The crystalline structure is overlaid with a ball-and-stick image obtained by diffraction from the crystal. Direct imaging resolves the crystal structure the same way that diffraction (indirect imaging) does; it does not resolve the amorphous structure, but amorphous structure can be determined by diffraction [4].

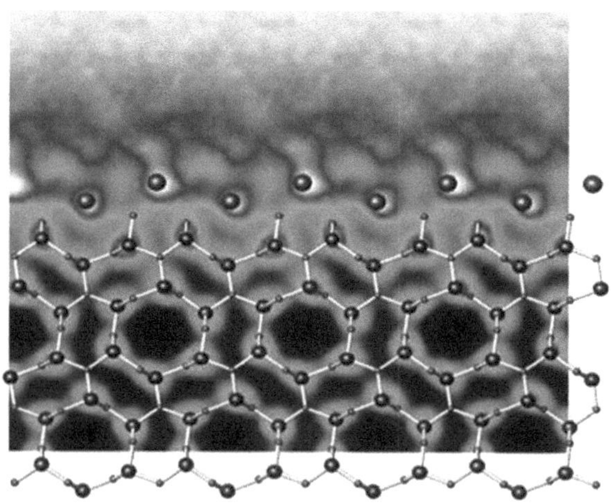

Fig. 2 Image courtesy of David Cockayne (Oxford University)

The current extreme example of indirect imaging is probably the reconstruction of three-dimensional images of single nanoparticles from coherent X-ray diffraction.

In this book we have intentionally taken a broad view of neutron imaging to include any process by which a picture or image of an object, or part of an object, can be produced based on the interaction with a neutron beam. These methods and their applications include

- Direct imaging methods such as neutron radiography being used to study operating fuel cells
- Three-dimensional tomographical reconstructions of mechanical objects
- The use of indirect methods such as diffraction or small-angle scattering to image strain patterns in materials
- Indirect imaging of the shapes of biological molecules by reconstruction from small-angle neutron scattering data.

Clearly, neutron imaging is a less well-known technique than X-ray imaging; most people know of the simple medical applications of X-ray radiography and the more recent extension to tomography (CAT or computed axial tomography scanning). This is largely due to the fact that it is simpler, and less costly, to generate and manipulate high-intensity sources of X rays than of neutrons. Hence the applications highlighted in this book rely heavily on the distinctive properties of neutron beams, which allow useful and often unique information to be derived from the image.

- Neutrons are weakly interacting neutral particles that penetrate deeply into most materials, so they can be used to internally image large objects, e.g., a full-size operating internal combustion engine, non-destructively.
- The amount of scattering or absorption of neutrons by atomic nuclei varies in an apparently random fashion through the periodic table. Hydrogen in particular has a very large scattering cross-section. Neutrons can therefore provide good contrast for light atoms in the presence of heavy atoms, e.g., the "classical" neutron image of a rose inside a lead flask (Fig. 3). This makes neutron imaging highly complementary to X-ray imaging (Fig. 1).
- The amount of scattering or absorption can also vary significantly between isotopes of the same chemical element; e.g., hydrogen has a very different scattering cross-section from that of its isotope deuterium. The contrast of particular elements/materials in an image can therefore be enhanced by substituting one isotope for another (Fig. 1).
- "Thermal" neutrons have wavelengths similar to inter-atomic distances, so mechanisms such as refraction or diffraction can be used to enhance images or to produce indirect images.
- Neutrons have a magnetic moment and a magnetic scattering cross-section that is comparable to the nuclear cross-section for many atoms. They can therefore be used to image magnetic structures.

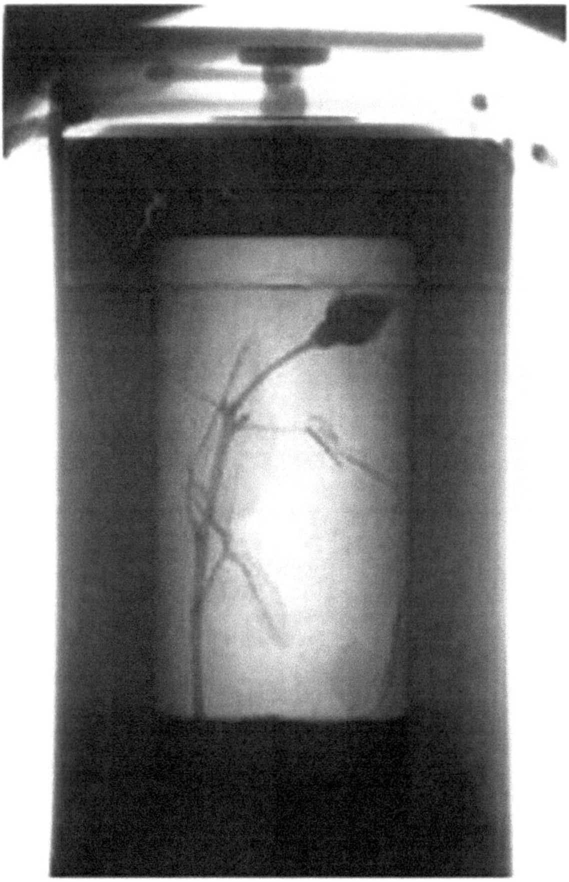

Fig. 3 Neutron radiograph of a rose in a lead flask [5]

Hence the intrinsic properties of neutrons allow a wide range of objects to be imaged, ranging from massive structures such as helicopter blades to the fine details of crystal structures and passing through the delicate composition of biological organisms and plants. Furthermore, these intrinsic properties provide for an extensive range of contrast enhancement mechanisms including absorption, scattering, diffraction, refraction, magnetic interactions, and, potentially, vibrations. These mechanisms can be used to determine the elemental compositions of objects, which may even be hidden, buried, or encapsulated within an impenetrable environment. The possibility of studying objects in situ, or in real operational environments, is promising for a range of industrial and academic applications.

Neutron imaging techniques have a huge potential but in the past, applications have been slow to develop, mainly because of the weakness of the source itself. Even the most powerful neutron sources in existence today have a source brightness that is comparable to a simple X-ray tube and many orders of magnitude lower than a third-generation synchrotron X-ray source. Hence, while synchrotron X-ray sources provide the capability of imaging single nano-particles with nanometer resolution, or dynamic images of larger objects with micrometer spatial resolution and microsecond time resolution, neutrons are presently limited to static images with spatial resolutions of the order of tens of microns, or dynamic images of 100 microns and microsecond exposure times for stroboscopic processes. Neutron imaging of smaller objects can only be achieved indirectly, using scattering techniques from an ensemble of particles.

Despite these limitations, the following chapters give a flavor of the wide range of applications that presently (or will potentially) benefit from neutron imaging techniques. The book is organized into three major sections.

Section A provides a comprehensive overview of basic neutron techniques aimed more specifically at a non-specialist audience. Frequent reference is made to the two introductory chapters in the first book of this series [2] by Roger Pynn (Neutron Scattering – a Non-Destructive Microscope for Seeing Inside Matter), and Helmut Schober (Neutron Scattering Instrumentation). Both these chapters are freely available at www.springerlink.com. In Chapter 1 Kenneth Herwig summarizes the essential neutron properties and techniques which are relevant to the majority of neutron imaging applications. Masatoshi Arai and Kent Crawford provide, in Chapter 2, an excellent survey of the different types and characteristics of neutron sources, including nuclear reactors, high-power spallation sources, and portable generators, which are typically used nowadays for neutron imaging. Although there is considerable overlap in the use of these sources, each type of source has specific advantages for certain types of applications. Due to the low-intrinsic brightness of neutron sources, efficient optical systems are imperative, so Ken Andersen's chapter (Chapter 3) presents the basic concepts of the neutron optics that are typically used on imaging beam lines. Finally in this section, Lowell Crow (Chapter 4) examines modern neutron detection methods for imaging. The first section reviews neutron capture converters which form the basis for thermal neutron detection, and the following sections examine detector systems with an emphasis on imaging applications.

Section B focuses on the neutron beam implementation of some well-known imaging techniques. Arthur Heller and Jack Brenizer (Chapter 5) present a summary of the history, methods, and related variations of neutron radiography techniques including a section on the application of standards. Even today, conventional film radiography remains the mainstay of high-resolution, large field-of-view, neutron imaging. In Chapter 6, Wolfgang Treimer extends the basic theories and applications to include three-dimensional tomography and introduces some of the newer methods for enhancing contrast such as wavelength dependent ("Bragg edge") imaging and small-angle scattering. Kenneth Tobin et al. (Chapter 7) provide a review of neutron image formation, resolution analysis concepts, and methods for both the design and characterization of radiography systems and conclude with a discussion of volumetric reconstruction techniques using analytic or iterative computed tomography algorithms.

The next two chapters describe techniques that depend intrinsically on the wave nature of the neutron. Franz Pfeiffer (Chapter 8) provides a fascinating overview of neutron phase imaging and its natural extension to neutron phase tomography. This technique offers the potential to image fundamental quantum mechanical interactions. Bhaskar Sur et al. (Chapter 9) show initial results from neutron holography experiments using both the internal and the external source approaches. Although in its infancy, neutron holography has the potential to resolve to atomic resolution the structures of materials, which are difficult to crystallize. Finally in this section Nikolay Kardjilov et al. (Chapter 10) describe some novel imaging techniques using the magnetic properties of the neutron. After demonstrating the power of neutrons to image magnetic fields in and around objects, they go on to describe theoretically some tantalizing but challenging potential applications of spin contrast imaging and neutron-based magnetic resonance imaging.

Section C provides the reader with an excellent, though non-exhaustive, overview of some specific applications of neutron imaging in diverse fields of research. Muhammad Arif et al. (Chapter 11) describe in situ neutron radiography and tomography studies of operating fuel cells and hydrogen storage systems, undoubtedly a high priority global research field for the foreseeable future, where the ability of neutrons to "see" hydrogen provides essential information to improve the practical chemical and mechanical engineering design.

Dayakar Penumadu (Chapter 12) provides an overview of some recent applications of neutron imaging methods to broad classes of materials science and engineering studies including metal casting, strain imaging, and characterization of discrete particle systems.

Carla Andreani et al. (Chapter 13) discuss the growing use of neutrons to image artifacts of interest in the domain of cultural heritage. Novel characterization methods allow determination of the provenance of ancient objects and can shed light on the methods and tools used at the time of manufacture.

In Chapter 14 Kenneth Watkin et al. summarize past and recent research efforts to apply neutron radiography to biological specimens, in the expectation that clinical and medical research, as well as forensic science, may benefit from advanced neutron imaging methods.

Moving on to prospective techniques for live imaging, Anuj Kapadia describes (in Chapter 15) the development of a tomographic technique that uses neutron inelastic scatter interactions to quantitatively identify the spatial distribution of elements in the body. The technique, called Neutron Stimulated Emission Computed Tomography (NSECT), uses a beam of fast neutrons to excite stable isotopes of elements in the body to determine their concentration and spatial distribution within the body. It has the potential to diagnose several element-related disorders in humans that are characterized by a change in element concentration in the diseased tissue.

In Chapter 16 William Heller and Gary Baker describe how small-angle neutron scattering measurements when combined with advanced computer modeling and contrast variation methods enable visualization of the structure and function of biological macromolecules.

Chapter 17 by Tomoko Nakanishi shows some fascinating examples of neutron imaging applied to plant physiology which allow in situ studies of vital processes in plant growth to be visualized.

Finally in Chapter 18 Dick Lanza summarizes security-related applications of neutron imaging, e.g., for the detection of illicit materials such as explosives or nuclear materials. One particular technique, neutron resonance radiography, is discussed in detail, as it illustrates many of the issues connected with imaging and detection.

We have attempted to provide a broad overview of the potential of neutron imaging methods. It is evident that in the process we will have omitted some important areas of application. Nevertheless we hope that you enjoy this edition and will be stimulated to read further.

The assistance of the Springer team during the various stages of the development of this book is greatly appreciated.

Oak Ridge, TN — Ian Anderson
Oak Ridge, TN — Hassina Bilheux
Oxfordshine, UK — Robert McGreevy

References

1. H. Kallman, *Research* **1**, 254 (1947).
2. http: //www.springer.com/series/8141, this series of books.
3. O. Peter, *Naturforsch.* **I**(10), 557 (1946).
4. G. B. Winkelman, C. Dwyer, T.S. Hudson, D. Nguyen-Manh, M. Döblinger, R.L. Satet, M.J. Hoffmann, D.J.H. Cockayne, *Phil. Mag. Lett.* **84**, 755–62 (2004).

Contents

Contributors

K. Andersen Institut Laue-Langevin, 6 rue Jules Horowitz, B.P. 156, 38042 Grenoble cedex 9, France

C. Andreani University of Milano Bicocca, Department of Materials Science, Via Roberto Cozzi 53, 20125 Milano, Italy

V. N. P. Anghel Atomic Energy of Canada Limited, Chalk River Laboratories, Chalk River, Ontario, Canada

J. F. Ankner Oak Ridge National Laboratory, Spallation Neutron Source, Oak Ridge, TN, USA

M. Arai J-PARC Center, JAEA, 2-4 Shirakata-Shirame, Tokai Ibaraki, Japan

M. Arif National Institute of Standards and Technology, Gaithersburg, MD, USA

G. A. Baker Oak Ridge National Laboratory, Oak Ridge, TN, USA

H. Z. Bilheux Oak Ridge National Laboratory, Spallation Neutron Source, Oak Ridge, TN, bilheuxhn@ornl.gov

P. R. Bingham Image Science and Machine Vision Group, Oak Ridge National Laboratory, Oak Ridge, TN, USA

J. S. Brenizer Pennsylvania State University, University Park, PA, USA

K. Crawford Spallation Neutron Source, Ok Ridge National Laboratory, Oak Ridge, TN, USA

L. Crow Neutron Facilities Development Division, Oak Ridge National Laboratory, Oak Ridge, TN, USA

G. Gorini University of Milano-Bicocca–Piazza dell'Ateneo Nuova, Milano, Italy

G. E. Granroth Neutron Scattering Sciences Division, Oak Ridge National Laboratory, Oak Ridge, TN, USA

J. Gregor Department of Electrical Engineering and Computer Science, University of Tennessee, Knoxville, TN, USA

A. K. Heller Department of Mechanical and Nuclear Engineering, The Pennsylvania State University University Park, PA, USA

W. T. Heller Center for Structural Molecular Biology, Oak Ridge National Laboratory, P.O. Box 2008, Oak Ridge, TN 37931, USA

K. W. Herwig Oak Ridge National Laboratory, Oak Ridge, TN, USA

D. S. Hussey National Institute of Standards and Technology, Gaithersburg, MD, USA

D. L. Jacobson National Institute of Standards and Technology, Gaithersburg, MD, USA

A. J. Kapadia Duke Advanced Imaging Laboratories, Department of Radiology, Duke University Medical Center, Durham, NC, USA

N. Kardjilov Helmholtz Centre Berlin for Materials and Energy, Berlin, Germany

J. Katsaras National Research Council, Chalk River Laboratories, Chalk River, ON, K0J 1J0, Canada

R. C. Lanza MIT Department of Nuclear Science and Engineering, Cambridge, MA, USA

W. T. H. Lee Neutron Scattering Sciences Division, Oak Ridge National Laboratory, Oak Ridge, TN, USA

T. Materna Institut für Kernphysik, Universität zu Köln, Köln, Germany

T. M. Nakanishi Graduate School of Agricultural and Life Sciences, The University of Tokyo, Yayoi, Bunkyo-ku, Tokyo, Japan

D. Penumadu University of Tennessee-Knoxville, Joint Institute for Advanced Materials (JIAM), Department of Civil and Environmental Engineering, Knoxville, TN, USA, dpenumad@utk.edu

F. Pfeiffer Department of Physics (E17), Technical University Munich, Munich, Germany

R. B. Rogge National Research Council of Canada, Chalk River Laboratories, Chalk River, Ontario, Canada

B. Sur Atomic Energy of Canada Limited, Chalk River Laboratories, Chalk River, Ontario, Canada

K. W. Tobin Image Science and Machine Vision Group, Oak Ridge National Laboratory, Oak Ridge, TN, USA

W. Treimer Department of Mathematics, Physics and Chemistry, University of Applied Sciences (TFH) Berlin, Luxemburger Str., Berlin, Germany, treimer@tfh-berlin.de; Helmholtz Center Berlin for Materials and Energy, Department SF1, Glienicker Str., Berlin, Germany, treimer@helmholtz-berlin.de

K. L. Watkin University of Illinois, College of Applied Health Science, 901 So. Sixth, Champaign, IL, USA

Section A

Chapter 1
Introduction to the Neutron

K.W. Herwig

Abstract As a fundamental particle, the neutron has many unique attributes that provide a variety of contrast mechanisms enabling many imaging techniques that will be discussed throughout this book. This chapter introduces the neutron with a short summary of its discovery and continues with a presentation of its fundamental properties. A physical picture of the attenuation of a primary beam of particles and the mathematical formulae that govern that attenuation are presented. This chapter concludes with a discussion of neutron interactions with matter that provide the physical basis underlying neutron imaging.

Keywords Neutron · Scattering · Absorption · Cross section

1.1 Introduction

As one of the constituents of the atom, the neutron was discovered well after its electrically charged companions the electron, by J. J. Thomson in 1897 [1], and the proton, for which there is some debate over any individual discoverer. A close combination of an electron and a proton was proposed as a neutral doublet by E. Rutherford in 1920, which he postulated would have "novel properties" including the ability to "move freely through matter," be difficult to detect, and perhaps impossible to contain in a vessel [2]. The term "neutron" was apparently first recorded in the literature in 1921 by W. D. Harkins in a discussion on classifying isotopes [3]. However, conclusive evidence for the existence of the neutron waited until 1932 when J. Chadwick reported first the "possible existence" [4] and shortly thereafter "the existence" of the neutron [5]. Chadwick was recognized for his discovery by receiving the 1935 Nobel Prize in Physics quite simply "for

K.W. Herwig (✉)
Oak Ridge National Laboratory, P.O. Box 2909, Oak Ridge, TN 37831, USA
e-mail: herwigkw@ornl.gov

I.S. Anderson et al. (eds.), *Neutron Imaging and Applications*,
Neutron Scattering Applications and Techniques, DOI 10.1007/978-0-387-78693-3_1,

the discovery of the neutron." Electrical neutrality is perhaps the signature fundamental property of the neutron that makes it so useful in both neutron scattering and in neutron imaging applications. As Rutherford anticipated, because the neutron is uncharged, it interacts primarily with nuclei; it is highly penetrating and well able to investigate the interior of large assemblies; and as a significant additional benefit, it does so non-destructively.

Another fundamental property of the neutron is its mass, m_n = 1.6749×10^{-27} kg [6], which gives the neutron a de Broglie wavelength on the order of atomic distances for energies comparable to the excitations in condensed matter at room temperature. Thus the neutron is simultaneously sensitive to both length and time scales relevant to condensed matter. This is particularly relevant to neutron scattering applications; but the microscopic response to length scales also plays a significant role in imaging applications, particularly for polycrystalline materials in which the primary attenuation is due to Bragg scattering from the crystal lattice. The de Broglie wavelength, λ in units of nm, is given by

$$\lambda = \frac{h}{m_n v} = \frac{395.6}{v} \tag{1.1}$$

where $h = 6.6261 \times 10^{-34}$ J × s is Planck's constant and v is the neutron velocity in m s^{-1}. The neutron energy, E in units of meV, is given by

$$E = \frac{1}{2} m_n v^2 = 5.2270 \times 10^{-6} v^2 \quad . \tag{1.2}$$

A neutron with an energy characteristic of room temperature, 300 K or 25.85 meV, will have a velocity of 2224 m s^{-1} and a wavelength of 0.18 nm. The neutron carries momentum as expressed by its wave vector $\vec{k}$. The direction of $\vec{k}$ is that of the neutron, while its magnitude $|\vec{k}| = 2\pi/\lambda$. Other basic properties of the neutron include its intrinsic spin of ½ and an associated nuclear magnetic moment of -0.9662×10^{-26} J T^{-1}. Thus in addition to nuclear interactions, the neutron will respond to external magnetic fields and interact with the magnetic moments of unpaired electrons in matter.

Neutrons are typically produced either by fission in nuclear reactors or by spallation, in which a high-energy proton beam impinges a heavy metal target and neutrons are ejected from the resulting excited nuclei. In both of these cases, the neutrons have spectrums of energies that are too high to be useful for either neutron scattering or imaging. Moderators are employed to cool the neutron spectrum to lower energies that depend on the temperature of the moderating medium, often hydrogen or hydrogenous materials. Neutrons are rather loosely assigned to bands of energies associated with characteristic moderator temperatures, as shown in Table 1.1, taken from [7]. See Chapter 2 for more details on sources and moderators.

Table 1.1 Neutron characteristics at various energy ranges

Neutron classification	Energy (meV)	Velocity (m/s)	λ (nm)
Ultra-cold	0.00025	6.9	57
Cold	1	437	0.9
Thermal	25	2187	0.18
Epithermal	1000	13,832	0.029

1.2 Interactions with Matter and Cross Sections

In neutron imaging applications, one is primarily interested in neutron interactions that attenuate a well-defined incident beam of neutrons. Neutrons can be removed from the incident beam either by absorption or by a change in direction as they interact with material in the beam. For the moment, consider the neutron as a classical point-size particle and the atoms in a sample as classical particles of fixed radius, *r*. Figure 1.1 illustrates the attenuation of a (well-collimated) neutron beam incident perpendicularly on a thin sample element of thickness *dx* located a distance *x* cm from the surface of the sample. Let *dx* be thin enough that every atom in this sample element can be sampled by the incident neutron beam (no shadowing). Define $I(x)$ as the incident neutron flux and $I(x + dx)$ as the transmitted flux, both in units of neutrons $cm^{-2}s^{-1}$. Let *n* be the number density of atoms in the sample in units of atoms cm^{-3}. The

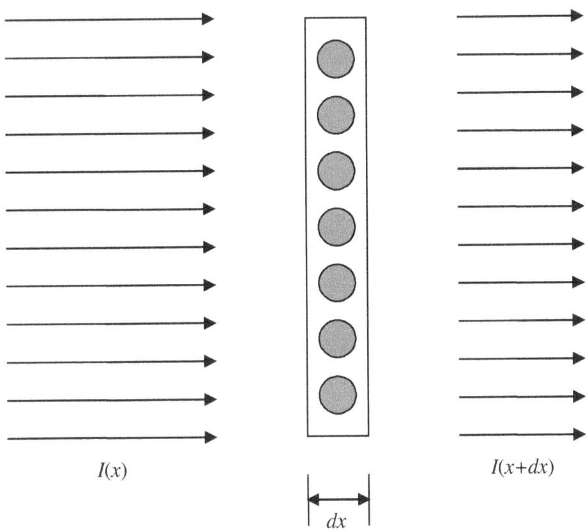

Fig. 1.1 Illustration of the attenuation of an incident beam of idealized neutrons by a thin sample

attenuation of the incident neutron beam is then given by the fractional area occupied by the classical atoms, $dx\, n\, \pi\, r^2$ and

$$I(x+dx) = I(x)\left(1 - dx\, n\left(\pi\, r^2\right)\right) = I(x)(1 - dx\, n\sigma) \ . \tag{1.3}$$

In this classical picture, the microscopic cross section, σ, is just the cross sectional area of the individual sample particles, $\pi\, r^2$, and has units of cm^2. More generally, the microscopic cross section is a measure of the effective interaction area for the neutron with a nucleus.

From Eq. (1.3), the rate of change of $I(x)$ is then

$$\frac{dI(x)}{dx} = -n\,\sigma I\,(x) \ , \tag{1.4}$$

and the solution for $I(x)$ is

$$I(x) = I_i \exp(-n\,\sigma\, x) \ , \tag{1.5}$$

where I_i is the flux of neutrons incident on the sample. The product of $n\sigma$ is called the macroscopic cross section, Σ. For real samples with multiple elements and isotopes, a sum over each species and all its associated cross sections must be performed in order to calculate the sample transmission

$$\Sigma_{tot} = \sum_j n_j \sigma_j = \sum_j \Sigma_j \ , \tag{1.6}$$

where the sum is over all isotopes in the sample and the transmitted neutron flux may be written as

$$I(x) = I_i \exp(-\Sigma_{tot} x) \ . \tag{1.7}$$

The interaction of neutrons with nuclei is governed by quantum mechanics and not the laws of classical physics; nonetheless, it is illustrative to explore the magnitude of the microscopic cross section in a classical context. The typical units for σ are barns, and 1 barn $= 10^{-24}$ cm^2. (The whimsical name of the unit originated with the Manhattan Project and was classified information until after World War II. The typical effective area, $\sim 10^{-24}$ cm^2, of a nuclear particle during a collision was a relatively large interaction cross section and hence was as "big as a barn" [8].) Consider a typical inter-atomic separation of 0.25 nm and the radius determined from a cross section of 1 barn. The ratio of the inter-atomic spacing to effective interaction radius is $\sim$ 44,000 and, as seen by a neutron, there is a vast amount of effective space between scattering centers, resulting in a correspondingly high penetration depth of neutrons into most materials.

A neutron may interact with a nucleus by either scattering or absorption and it is typical to write a total microscopic cross section, σ_t, as the sum of these terms

$$\sigma_t = \sigma_s + \sigma_a \tag{1.8}$$

where σ_s and σ_a are the scattering and absorption cross sections, respectively. An in-depth introduction to neutron scattering is provided by R. Pynn [9], so only a few relevant terms are presented here. Because the neutron interaction is with the nucleus, the strength of the scattering cross section varies non-uniformly across the periodic table and depends not only on the individual element but also on the particular isotope of that element. Figure 1.2 shows the total neutron scattering and absorption cross sections for bound atoms across the periodic table as derived from the tabulation in [10]. In practice, the scattering cross sections are typically taken as constant in the epithermal, thermal, and into the cold neutron energy ranges. In general, however, these cross sections have an energy dependence with gadolinium (due to the close proximity of a nuclear resonance) and bound hydrogen (e.g., in the case of water) that exhibits significant variation. (See [11] for a short discussion of this effect.)

The neutron scattering cross section can be further broken down into coherent and incoherent components. In a coherent scattering process, neutron waves that are scattered by different nuclei combine with one another to produce an interference pattern that depends on the relative locations of the atoms in the material. Incoherent scattering may arise when more than one isotope is present in the sample (whether the same element or not) or when a given isotope has a non-zero nuclear spin [12]. In

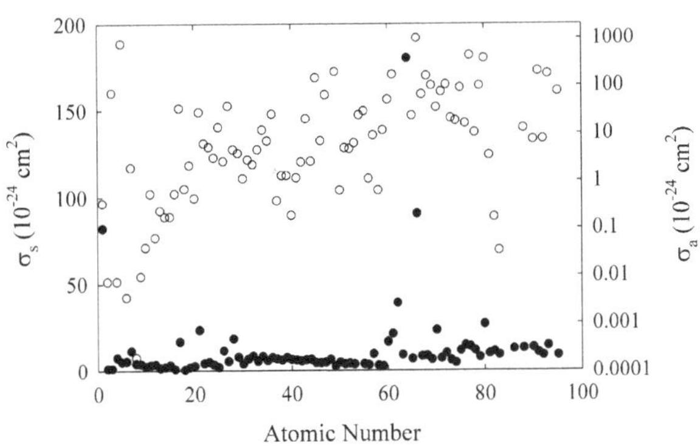

Fig. 1.2 Neutron scattering and adsorption cross sections for thermal neutrons ($v = 2200$ m/s) as tabulated in [10]. *Open circles* are adsorption cross sections and are plotted on the logarithmic scale of the right y-axis. *Closed circles* are the bound atom scattering cross sections plotted on the linear scale of the left y-axis

these cases, there is a randomization of the strength of the scattering cross section, which results in a component of the neutron scattering that depends only on the scattering from individual atoms and does not give rise to an interference pattern. For each of these scattering processes, the neutron may scatter elastically, exchanging no energy with sample atoms, or inelastically, either gaining or losing energy. During the scattering process, the neutron may also exchange momentum with the atom, $\vec{Q} = \vec{k_f} - \vec{k_i}$, where $\vec{k_f}$ and $\vec{k_i}$ are the final and initial neutron wave vectors, respectively. Elastic coherent scattering gives rise to structural information on the relative locations of atoms in the sample; in crystalline materials, it is called "diffraction." Inelastic coherent scattering yields information on collective excitations in the sample, such as phonons and spin waves. Elastic incoherent scattering sums over the intensity from individual atoms, whereas inelastic incoherent scattering conveys information on single-particle excitations such as diffusion, methyl group rotations and, at high energy and momentum transfers, momentum distributions.

All of these scattering processes attenuate the incident neutron beam and give rise to contrast, which can be used for real-space imaging. As an extreme but very useful example, consider the transmission through a sample of water. Hydrogen has a very large incoherent scattering cross section of approximately 80 barns for bound atoms and thermal neutrons, 10–30 times greater than typical values for other elements. At a density ρ of 1 g cm^{-3}, the hydrogen number density may be calculated as

$$n = \frac{2\rho}{M} N_A = 6.691 \times 10^{22} \, \frac{\text{H} - \text{atoms}}{\text{cm}^3} \quad , \tag{1.9}$$

where M is the molecular weight (18 g mol^{-1} for water), 2 is for the two hydrogen atoms/water molecule, and N_A is Avogadro's number, 6.022×10^{23} mol^{-1}. The macroscopic cross section due to scattering from hydrogen in water is $\Sigma_H = 5.353$ cm^{-1}. In comparison, the macroscopic scattering cross sections are $\Sigma_{Fe} = 0.99$ $cm^{-1}$and $\Sigma_{Al} = 0.091$ cm^{-1} for iron and aluminum, respectively. Figure 1.3 shows the attenuation of a neutron beam due to these scattering cross sections as a function of depth x into a sample. What is very clear from Fig. 1.3 is that the presence of hydrogen in a specimen gives rise to a very rapid attenuation of the neutron beam. This has been used to great advantage recently in real-time identification of the formation and motion of water in operating proton exchange membrane fuel cells (see Chapter 11 and, e.g., [13]). The high penetration depth of the neutron in aluminum and iron is also illustrated (Fig. 1.1 of [9]).

Real materials are much more complicated than the cases determined above for simple attenuation; this is particularly evident in the energy dependence of the total scattering cross section for polycrystalline materials. Figure 1.4 shows the measured elastic cross section for polycrystalline iron [14]. The coherent nature of the scattering and the regular positions of the iron atoms in the

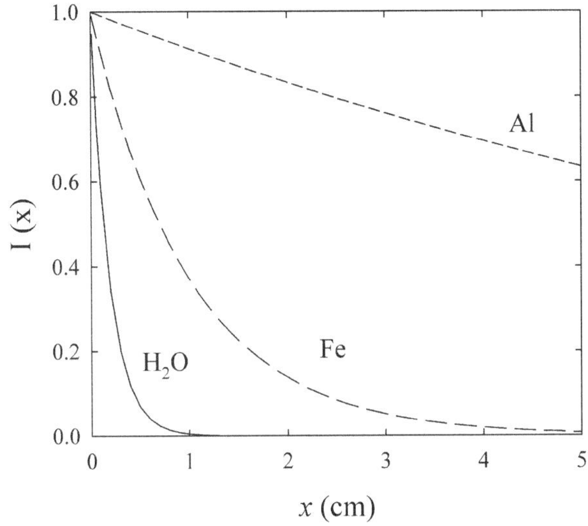

Fig. 1.3 Attenuation as a function of depth into samples of water, iron, or aluminum due to neutron scattering

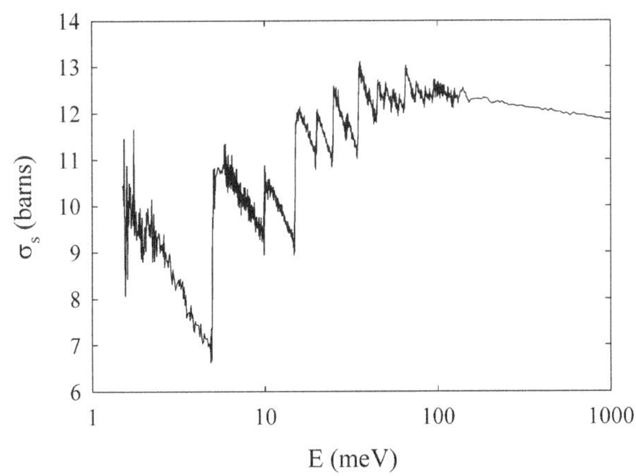

Fig. 1.4 The total neutron scattering cross section for polycrystalline iron

crystalline lattice give rise to abrupt changes in the cross section and associated neutron transmission. Associated with a given spacing between planes of atoms in the material, d is a Bragg reflection whose scattered intensity is determined by the local arrangement of atoms within a unit cell. The condition for neutron scattering is determined by Bragg's law

$$n\lambda = 2d\sin(\theta) \quad , \tag{1.10}$$

where θ is half the angular separation between the incident and scattered neutron beams. There is clearly a maximum neutron wavelength λ_{max} for any given d at which the condition of Eq. (1.10) is satisfied, namely $\lambda_{max} = 2d$ at $\theta = 90°$. For shorter wavelengths and lower scattering angles, there will also be neutron scattering that satisfies this condition. Thus there is a sharp edge at λ_{max} for each family of reflections allowed by the crystal structure, followed by a decreasing tail at shorter λ as seen in Fig. 1.4. This wavelength dependence of the neutron cross section and associated transmission can be used to great advantage in optimizing the conditions for high contrast from various materials in a sample (Neutron Tomography (Chapter 6) and Material Science and Engineering with Neutron Imaging (Chapter 12) and, e.g., [15]). Radiographs at different neutron wavelengths can be obtained through either time-of-flight means (spallation neutron sources) or monochromatization (reactor sources).

Neutron absorption, another neutron–nuclei interaction that attenuates an incident neutron beam, is a mechanism whereby contrast can be obtained. In this interaction, a neutron is absorbed by a nucleus, often resulting in an unstable, radioactive nucleus that will decay with a particular half-life. The half-life is the time required for the radioactivity to decay to half of its original value. As the nucleus decays, it may emit a variety of secondary radiation, including an α-particle, which is identical to a helium nucleus and carries two units of positive charge; a β-particle, which is an energetic electron or positron; γ-rays, which are high-energy photons originating in the nucleus; and others. An α-particle interacts strongly with matter and is rapidly attenuated even in air, whereas the latter two can penetrate relatively deeply into matter. Emitted γ–rays have a well-defined and catalogued energy spectrum that can act as a signature for the decay of particular isotopes. One mature technique is neutron activation analysis, whereby decay radiation, either delayed or prompt, is detected from nuclei that have absorbed a neutron. Prompt decay gives rise to an imaging paradigm whereby a sample rastered through a neutron beam can enable a quantitative measure of the presence of particular isotopes in the volume sampled by the neutron beam (Novel Neutron Imaging Techniques for cultural Heritage Objects (Chapter 13)). An energy sensitive γ-detector can register the energy of the promptly emitted γ, yielding a fingerprint of the isotope. In the low epithermal through cold neutron energy ranges, these absorption processes follow a $1/v$ probability law for most nuclei. The value varies across the periodic table as demonstrated in Fig. 1.2 for a neutron velocity of 2200 m/s. The probability of absorption then increases linearly with increasing neutron wavelength. Exceptions include gadolinium, which has many resonances; the lowest occurs at approximately 2 eV, with a corresponding nonlinear response in the scattering cross section as well.

The half-life of the activated isotopes may be long enough that it is practical to remove an object exposed to a neutron beam and use its residual activity to expose either a photographic film or an image plate. The radiation emitted from particular isotopes will be present to a lesser or greater degree after neutron irradiation, depending on their respective half-lives and the intervening time before the film or image plate is exposed by the sample. Multiple exposures at subsequent times will be more sensitive to the longer-lived isotopes as the shorter-lived ones decay to low levels of activity, giving rise to an additional means of obtaining isotopic contrast.

An additional neutron–nuclei interaction involves a high-energy neutron (~MeV) that scatters inelastically, imparting a relatively large energy to the nucleus. The excited nucleus may give up the energy in a quantized fashion through γ-emission. Similarly to prompt γ detection, an energy-sensitive detector can be used to identify the signatures of particular isotopes. This is the underlying principle of neutron stimulated emission computed tomography, described in Chapter 15 of this book.

Finally, to this point, only neutron interactions with nuclei have been considered; but, as mentioned, the neutron also carries a magnetic moment that can interact with magnetic fields. This provides an additional means by which neutrons can be scattered out of an incident beam as they interact with the magnetic fields produced by unpaired electrons in materials. This interaction may be elastic, giving rise to magnetic Bragg peaks that yield information on the magnetic order in a sample, or inelastic, giving insight into fluctuations. A discussion of the use of magnetic scattering and the production and use of polarized neutron beams is given by R. Pynn in his presentation on neutron scattering techniques [9]. Some examples of applications to neutron imaging are given in Chapter 10.

In conclusion, the neutron has several intrinsic properties that make it very useful for imaging purposes and result in a variety of contrast mechanisms. It is uncharged and is able to penetrate deeply into matter in a non-destructive fashion. The neutron cross sections for interactions with nuclei are both isotope and neutron energy dependent and include processes that attenuate a neutron beam by either scatter or absorption. It has a wavelength/energy relationship such that the neutron simultaneously responds to both length and time scales relevant to condensed matter at room temperatures. Finally, the activation of a nucleus upon neutron absorption can yield emission of γ-rays having energies that are isotope dependent.

References

1. J.J. Thomson, Cathode Rays, *The Electrician* **39**, 104–109 (1897).
2. E. Rutherford, *Proc. Roy. Soc. A* **97**, 374–400 (1920).
3. W.D. Harkins, Natural systems for the classification of isotopes, and the atomic weights of pure atomic species as related to nuclear stability, *J. Am. Chem. Soc.* **43**, 1038–1060 (1921).
4. J. Chadwick, Possible existence of a neutron, *Nature* **129**, 312 (1932).
5. J. Chadwick, The Existence of a neutron, *Proc. Roy. Soc. A* **136**, 692–708 (1932).
6. All values for fundamental constants taken from the NIST Reference on Constants, Units, and Uncertainty, available at http://physics.nist.gov/cuu/Constants/.
7. D.L. Price and K. Sköld in N*eutron Scattering*, edited by K. Sköld and D.L. Price (Academic Press, Orlando, Fl., 1986), Vol. A.

8. M. Perricone, *Symmetry* **3**, 4 (2006), available on-line at http://symmetrymagazine.org.
9. http://www.springer.com/series/8141, this series of books.
10. V.F. Sears, *Neutron News* **3**, 29–37 (1992).
11. J.R.D. Copley, *Neutron News* **18**, 30–31 (2007).
12. See the following for discussions on the origin of coherent and incoherent scattering cross-sections R.-J. Roe, *Methods of X-Ray and Neutron Scattering in Polymer Science* (Oxford University Press, New York, 2000), G.L. Squires, *Introduction to the Theory of Thermal Neutron Scattering* (Cambridge University Press, Cambridge, 1978), and V.F. Sears, *Neutron Optics: An Introduction to the Theory of Neutron Optical Phenomena and their Applications* (Oxford University Press, Oxford, 1989).
13. M.A. Hickner, N.P. Siegel, K.S. Chen, D. s. Hussey, D.L. Jacobson, and M. Arif, *J. Electrochem. Soc.* **155**, B294–B302 (2008).
14. J.A. Harvey, H.A. Mook, N.W. Hill, and O. Shahal in *Nuclear Data for Science and Technology: Proceedings of the International Conference, Antwerp, 6–10 September 1982*, edited by K.H. Bhockhoff (D. Reidel, Boston, 1983), p. 961, and available at http://www.nndc.bnl.gov/exfor/exfor00.htm along with many other measured cross-sections.
15. W. Kockelmann, G. Frei, E.H. Lehmann, P. Vontobel, and J.R. Santisteban, Energy-selective neutron transmission imaging at a pulsed source, *Nucl. Instrum. Meth. Phys. Res. Sect. A* **578**, 421–434 (2007).

Chapter 2
Neutron Sources and Facilities

M. Arai and K. Crawford

Abstract This chapter provides a brief survey of the types of neutron sources that are available for imaging applications, primarily focusing on high-flux sources such as reactors or spallation sources but also including smaller or portable sources based on radioactive decay or small accelerators. Although nearly all neutron imaging studies performed so far have used portable or reactor-based sources, spallation sources will be of increasing importance in the future. Their designs are less familiar and somewhat more complex, so the chapter concentrates on them.

Keywords Reactor source · Spallation source · Portable source · Neutron energy · Pulse · Pulse width · Thermal neutron · Moderators · Reflectors · Neutron production · Fission · Spallation · Proton · Accelerator · Instrument · Time-of-flight · Flux

2.1 Introduction

One might think it is not necessary to have any detailed knowledge of the neutron source in order to use the techniques of neutron scattering or imaging. To an extent this is true. However, the distribution of neutrons from the source in terms of energy and time, and the distribution of "background" (fast neutrons, delayed neutrons, gammas, ...), has a direct bearing on the design of neutron scattering and imaging instruments and their performance, on how measured "raw" data must be corrected to make them scientifically meaningful, and on the types of measurements that can be undertaken. Even if the average user does not design the instrument, the user needs to choose the source/instrument to be used for a particular experiment or measurement, so

---

M. Arai (✉)
Japan Proton Accelerator Research Complex Center, Japan Atomic Energy Agency, Japan
e-mail: masatoshi.arai@j-parc.jp

I.S. Anderson et al. (eds.), *Neutron Imaging and Applications*,
Neutron Scattering Applications and Techniques, DOI 10.1007/978-0-387-78693-3_2,

some knowledge of the different types of sources is advisable. There are two kinds of neutron sources for neutron scattering and high-resolution/high-flux imaging facilities: (1) reactor sources and (2) large accelerator-based neutron sources. However, low-resolution/low-flux imaging can also be carried out with small radioactive or accelerator-based sources, which can be fixed or portable.

Nuclear reactors use the fission process to produce neutrons. Most of the current reactor sources for scattering applications were built in the 1960s and 1970s and were primarily designed for materials testing for the nuclear industry, providing medium flux. The best reactor source, optimized for neutron scattering applications, is still the High-Flux Reactor (HFR) at the Institut Laue-Langevin (ILL), built in 1972 in Grenoble, France [1, 2]. Reactors coming into operation more recently include JRR3 at the Japan Atomic Energy Agency (1990) [3], HANARO at the Korea Atomic Energy Research Institute (1997) [4], FRM-II in Munich, Germany (2004) [5], OPAL at the Australian Nuclear Science and Technology Oganisation (2006) [6], and the China Advanced Research Reactor in Beijing, China (potential operation in 2008) [7]. With the exception of the HFR, these are all medium-sized research reactor sources (typical power 10–20 MW) built with advanced technology.

Electron accelerators can produce neutrons in a target material using the Bremsstrahlung photo-neutron reaction. Although electron accelerators are relatively inexpensive to construct, the large amount of heat dissipated in the target per neutron produced severe limits on the potential performance.

Neutrons can also be produced by the spallation process, in which high-energy protons strike a solid target. The development of proton accelerator technology, driven by other applications such as particle physics, helped to increase the potential power and hence the neutron flux of spallation sources. The pulsed nature of most accelerator-based neutron sources can offer a significant advantage in experiments using the time-of-flight (TOF) method, in which the speed of the neutron is measured by timing its flight from the source to the detector. Pulsed proton-driven neutron sources recently completed or under construction include the Spallation Neutron Source (SNS) in the United States [8, 9], the Japanese Spallation Neutron Source (JSNS) of the Japan Proton Accelerator Research Complex (J-PARC) project in Japan (2008) [8, 9], and the China Spallation Neutron Source (CSNS) in China (under construction) [10, 11].

Both fission and spallation produce neutrons in the megaelectron volt energy range as discussed in Section 2.2. However, neutron scattering and most neutron imaging applications require neutrons at electron volt or lower energies. Moderators are used to slow the neutrons to these energies, as described in the section "Moderation Mechanisms." Reactor technology is highly developed and has not changed significantly for many years. The technologies for accelerator-driven sources offer many more parameters that can be varied to optimize the source for different purposes, so more space will be devoted to describe these options in this chapter.

2.2 Neutron Production

2.2.1 *Reactors*

At research nuclear reactors, neutrons are produced by the well-known fission process

$$\begin{aligned} &^{235}\mathrm{U} + \mathrm{n} \rightarrow \mathrm{X} + \mathrm{Y} + 2.5\,\mathrm{n}\,, \\ &(\sim 200\,\mathrm{MeV}\ \text{total energy release}, \sim 2\,\mathrm{MeV}\ \ \text{per neutron}) \end{aligned} \tag{2.1}$$

where X and Y are fission fragments or atoms of smaller atomic weight. Reactors are designed and optimized for different purposes. The exact design is dependent on a number of features—the shape and size of the core, the arrangement and type of the fuel elements, control rods, coolant, moderator, reflectors, and beam tubes. Power reactors are optimized for heat extraction and efficient use of fuel, so they have quite a different design from research reactors that are optimized for high (external) thermal neutron flux. In a typical research reactor design, one of the neutrons produced per fission is needed to sustain the chain reaction, ~0.5 is lost, and one is available for external use (i.e., ~200 MeV of heat is produced for each available neutron).

Recent design innovations have made a compact reactor core with high enrichment capabilities practical, and this in turn produces very high neutron fluxes outside the core for beam tube applications, as is the case at the FRM-II reactor in Munich (Fig. 2.1). The highly optimized combination of core design and moderator arrangement makes the thermal neutron flux available for experiments comparable to that of the world's preeminent research reactor facility for neutron scattering at the ILL I in Grenoble—8×10^{14} $\mathrm{ncm^{-2}s^{-1}}$for 20-MW reactor power compared with 1.5×10^{15} $\mathrm{ncm^{-2}s^{-1}}$for the ILL power of 58 MW. Table 2.1 shows the reactor power and source flux for the operating world-class research reactor sources (for scattering applications) along with some other parameters that will be discussed later in the chapter.

The 1980s Advanced Neutron Source project in the United States attempted to design a high-flux reactor of significantly higher power than ILL. However, it became clear that the technical challenges, coupled with increasing safety regulation, made this effort economically unviable [13, 14]. The future development of higher-flux neutron sources will therefore be based on accelerators.

2.2.2 *Proton Accelerator-Based Sources*

High-energy protons can create large numbers of "spalled" neutrons from bombardment of heavy nuclei. For example, a 1-GeV proton is capable of producing approximately 25 neutrons from a lead target, with heat deposition in the target of about half of the proton beam power—meaning one order of

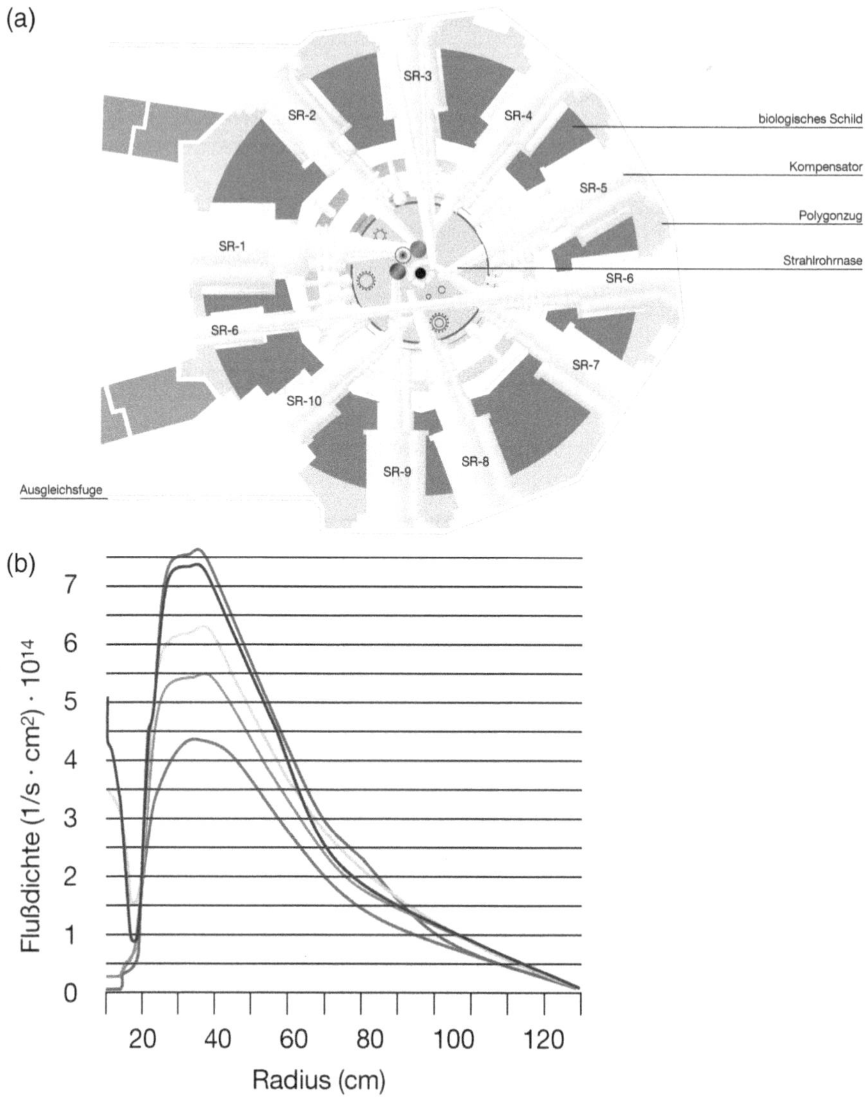

Fig. 2.1 (**a**) Layout of the reactor pool of FRM-II. The reactor core is very compact—24 cm in diameter—so that the maximum neutron flux is achieved at the moderator position, as shown in (**b**), giving a very high flux for instruments [5]

magnitude less heat must be dissipated than in a fission reaction producing the same time-averaged neutron flux. Spallation reactions occur for proton energies above 100 MeV. High-energy neutrons, pions, and spalled nuclei cause inter-nuclear cascades followed by low-energy neutron evaporation from the excited nuclei, as illustrated in Fig. 2.2.

Table 2.1 Existing medium- and high-flux reactor sources and their respective parameters

Country	United States	United States	Canada	France	France	Germany	Germany	Australia	Korea	Japan
Neutron source	HFIR	NBSR	NRU	HFR	ORPHEE	BENSC	FRM-II	OPAL	HANARO	JRR-3 M
Organization	Oak Ridge National Laboratory	National Institute of Standards and Technology	Atomic Energy of Canada Limited	Institut Laue-Langevin	Laboratoire Léon Brillouin	Helmholtz-Zentrum Berlin	Technische Universitat Munchen	Australian Nuclear Science and Technology Organization	Korea Atomic Energy Research Institute	Japan Atomic Energy Agency
Power (MW)	85	20	120	58	14	10	20	20	24 (present) 30 (designed)	20
Flux ($n \cdot cm^{-2} \cdot s^{-1}$)	1.5×10^{15}	3×10^{14}	3×10^{14}	1.5×10^{15}	3×10^{14}	2×10^{14}	8×10^{14}	3×10^{14}	2×10^{14}	3×10^{14}
Number of cold/hot sources	1/0	1/0	0/0	2/1	1/1	1/0	1/1	1/0	1(planned)/0	1/0
Number of instruments	9(present) +6 (planned by 2012)	24	5	26	22	22	20 (present) + 10 (under construction)	6	6	24
Existing neutron imaging instrument		BT-2 [14]			[15]	CONRAD [16]	ANTARES [17]		NR-port [18]	TNRF [19] And TNRF-2 [20]
Facility operating since	1967	1970	1957	1972 (refurbished 1993)	1980	1973	2004	2006	1997	1990

HFIR: High-Flux Isotope Reactor [21]; NBSR: National Bureau of Standards Reactor [22]; NRU: National Research Universal Reactor, Chalk River, Canada [23]; HFR: High-Flux Reactor at ILL [1, 2]; ORPHEE: reactor at LLB [24]; BENSC: Berlin Neutron Scattering Centre [25]; FRM-II: Forschungsneutronenquelle Heinz Maier-Leibnitz [5]; OPAL: Open Pool Australian Light-water Reactor [6]; HANARO: High-flux Advanced Application Reactor [4]; JRR-3 M: Japan Research Reactor No. 3 Modified [3]. Consult the web sites for these facilities to obtain additional information and current details. A number of smaller research reactors, primarily at universities, are not listed here.

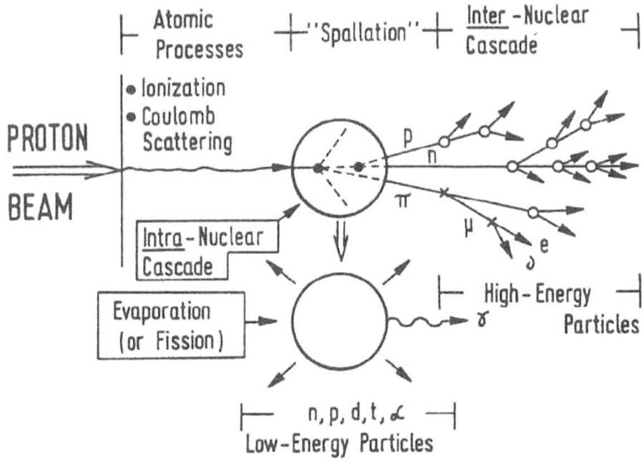

Fig. 2.2 Nuclear spallation by high-energy proton bombardment [26]

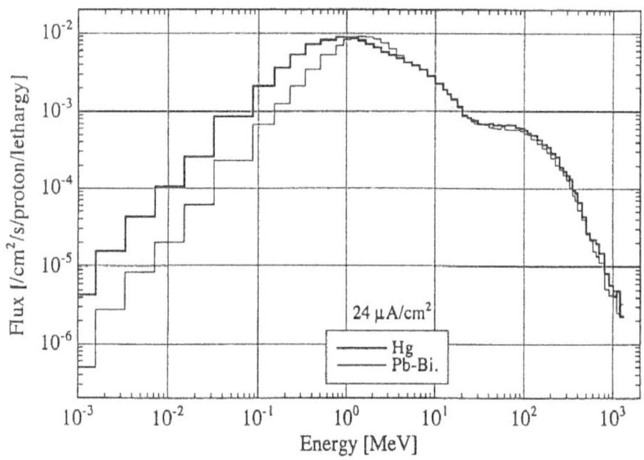

Fig. 2.3 Neutron energy spectrum from mercury and PbBi alloy targets [27]

The energy of a small fraction of the neutrons produced in spallation processes can be as high as the incident proton energy (these neutrons require very thick shielding), but the spectrum reaches a maximum of around 2 MeV for the evaporating neutrons, as shown in Fig. 2.3.

For efficient neutron production, as many protons as possible should undergo high-energy collisions with nuclei rather than gradually losing energy through other processes. The proton mean free path, dominated at high energies by energy-independent nuclear collision cross sections is approximately 200 g cm^{-2} (dividing by the selected target material density gives units of length). The proton stopping length, dominated at low energies by electron excitation energy loss, depends on the material and the energy but is roughly 600 g cm^{-2} (for lead and other heavy elements for 1-GeV protons). When the stopping length of a proton is greater than three times its mean free path, neutron production efficiency becomes close to 100%, so the proton energy should be as high as 1 GeV or greater.

In the early days of design work on spallation sources, proton energies were less than 1 GeV because of the lack of experimental experience at higher energies. However, more sophisticated codes and experiments in the 1990s demonstrated that the neutron production rate is almost proportional to the accelerator power even at 12 GeV [28, 29], as shown in Fig. 2.4. This resulted in flexibility in optimizing accelerator and neutron target design. Proton current and energy are equally optimized for beam experiments. Current and energy for spallation sources are listed in Table 2.2 along with some other parameters to be discussed.

For a short-pulse neutron source, the repetition rate of the proton acceleration is an important parameter to be considered. When a long neutron flight path is used for improved TOF resolution, a slow repetition rate is important to minimize frame overlap (where the fast neutrons from one pulse overlap with the slow neutrons from the previous pulse). However, if the same time-averaged power is maintained, a lower repetition rate requires more power per pulse, creating a more difficult engineering problem for the accelerator and target.

Most accelerator-based neutron sources are pulsed, and heat is produced in the target only during the pulses. This allows the heat to dissipate slowly in the period between pulses, so the instantaneous power and neutron flux can be very high. However, thermal shock in the target remains a problem to be overcome at the highest levels of proton power. Building spallation neutron sources

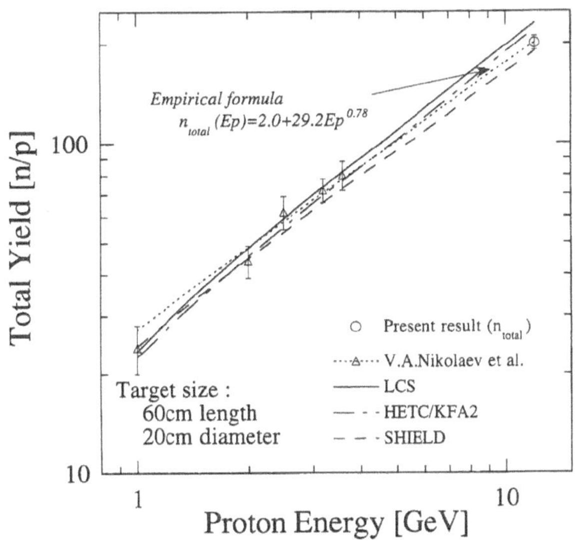

Fig. 2.4 Neutron yield from a lead target as a function of proton energy. Neutron yield is almost proportional to proton energy at up to 12 GeV. Here the *symbols* show experimental results, whereas the *lines* show the results of various calculations [28, 29]

Table 2.2 Past, existing, and future spallation source and their respective parameters

Country	United States	United States	United States	U.K.	Switzerland	China	Europe	Japan	Japan
Neutron source	IPNS	LANSCE	SNS	ISIS	SINQ	CSNS	ESS	KENS	JSNS
Organization	Argonne National Laboratory	Los Alamos National Laboratory	Oak Ridge National Laboratory	Rutherford Appleton Laboratory	Paul Scherrer Institute	Institute of High Energy Physics	Undecided	High Energy Accelerator Research Organization	Japan Atomic Energy Agency
Proton energy (MeV)/ Current (μA)	450/15	800/70	1000/1400	800/200	590/1500	1600	1333/7500	500/9	3000/333
Proton beam power	7 kW	56 kW	1.4 MW	160 kW	1 MW	100 kW	5 MW	4.5 kW	1 MW
Repetition rate (Hz)	30	20	60	50/10 (2 targets)	Continuous	25	16 (long pulse)	20	25
Target material	Depleted Uranium	Tungsten	Mercury	Tantalum	Zircaloy	Tungsten	Mercury	Tungsten	Mercury
Moderator	S-CH_4/L-CH_4	L-H_2/H_2O	L-H_2/H_2O	L-H_2/L-CH_4/H_2O	L-D_2/D_2O	H_2OL-CH_4L-H_2	L-H_2	S-CH_4/H_2O	L-H_2
Number of instruments	12	7	24 (beam ports)	22 (TS1) 7 (TS2)	15		20 (beam ports)	15	23 (beam ports)
Existing neutron imaging instrument					NEUTRA [30] and ICON [31]				
Facility operating since or planned to operate in	1981 (closed 2008)	1983	2006	1985 (TS1) 2008 (TS2)	1996	2014	Under planning	1980 (closed 2005)	2008

IPNS: Intense Pulsed Neutron Source [32]; LANSCE: Los Alamos Neutron Science Center [33]; SNS: Spallation Neutron Source [8, 9]; ISIS: [34, 35]; SINQ: Swiss Spallation Neutron Source [36, 37]; CSNS: Chinese Spallation Neutron Source [10, 11]; ESS: European Spallation Source [38, 39]; KENS: Koh-Energy-ken Neutron Source [40, 41]; JSNS: Japanese Spallation Neutron Source [8, 9]. Consult the websites for these facilities to obtain additional information and current details.

instead of reactors is therefore becoming a worldwide trend, as demonstrated by the SNS in the United States [8, 9] and the J-PARC project in Japan [42]. China has started construction of the CSNS [10, 11], and European countries are planning a long-pulse neutron source, the European Spallation Source (ESS) [38, 39], in the near future.

So far, short-pulse spallation neutron sources, typically delivering 1-μs proton pulse widths, have predominated because of the good timing resolution provided for TOF measurements of the neutron energy. Ring structures (synchrotrons or accumulator/storage rings) in the accelerator system are required to provide high proton intensities in such short pulses. Figure 2.5 shows an example of one such accelerator system using an accumulator ring.

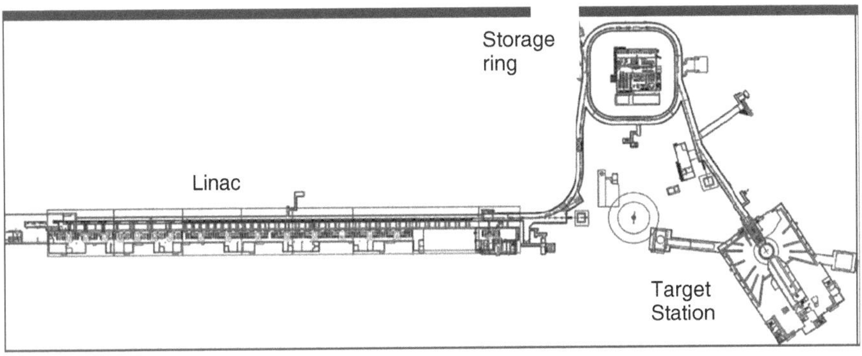

Fig. 2.5 A typical short-pulse spallation neutron source, the SNS facility, (Oak Ridge National Laboratory, United States) [8, 9]

Long-pulse sources, typically having 1-ms proton pulse widths, are another option rapidly gaining favor for neutron sources not requiring such high timing resolution. No accumulator ring is required for such sources, so a very high-intensity direct proton beam can be delivered from the linac to the neutron target [43]. Continuous cyclotron-based spallation sources, such as that at the SINQ at the Paul Scherrer Institute in Switzerland [36, 37], produce continuous neutron beams; their performance is similar to that of a medium-flux reactor. Table 2.2 shows spallation neutron sources either in operation or being planned.

2.3 Moderation Mechanisms for Reactors and Spallation Sources

2.3.1 Reactor Neutron Sources

The cross section for neutron-induced fission is much higher for thermal neutrons (100 meV) than for the fast neutrons (1–2 MeV) that are produced. For a reactor to achieve a self-sustaining chain reaction from a small mass of fissile material, and to obtain suitable neutrons for neutron scattering, the fast neutrons within the core must be "slowed down." This is done using a moderator, which is usually also the coolant. A moderator reduces the neutron energy by inelastic scattering and so is preferably a material containing light elements, e.g., H_2O or D_2O. Hydrogen has a large scattering cross section and a large absorption cross section, so when H_2O coolant is used, the core needs to be relatively compact to achieve a high-flux density at the neutron beam tubes. These beam lines are directed tangentially to the cylinder of maximum thermal flux in order to reduce the background of high-energy neutrons and gammas, as illustrated in Fig. 2.1. With D_2O, a larger core can be used (meaning easier cooling because heat can be dissipated more easily). Surrounding the moderator there may be "reflector" materials that do not provide much moderation but scatter or reflect some of the fast neutrons back into the moderator to enhance the thermalized neutron flux output.

The neutron energy distribution can be altered from that produced by the reactor coolant/moderator or reflector/moderator materials and design by strategic placement of special moderators, shifting the neutron energy spectrum to either slightly lower energies (a "cold source"—e.g., liquid hydrogen) or higher energies (a "hot source"—e.g., carbon heated by gamma radiation from the reactor). Such special moderators expand the range of usable neutron energies and hence expand the research capabilities of the reactor facility. Table 2.1 shows which of the research reactor facilities provide such capabilities.

2.3.2 Pulsed Spallation Neutron Sources

In designing spallation sources, as much consideration should be paid to the moderators as to the accelerator performance. The design of moderators is strongly dependent on the kinds of instruments, resolution, and intensity that are required. Among the most important parameters to be optimized for moderators are (1) temperature, (2) neutronic structure, and (3) materials.

Neutrons in moderators reach thermal equilibrium after multiple scattering events and have, in the case of sufficiently thick moderators, a Maxwellian distribution in energy around the temperature of the moderator. On the other hand, for pulsed sources, the time that neutrons spend in the moderator broadens the pulse. Short-pulse spallation sources need to have a sharp pulse structure of thermalized neutrons, so the moderator dimensions need to be small and are optimized at around $10\times10\times10\,cm^3$. The under-moderated neutrons result in a rich "epithermal" flux, proportional to $1/E$ where E is the energy of the neutrons, in the higher energy "slowing down" region above the Maxwellian distribution (Fig. 2.6). Most pulsed neutron sources utilize moderators in a "wing" geometry so that the beam tubes are directed tangentially to the target to minimize the flux of high-energy neutrons and gammas in the neutron beams. The "flux trap" geometry is also effective, as is demonstrated by the Los Alamos Neutron Science Center [44]. Low-temperature moderators can extend the slowing-down region to shift the Maxwell distribution to lower energy, although flux is sacrificed in the thermal energy region.

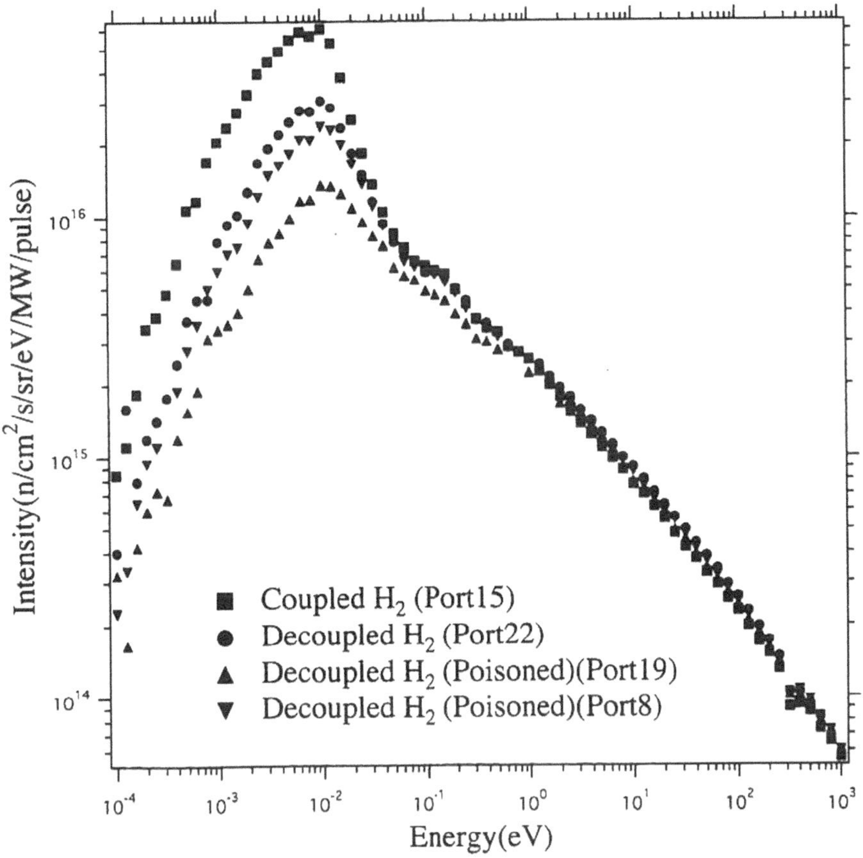

Fig. 2.6 The neutron energy distribution (flux) of the J-PARC neutron source for coupled, decoupled, and poisoned decoupled moderators. The flux consists of a Maxwell distribution at low energies and a 1/E region at higher energies [45]

The pulse width is a key parameter for short-pulse spallation sources and is directly influential on neutron beam instrument performance. It is almost proportional to the neutron wavelength in the 1/E region of flux, is broadened in the thermal equilibrium region, and then saturates in the very low-energy region. The broadening starts to occur at about 300 meV (neutron wavelength of 0.5 Å) for an ambient-temperature moderator and at about 15 meV (2.5 Å) for a methane moderator at 20 K. This is clearly seen in Fig. 2.7. In the 1/E region for each moderator, the pulse width, Δt, is proportional to wavelength, λ, as [46]

$$\Delta t[\mu s] \sim \frac{2}{\sqrt{E[eV]}} \sim 7\lambda[\mathring{A}] . \tag{2.2}$$

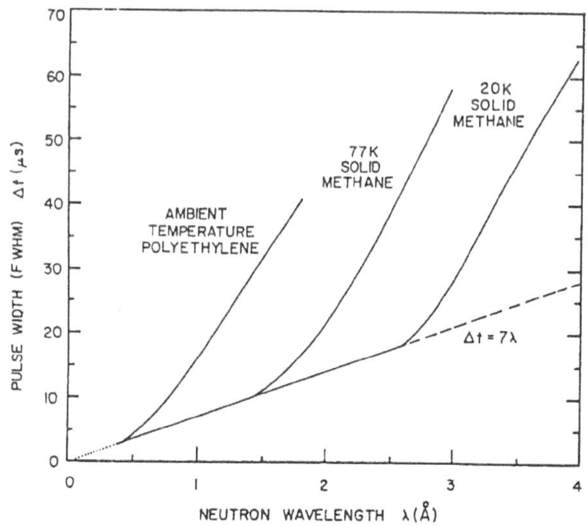

Fig. 2.7 Pulse widths for typical moderators [45]

This proportionality between the pulse width and wavelength is of great importance for high-resolution instruments, as will be discussed in a later section.

Pulse widths can also be reduced by surrounding the moderator with an absorbing material (decoupler) such as cadmium on all sides except the side from which the neutron beam emerges. This prevents the neutrons slowed down in the reflector from entering the moderator and emerging as part of the neutron beam. Since those neutrons would typically emerge into the beam later than would the neutrons slowed down in the moderator, such decoupling prevents the extra broadening of the neutron pulse that would result from such neutrons. However, this decoupling also results in a reduced total intensity in the neutron pulse. Still further reductions in pulse width can be achieved by placing an absorbing material (cadmium or gadolinium) as "poisoning" in the moderator, effectively reducing the moderator size for low-energy neutrons, but this again results in a penalty in intensity. If no such absorbing materials are used, the moderator is fully "coupled" to its surroundings and produces the highest intensity but with relatively broad pulse widths.

The neutrons from small moderators are empirically understood to consist of two components: (1) the thermal equilibrium component, the Maxwellian or so-called storage component, and (2) the pre-equilibrium component caused when neutrons undergoing a small number of scattering processes escape the moderator before thermalizing, the so-called slowing down component [47]. The former component broadens the pulse width (Fig. 2.8) giving intense low-energy flux as indicated in Fig. 2.6.

Moderator materials should be chosen that are suitable for thermalization of neutrons with good neutron cross sections in the low-energy range.

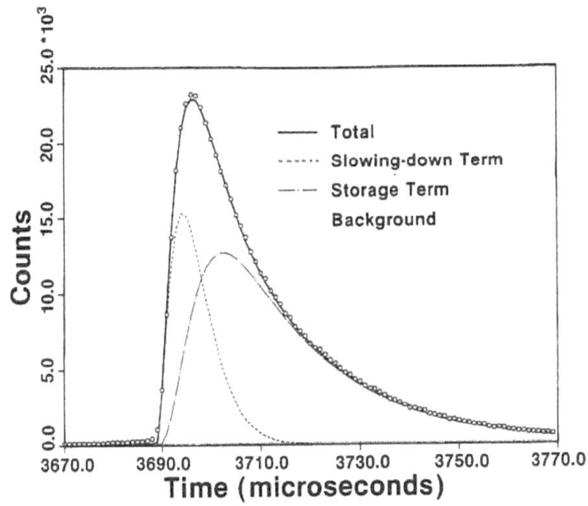

Fig. 2.8 Typical pulse structure from a short-pulse spallation source. The sharp rising edge comes from the slowing down component followed by the storage component with a long time tail [47]

Cold coupled moderators have been extensively optimized for pulsed sources by Watanabe and Kiyanagi, greatly enhancing neutron flux from the cold neutron moderator while minimizing heat deposition in the moderator (Fig. 2.9) [48].

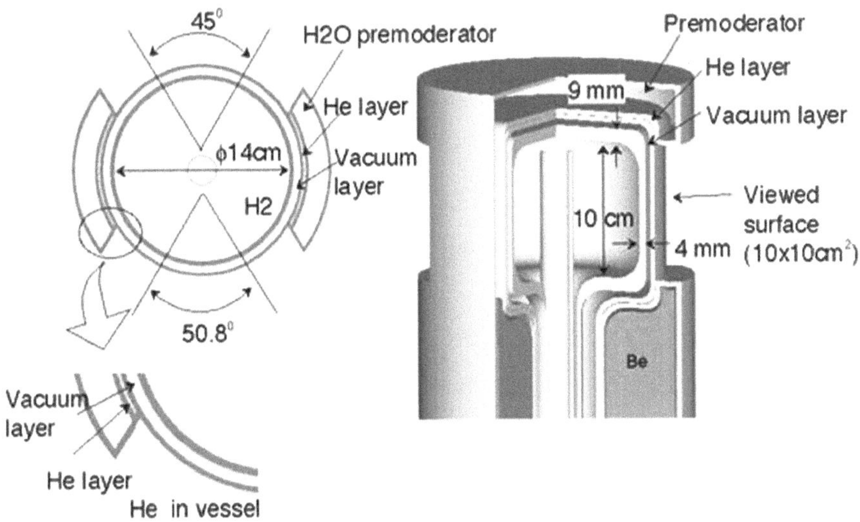

Fig. 2.9 Design of a coupled moderator for J-PARC. The hydrogen moderator is surrounded by a water premoderator, which reduces neutron energy in the first stage and removes heat before neutrons go into the hydrogen moderator [48]

2.4 Comparison of Source Types

We can consider the case of crystalline diffraction to illustrate the differences between reactor sources and pulsed spallation sources and the moderator choice considerations for a pulsed source (this would also apply to, for example, Bragg edge imaging, see Chapters 6 and12). When a diffraction measurement is made at a reactor, the neutron energy is typically monochromatized by using Bragg scattering from a single crystal, and neutrons scattered from the sample are recorded as a function of angle. At a pulsed neutron source, the TOF method naturally is used. Neutrons are counted as a function of flight time, starting at neutron emission resulting from the proton pulse on the target. Neutrons propagate along the primary flight path, L1; are scattered by a sample; and are detected by a detector at a certain scattering angle and distance, L2, as shown in Fig. 2.10. The peak pulse width is maintained during the propagation from source to detector. Consequently, a sharp peak width from a decoupled moderator is preferable for a high-resolution measurement with a long flight path, leading to better separation of peaks at the detector. Alternatively, a coupled moderator could be used to obtain higher flux, if a much longer flight path could be used to maintain the resolution. Hence, essentially in a pulsed spallation source, the longer the flight path the higher the resolution. The transmission of modern neutron guides in the relevant wavelength range is sufficiently good that flight paths of up to 250 m might be considered, but the cost of the guides and associated shielding is significant.

Continuous reactor sources have the advantage of higher time-averaged intensity. For instance, neither SNS nor JSNS will exceed the average neutron

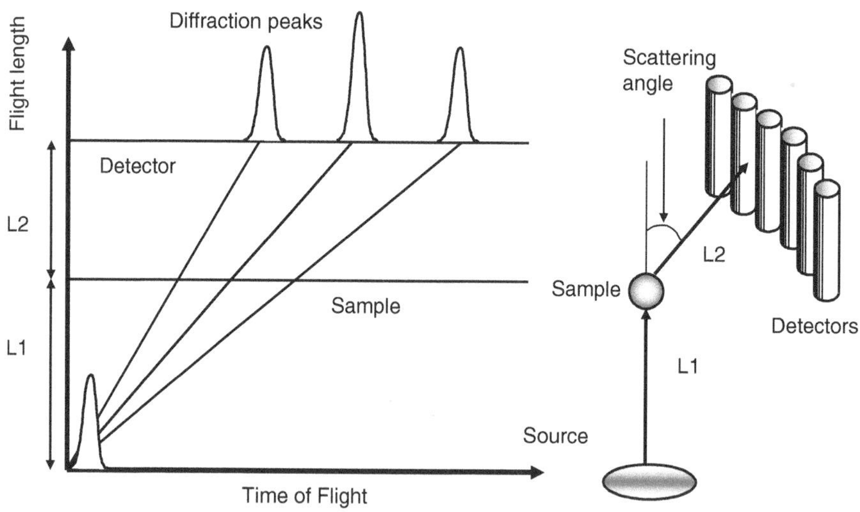

Fig. 2.10 Flight-length/time-of-flight diagram for a diffraction measurement

flux of high-performance reactors such as the High-Flux Isotope Reactor at Oak Ridge National Laboratory, the HFR, or FRM II. Furthermore, reactor neutron sources offer high availability; for instance, FRM II operates 260 full-power days per year. However, spallation neutron sources deliver the highest peak flux, which, coupled with the advantages of instrumentation based on the TOF principle (see Chapter 3 for further details), becomes the effective figure of merit for many applications. Also, as noted earlier, spallation sources offer the only serious opportunities for improvement of source intensities in the future.

2.5 Neutron Facilities

Most neutron facilities are user facilities (i.e., open to external scientists). Prospective users submit experimental proposals, which are evaluated by review committees for approval. To attract a broad range of users, facilities should offer a suitably wide variety of instruments, ease of access, good maintenance, and high performance. The number of experiments/users that a facility can accommodate depends on the power of the source (i.e., the speed of individual measurements), the number of days of source operation, and the number of instruments. Apart from relatively routine measurements that can be automated, the minimum practical turnaround time for experiments tends to be 1–2 days, so the capacity saturates regardless of the source power. However, higher-power sources enable significantly more complex experiments to be carried out in the same time (e.g., imaging at higher resolution, tomography, or the use of contrast enhancement techniques), as described in Chapters 6 and 12.

Most modern instruments are equipped with a large number of detector segments and easily produce nearly 1 GB or more of data in a single measurement; analysis and storage of these large amounts of data are becoming a limiting factor. Drastic improvements in accelerator or reactor performance are not easy, technologically or financially. However, improvements in neutron optics, detectors, and instruments have the potential to dramatically increase the flux of useful neutrons for experiments, as has been the exemplary experience at ILL. Maintaining a suitable balance of effort among the accelerator, target, instruments, and support activities (e.g., sample preparation, sample environment, data analysis) is therefore an important factor in maintaining facility capabilities.

2.6 Smaller Neutron Sources for Imaging and Other Applications

Neutrons can be produced by spontaneous fission, although this is unusual. For example, ^{252}Cf can decay by α-decay with a half-life of 2.65 years (this produces helium gas, which creates internal pressure in the source) and by n-decay with a half-life of 85.5 years. The average neutron energy is 2.14 MeV, and the rate is 2.34×10^{12} ns^{-1} g^{-1}. Such sources are produced by irradiation in a high-flux reactor.

A more common method is neutron production as a secondary process, as used by Chadwick when he first discovered the neutron. For example, in an α-n source such as ^{241}Am/Be, ^{241}Am undergoes α-decay; the α-particle can be absorbed by a light element such as beryllium, which then decays by neutron emission. This can be written as

$$\begin{aligned} &^{241}\mathrm{Am} \rightarrow {}^{237}\mathrm{Np} + {}^{4}\mathrm{He}\,(5.6\,\mathrm{MeV})\ \text{followed by} \\ &^{9}\mathrm{Be} + {}^{4}\mathrm{He} \rightarrow {}^{12}\mathrm{C} + \mathrm{n}\,(\text{few MeV})\,. \end{aligned} \tag{2.3}$$

The half-life is 433 years, and these sources can produce 10^6–10^8 n s^{-1} g^{-1} Am. An alternative is a γ-n source, for example

$$\begin{aligned} &^{124}\mathrm{Sb} \rightarrow {}^{124}\mathrm{Te} + \mathrm{h}\nu(1.7\,\mathrm{MeV})\ \text{followed by}\ {}^{9}\mathrm{Be} + \\ &\mathrm{h}\nu \rightarrow 2\,{}^{4}\mathrm{He} + \mathrm{n}\,(\text{few MeV})\,. \end{aligned} \tag{2.4}$$

The half-life is 60 days, and such sources can produce 10^9–10^{10} n s^{-1}.

Radioactive decay sources have the advantage of being small and highly portable, but they have low intensity and are always "on." They can be used for testing (e.g., of neutron detectors), in medicine (e.g., activation analysis, cancer treatment with ^{252}Cf needles), and for low-resolution/low-flux radiography.

Higher intensities can be produced by small accelerator-based neutron sources. Over the years, these have evolved sufficiently that compact portable sources are now commercially available from a range of vendors. They are normally based on the "D-T" reaction:

$$^{2}\mathrm{H}(\sim 150\,\mathrm{keV}) + {}^{3}\mathrm{H} \rightarrow {}^{4}\mathrm{He} + \mathrm{n}\ (14.2\,\mathrm{MeV})\,. \tag{2.5}$$

Sealed tube sources with a typical length of 1 m and diameter of 10 cm, operating at a power of 0.5 kW, can produce up to 3×10^{10} n s^{-1}. At a distance of 1 m, this gives a flux on the order of 2×10^5 n cm^{-2} s^{-1}. Note that these are fast neutrons; if they were moderated to be used for thermal neutron imaging, then the flux would be considerably lower. They are widely used for industrial and security applications based on fast neutron radiography (Chapter 18) and prompt gamma activation analysis.

The Low Energy Neutron Source at Indiana University [49] is a small cyclotron-based source that uses the reaction

$$\mathrm{p}(7 - 13\,\mathrm{MeV}) + {}^{9}\mathrm{Be} \rightarrow {}^{9}\mathrm{B} + \mathrm{n}(5 - 11\,\mathrm{MeV})\,. \tag{2.6}$$

This source can produce a flux of moderated thermal neutrons comparable to that produced by a small spallation source and is therefore suitable for both scattering and imaging applications.

References

1. W.J. Drexel, *IEEE Trans. Nucl. Sci.* **NS-29**, 123–126 (1982).
2. C.J. Carlile, *Physica B* **385–386**, 961–965 (2006).
3. Fumio Sakurai, Yoji Horiguchi, Shinsho Kobayashi, and Masaji Takayanagi, *Physica B* **311**, 7–13 (2002).
4. Chang-Hee Lee, Young-Hwan Kang, and Il-Hiun Kuk, *ICANS-XV Proceedings of the 15th Meeting of the International Collaboration on Advanced Neutron Sources*, Tsukuba, Japan, 146–154 (2000).
5. J. Neuhaus and W. Petry, *Neutron News* **18**:2, 13–15 (2007).
6. Shane J. Kennedy, *Physica B* **385–386**, 949–954 (2006).
7. D.F. Chen, Y.T. Liu, C. Gou, and C.T. Ye, *Physica B* **385–386**, 966–967 (2006).
8. T.E. Mason, *et al.*, *Physica B* **385–386**, 955–960 (2006).
9. Tony A. Gabriel, John R. Haines, and Thomas J. McManamy, *J. Nucl Mater.* **318**, 1–13 (2003).
10. J. Zhang, Q.W. Yan, C. Zhang, P.L. Zhang, S.N. Fu, F.W. Wang, Z. Zhang, and S.X. Fang, *J. Neutron Res.* **13**:1, 11–14 (2005).
11. Q.W. Yan, W. Yin, and B.L. Yu, *J. Nucl. Mater.* **343**, 45–52 (2005).
12. Colin D. West, *Physica B* **174**, 430–437 (1991).
13. A. Lawler, *Science* **279**, 470–471 (1998).
14. D. Hussey et al., *Nucl. Instrum. Methods Phys. Res. A* **542** (2005) 9–15.
15. http://www-llb.cea.fr/neutrono/nr1.html.
16. A. Hilger et al., *Physica B* **385–386** (2006) 1213–1215.
17. E. Calzaba et al., *Nucl. Instrum. Methods Phys. Res. A* **542** (2005) 38–44.
18. I.-C. Lim, M.-S. Kim, K.-Y. Nam, C.-M. Sim, B.-C. Lee, H.-Y. Choi, and S.-Y.Hwang, Characteristics of neutron beam from IR beam port of HANARO for its application to dynamic neutron radiography, in *Proceedings of the 2002 ANS RPD Topical Meeting, Int. Conf. New Frontiers of Nuclear Technology*, Seoul, Korea, Oct. 2002.
19. M. Matsubayashi and A. Tsuruno, JRR-3 Neutron Radiography Facility, *Proceedings of the 4th World onf. Neutron Radiography*, San Francisco, California, May 10–16, 1992, p. 415, Gordon & Breach Science Publishers.
20. M. Matsubayashi et al., *Nucl. Technol.* **132** (2000) 309.
21. S.E. Nagler and H.A. Mook, *Neutron News* **19**:2, 12–13 (2008); also other articles this issue.
22. R.L. Cappelletti, C.J. Glinka, S. Krueger, R.A. Lindstrom, J.W. Lynn, H.J. Prask, E. Prince, J.J. Rush, J.M. Rowe, S.K. Satija, B.H. Toby, A. Tsai, and T.J. Udovic, *J. Res. Natl. Instrum. Stand. Technol.* **106**, 187–230 (2001).
23. J. Root, *Neutron News* **14**:2, 12–13 (2003).
24. P. Monceau and M. Alba, *Neutron News* **14**:3, 10–12 (2003).
25. R. Michaelsen, *Neutron News* **12**:3, 7–9 (2001).
26. N. Watanabe, *Rep. Prog. Phys.* **66**:339–381 (2003).
27. M. Teshigawara et al., *JAERI-Res.* 99–0101 (1999).
28. M. Arai et al., J. *Neutron Res.* **8**, 71–83 (1999).
29. J.M. Carpenter, et al., Physica B **270**, 272–279 (1999).
30. E. Lehmann, et al., *Nucl. Instrum. Meth. A* **377** (1996) 11.
31. G. Kuehne et al., *Nucl. Instrum. Methods Phys. Res. A* **542** (2005) 264–270.

32. Raymond G. Teller, James W. Richardson, and John M. Carpenter, *Neutron News* **15**:3, 9–13 (2004).
33. J.A. Roberts, *Neutron News* **10**:4, 11–14 (1999).
34. R. Eccleston and C. Wilson, *Neutron News* **15**:1, 15–18 (2004).
35. Penfold, *Neutron News* **15**:2, 9–12 (2004).
36. W. Wagner, J. Mesot, P. Allenspach, G. Kuehne, and H.M. Rønnow, *Physica B* **385–386**, 968–971 (2006).
37. W. Wagner and H. Hayck, *Proc. ICANS-XVI*, ISSN 1433-559X, 53–66 (2003).
38. E. Cartlidge, *Phys. World* **10**, June, 10 (2007).
39. D. Richter, *Appl. Phys.* **A 74** [Supl.], S18–S22 (2002).
40. Ishikawa, *ICANS-IV Proceedings of the 4th Meeting of the International Collaboration on Advanced Neutron Sources*, Tsukuba, Japan, 89–101 (1980).
41. Ikeda, ICANS-XV *Proceedings of the 15th Meeting of the International Collaboration on Advanced Neutron Sources*, Tsukuba, Japan, 93–96 (2000).
42. Shoji Nagamiya, *J. Neutron Res.* **13**:1, 7–10 (2005).
43. F. Mezei, *Proceedings 12th meeting of International Collaboration on Advanced Neutron Sources ICANS-XII*, Abingdon, Oxfordshire, UK, I-377–384 (1993).
44. Takashi Ino, Motoki Ooi, Yoshiaki Kiyanagi, Yoshimi Kasugai, Fujio Maekawa, Hiroshi Takada, Guenter Muhrer, *Eric J. Pitcher, and Gary J. Russell, Nucl. Instrum. Meth. A* **525**, 496–510 (2004).
45. Tamura et al., *JAERI-Report 2003–008* (2003).
46. F.R. Mildner and R.N. Sinclair, *J. Nucl. Energy* **6**, 225 (1979).
47. Ikeda and J. Carpenter, *Nucl. Instrum. Meth. Phys. Res.* **A239** 536–544 (1985).
48. Watanabe et al., *Proceedings of the 10th meeting of International Collaboration on Advanced Neutron Sources ICANS-X*, Los Alamos, USA, pp.787–797 (1988).
49. Leuschner, D.V. Baxter, J.M. Cameron, V. Derenchuk, C. Lavelle, A. Lone, H. Nann, T. Rinckel, and W.M. Snow, *J. Res. Natl. Instrum. Stand. Technol.* **110**, 153–155 (2005).

Chapter 3
Neutron Optics

K. Andersen

Abstract An overview is given of neutron optics, aiming at a qualitative overview of the main concepts and components currently used for designing the optics of neutron instruments. The section on source optics shows the brightness curves of the Institu Laue-Langevin and ISIS cold sources and includes a short discussion of Liouville's theorem. Neutron guides and supermirrors are dealt with in the section of reflection optics. In the diffraction optics section, I describe the concepts of mosaic and bent-perfect crystals, giving reference numbers for their performance and also attempting to cover some basic focusing concepts. There is a very short section on refractive optics, covering both materials and magentic lenses. The final section on polarisation optics is more detailed, partly because of the current rate of progress in the field and partly because it is an area with which the general reader may be less familiar.

Keywords Neutron optics · Neutron polarisation · Neutron instumentation

3.1 Introduction

Throughout this chapter, I restrict the descriptions to the qualitative 'hand-waving' type. Neutron optics is a large field, which is covered in much more detail elsewhere [1, 2]. I have attempted here to give an overview of the field and have gone into more detail only where I felt that additional information was useful or that significant progress had been made since the publication of the Institut Laue-Langevin (ILL) Neutron Data Booklet [1]. In these cases, I have also indicated recent references for further study.

K. Andersen (✉)
Institut Laue-Langevin, 6 rue Jules Horowitz, B.P. 156, 38042 Grenoble Cedex 9, France
e-mail: andersen@ill.fr

I.S. Anderson et al. (eds.), *Neutron Imaging and Applications*,
Neutron Scattering Applications and Techniques, DOI 10.1007/978-0-387-78693-3_3,

3.2 Source Optics

Neutron sources are typically 10–20 cm in size in all directions. They consist of an efficient moderator material (usually containing hydrogen or deuterium) at a well-defined temperature to produce a quasi-Maxwellian wavelength spectrum. Their performance is characterised by their brightness (or brilliance) $d^2\Phi/d\Omega d\lambda$ (i.e., flux Φ per unit solid angle Ω and per unit wavelength λ). It is usually given in units of neutrons/sec/cm^2/sterad/Å. Brightness curves for the ILL vertical cold source are shown in Fig. 3.1. For a pulsed source, the brightness can be given as the peak brightness (i.e., the instantaneous brightness at its maximum in time) or integrated over the full time spectrum of the moderator. These curves are also shown in Fig. 3.1 for the liquid H_2 moderator on the first target station at ISIS.

The brightness is a property of the source and does not change with distance. The neutron flux at the sample is obtained by integrating over the source solid angle seen by the sample and over the desired wavelength range. It then needs to be scaled down by any imperfections in the transmission optics such as guide or monochromator reflectivities or chopper transmissions.

The efficiency of any neutron optical device is limited by Liouville's theorem, also known as the 'no free lunch' theorem, which states that phase space density in a conservative system is conserved. What this means is that any compression in one dimension, such as focusing the source onto a spot smaller than the source, is accompanied by an equivalent defocusing in another dimension, in this case the divergence. This can be verified by looking at the standard lens equation illustrated in Fig. 3.2: $w_1/L_1 = w_2/L_2$ (i.e., the size of the focal spot scales with the distance to the lens).

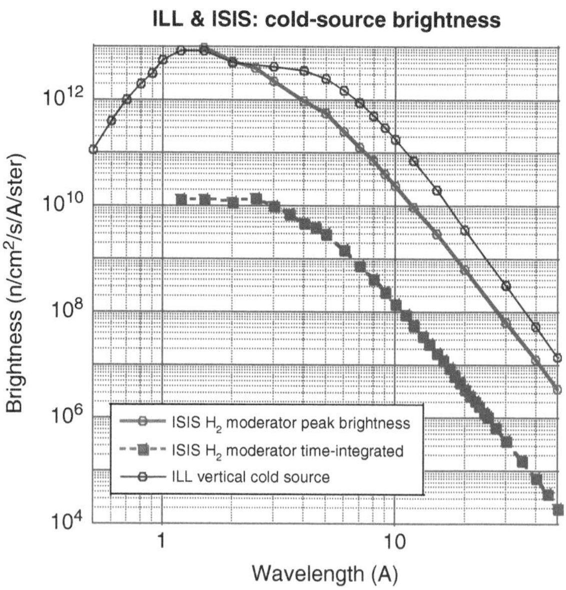

Fig. 3.1 Brightness of the ILL vertical cold source and the ISIS liquid H_2 moderator

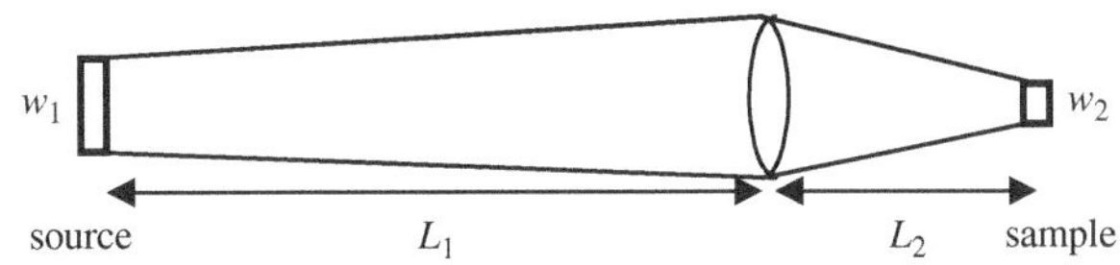

Fig. 3.2 Source–sample optics with focusing lens

It can easily be verified that the reduction in the size of the focal spot is accompanied by an increase in the divergence hitting the focal spot, in accordance with Liouville's theorem. Liouville's theorem should really be treated as a best-case scenario. No optics system is fully conservative: some neutrons are always lost from absorption, scattering, or other imperfections.

3.3 Reflection Optics: Neutron Guides

A neutron beam impinging on a flat surface can be treated as a plane wave hitting a potential step. The potential V is proportional to the scattering-length density Nb – the product of the atomic number density and the average coherent scattering length. Since the gradient of the potential is perpendicular to the surface, this potential step should be compared to the component of neutron kinetic energy which is perpendicular to the surface $K_\perp$. When $K_\perp < V$, the neutron beam is fully reflected. The condition $K_\perp = V$ corresponds to the critical angle for total reflection, which is given by

$$\sin\theta = \lambda\sqrt{Nb/\pi} \ . \tag{3.1}$$

For natural nickel, which has a scattering-length density of 9.4×10^{-6} Å^{-2}, it occurs at an incidence angle θ of 0.10° per Å neutron wavelength λ.

Total reflection from a flat surface enables neutron beams to be transported long distances from the source down glass guides with appropriate surface coatings. Nickel is a common choice for this surface coating, as it is relatively inexpensive and has a large scattering-length density. In order to significantly increase the critical angle beyond that of nickel, the guide surfaces can be coated with supermirrors. A neutron supermirror is an artificial multilayer consisting of thin layers of alternating materials with a large contrast in scattering-length density. Hundreds or even thousands of layers are deposited, usually by sputtering, to create a quasi-continuous range of d-spacings. The d-spacing is the combined thickness of two adjacent layers, which we can treat using Bragg's law:

$$\sin\theta = \lambda/2d \ . \tag{3.2}$$

A supermirror is usually made by first depositing the thin layers, gradually increasing the layer thicknesses in order to cover the range of d-spacings up to

$d = \sqrt{\pi/4Nb}$, where Nb is the scattering-length density of the high-Nb material. It is then finished off with a thick (several hundred Å) layer of the high-Nb material, to ensure total reflection at the small angles. The maximum critical angle is determined by the smallest d-spacing which is technically feasible. In practice that tends to be limited by substrate roughness, interdiffusion between the layers, and stress build-up, which can eventually destroy the periodicity of the mirror when too many layers are deposited. The critical angle of the supermirror is usually given relative to the critical angle of nickel, using the 'm' number: $m=1$ is nickel, $m=2$ is a supermirror with twice the critical angle of nickel, etc. Presently, neutron supermirrors are commercially available with critical angles of $m=4$, though values as high as $m=6$ have been reported by research groups.

The reflectivity of a surface describes the probability of specular (i.e., $\theta_{out}=\theta_{in}$) reflection as a function of incidence angle θ_{in}. Below the total-reflection angle of the material, the reflectivity can almost be perfect (i.e., greater than 99%), while in the Bragg-scattering regime of a supermirror it tends to be rather lower, typically around 80% at $m=3$.

Supermirror-coated glass guides (Fig. 3.3) can be used to transport neutrons far from the neutron source, which has several desirable effects. There is room for more instruments if they are not all crowded around the neutron source. Signal-to-background ratios can also be significantly improved by transporting the useful neutron beam far from the source. The flux of gamma rays and fast neutrons emitted from the source decreases approximately by $1/r^2$ (where r is the distance from the source). Gamma rays and fast neutrons can also be substantially attenuated by curving the neutron guide so as to avoid direct line-of-sight back to the source. This condition is given by

$$L > \sqrt{8wR_c}\ , \tag{3.3}$$

where L is the guide length, w is the guide width, and R_c is the radius of curvature. Curving the guide also preferentially reduces the transmission of short-wavelength neutrons. The characteristic wavelength λ^* (in Å) of a curved guide is given by

$$\lambda^* = \frac{575}{m}\sqrt{2w/R_c}\ , \tag{3.4}$$

and corresponds to two-third of the incident neutrons being transmitted. At shorter wavelengths, the transmission falls rapidly to zero.

Guides can also be used to improve the resolution of pulsed-source instruments. This is typically determined by the relative time uncertainty $\Delta t/t$, where Δt is mainly determined by the time-width of the pulse in the moderator and t is the total time-of-flight from source to detector. Hence by moving the instrument further from the source, the relative time uncertainty can be reduced, thus improving the instrumental resolution. The

Fig. 3.3 A supermirror guide on the ISIS second target station. The guide is viewed end-on, showing the multiple reflections from the side and top and bottom walls. Photo courtesy of the Science & Technology Facilities Council (UK)

length of the instrument is limited by the high cost of the supermirror guide and associated shielding and the need to avoid frame overlap. Frame overlap can occur when the slowest neutrons from one pulse are overtaken by the fastest neutrons from the subsequent pulse. Long instruments usually incorporate mechanical choppers which limit the wavelength bandwidth to avoid frame-overlap problems.

Finally, guides can be used to focus neutron beams to a smaller size, increasing the flux at the sample, while at the same time increasing the divergence, in accordance with Liouville's theorem. Most guides comprise straight segments, which can be placed so as to approximate a curved surface. However, in some cases, it is sometimes worth the extra complication and cost to build them with continuous curvature in elliptical or parabolic shapes.

3.4 Diffraction Optics

Diffraction optics are based on Bragg's law:

$$n\lambda = 2d\sin\theta \quad . \tag{3.5}$$

Monochromators are used to select the wavelength of neutrons incident on the sample, while analysers are used to select the wavelength after scattering. Both are much more prevalent at steady-state sources, as alternatives to the use of time-of-flight for labelling the neutron wavelength. For this chapter I will refer to both as monochromators for simplicity. Most monochromators for neutron instruments use the first order ($n = 1$) reflection since it is the most intense reflection. The higher-order reflections can be problematic as they contaminate the beam with undesired shorter-wavelength neutrons, which need to be dealt with by other means.

Monochromators usually consist of single crystals, but artificial multilayers can also be made periodic to produce efficient Bragg diffraction. However, these are more usually stacked aperiodically to produce supermirrors and are considered in the section on reflection optics. A perfect single crystal accepts only a very narrow range of wavelengths within a given range of scattering angles, given by the Darwin width, which is of the order of arc-seconds. Though the peak reflectivity can be very high, the integrated reflectivity is low, and to increase the counting rate on the instrument, two types of crystals are habitually used: bent perfect crystals and mosaic crystals.

When a perfect crystal is bent, the number of reflected neutrons increases dramatically. This can be qualitatively understood as the combination of two effects: when bending, the internal (concave) face is compressed in the direction of the face. To conserve density, the crystal must expand slightly in the other (radial) direction. The d-spacing is thus made to vary as a function of depth. The other effect is the variation of orientation of the crystal along the face due to the continuous curvature. The two effects can add to give a large increase in integrated intensity compared to the flat crystal. A mosaic crystal can be visualised as an assembly of small perfect crystals that are slightly misaligned with respect to each other. The misalignment is known as the mosaicity and is usually modelled by a Gaussian distribution of misorientations with the mosaic spread given by the FWHM (full width at half maximum) of the Gaussian.

A monochromator crystal is characterised by its peak reflectivity and its mosaic spread. For a high peak reflectivity, you need a large structure factor associated with the reflection and a large scattering cross section. It is reduced by extinction, absorption, parasitic reflections, and inelastic scattering. Extinction is caused by multiple first-order Bragg reflections either coherently within a perfect crystal or mosaic crystal block (primary extinction) or incoherently between mosaic blocks (secondary extinction). Parasitic reflections are predominantly due to other Bragg reflections, attenuating the diffracted beam as it leaves the crystal.

The peak reflectivity represents the reflectivity at the centre of the peak measured in a rocking-curve scan. It is strongly wavelength dependent. For a nonabsorbing crystal it increases with wavelength until the Bragg cutoff is reached at $\lambda = 2d$. The measured mosaic spread is usually given by the FWHM of the rocking curve. It is wavelength-dependent and differs from the true mosaic distribution of crystallites, primarily due to the finite thickness of the crystal. The behaviour of mosaic crystals is treated in [3].

The external surface of the crystal can be cut at any angle with respect to the crystal planes. Most monochromators are cut either in reflection (Bragg) geometry or in transmission (Laue) geometry, with the reflecting planes respectively parallel and perpendicular to the external surface. The neutron reflectivity is always higher in reflection geometry, but transmission geometry may be chosen for convenience, particularly when the Bragg angle is small. Table 3.1 gives typical performance numbers for commonly used monochromator crystals. The resolution of a single crystal monochromator can be studied by differentiating Bragg's law to give

$$\begin{aligned}\Delta\lambda^2 = \left(\frac{\partial\lambda}{\partial d}\Delta d\right)^2 + \left(\frac{\partial\lambda}{\partial\theta}\Delta\theta\right)^2 &= (2\sin\theta\Delta d)^2 + (2d\cos\theta\Delta\theta)^2 \\ &\Rightarrow \Delta\lambda/\lambda \approx \cot\theta\Delta\theta\end{aligned}, \quad (3.6)$$

where we have dropped the first term which is usually small. From here we can see that the diffracted wavelength spread is proportional to $\cot\theta$, which becomes very large for small take-off angles. This means that when the crystal is used to obtain wavelengths which are significantly shorter than the d-spacing, the wavelength resolution becomes quite poor, which can be useful if you wish to maximise flux at the sample at the expense of resolution. At the other end of

Table 3.1 Typical performance of some commonly used monochromator crystals and their reflections. Heusler is the Cu_2MnAl alloy, which provides a polarised beam. The peak reflectivity is given for a crystal in reflection geometry

Material	Reflection	d-spacing	Peak reflectivity at $\lambda = 2$ Å	Peak reflectivity at $\lambda = 4$ Å	Mosaic
Silicon	111	3.135 Å	80%	90%	bent perfect
Graphite	002	3.355 Å	70%	80%	0.5°
Germanium	111	3.266 Å	60%	80%	0.2°
Copper	111	2.087 Å	60%	80%	0.3°
Heusler	111	3.437 Å	25%	35%	0.3°

the range, when the Bragg angle approaches 90°, the $\cot\theta$ term becomes small and very high resolution measurements are possible.

Monochromators usually comprise several crystals assembled together to form a continuous surface, which can focus on to the sample. Large focusing monochromators can give very large flux gains at the sample by accepting large amounts of solid angle from the source. Figure 3.4 shows two large focusing monochromators built for the IN8 triple-axis spectrometer at the ILL [4].

Vertical focusing contributes little to instrumental resolution as the vertical tilt does not affect the Bragg angle to a first approximation. Horizontal focusing of a parallel beam invariably entails varying the Bragg angle across the surface of the monochromator and thus tends to degrade resolution to some extent. The resolution degradation can be strongly reduced by arranging the instrument optics in order to obtain focusing in Q or energy, as illustrated in Fig. 3.5.

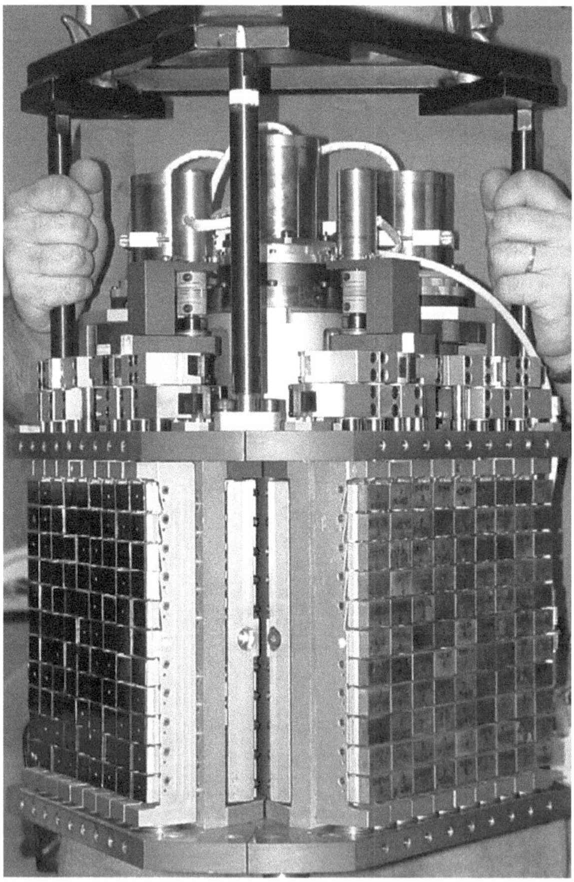

Fig. 3.4 Graphite (*left*) and copper crystals on the IN8 monochromator. Both faces consist of 99 mosaic crystals and have variable horizontal and vertical curvature

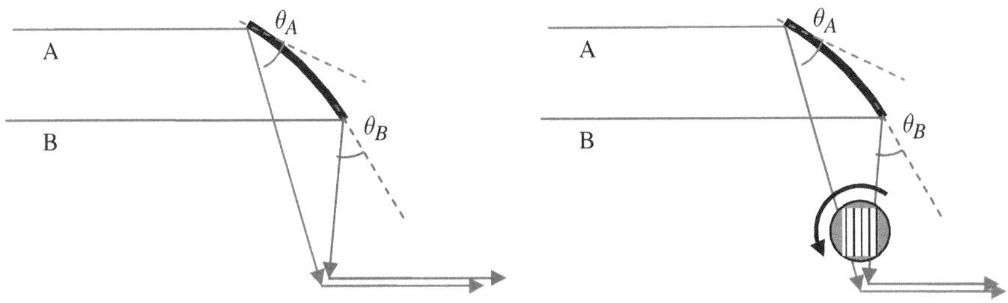

Fig. 3.5 Q-focusing and time-focusing. The *left-hand frame* shows Q-focusing as employed on powder diffractometers on continuous sources. *The right-hand frame* shows time-focusing as employed on time-of-flight spectrometers at continuous sources

In both cases, the monochromator focuses on the sample in space and the wavelength selected by the monochromator varies with position on the monochromator. As a result, the incident angle on the sample is strongly correlated with the wavelength. In the case of Q-focusing, the angular difference between the short- and long-wavelength neutrons coming from the monochromator is cancelled out by the Bragg scattering at the sample, and the result is a scattered beam which is as well collimated as the beam incident on the monochromator. Q-focusing works only when the direction of scattering at the sample is the opposite of that at the monochromator, and is most efficient when the d-spacing measured at the sample is similar to that of the monochromator.

In the case of time-focusing, a Fermi chopper (a rotating collimator) pulses the beam before the sample. It spins in the direction which lets through the slower neutrons (B) before the faster ones (A), so that if they transfer the same amount of energy at the sample, they arrive at the detector at the same time.

When Q- or time-focusing are properly optimised they can provide the same energy resolution as what would be obtained with a flat monochromator, but with a large gain in flux at the sample.

When using mosaic crystals, the reflected beam is somewhat smeared by the misorientation of the crystallites of the monochromator crystals. The left-hand panel of Fig. 3.4 illustrates Q-focusing starting from a parallel beam, which can be considered as a point-source at an infinite distance. In general, a focusing monochromator can be optimised for a point-source at any distance by adjusting the curvature. The finite size of the real source then needs to be optimised to match the mosaic spread of the crystals and the sample size. Neutrons originating outside this optimum source size will not hit the sample and therefore only contribute to background. Hence the concept of the 'virtual source' in which a diaphragm is used to reduce the beam size at the source, giving significant gains in signal-to-background ratios.

3.5 Refraction Optics

Neutron beams can be deviated by a change of the refractive index of the medium. This is associated with a change in the potential energy either from a magnetic field (the Zeeman energy) or from the refractive index n of the material:

$$n = 1 - Nb(\lambda^2/2\pi) \quad . \tag{3.7}$$

Since the refractive index is smaller than that of vacuum, a focusing lens must have a concave shape, unlike for the case of light.

A magnetic field gradient can also deviate the neutron trajectory:

$$\frac{\partial^2 r}{\partial t^2} = \pm \frac{\mu}{m} \nabla \left| \vec{\mathbf{B}} \right| \quad , \tag{3.8}$$

where μ is the magnetic moment of the neutron and m is its mass. The plus/minus sign refers to the two spin states of the neutron: antiparallel or parallel to the magnetic field, respectively.

Unfortunately for neutron optics purposes, the amount of focusing which can be achieved by refraction and magnetic-field gradients is usually quite small. Currently, used lens devices are restricted to small beam sizes and tend to suffer from long focal lengths (of the order of tens of metres) and chromatic aberrations, though work is underway to address these problems both for refractive [5] and for magnetic [6] lenses.

3.6 Polarisation Optics

Considerable technical advances have been made in polarisation optics over the last 10 years or so. For this reason, I go into a little more detail here than in the other sections. The reader is referred to [7] for a good overview of the technique.

Each individual neutron in a neutron beam has a spin of ½. When placed in a magnetic field, this defines a quantisation axis with respect to which there are only two possible orientations: parallel and antiparallel or ‘up’ and ‘down’. The beam polarisation can be expressed as a scalar P which is usually defined to vary between 0 and 1:

$$P = \frac{N_\uparrow - N_\downarrow}{N_\uparrow + N_\downarrow} \quad , \tag{3.9}$$

where there are $N_\uparrow$ ($N_\downarrow$) neutrons with spin up (down). The beam polarisation is often expressed in terms of the flipping ratio F given by

$$F = \frac{N_\uparrow}{N_\downarrow} = \frac{1+P}{1-P} \Rightarrow P = \frac{F-1}{F+1} \quad . \tag{3.10}$$

Bragg reflection from a single crystal can provide a well-polarised beam when the nuclear and magnetic structure factors are equal in magnitude, so that the combined structure factor is approximately zero for one of the two spin states. Several crystals have reflections which satisfy this criterion. The most commonly used are the 200 reflection of $Co_{0.92}Fe_{0.08}$ ($d = 1.76$ Å) and the 111 reflection of the Heusler alloy Cu_2MnAl ($d = 3.43$ Å). Of these, the Heusler alloy has the highest reflectivity and is available in good quality. Heusler crystals can be assembled into large focusing arrays, similar to nonpolarising crystals, provided they are designed so that a sufficiently large magnetic field (about 1,000 Oe [8]) can be applied to properly magnetise the crystals. A good-quality magnetised Heusler crystal will have a reflectivity for an unpolarised beam at $\lambda = 2$ Å of about 25%. Seeing as the reflection is polarising, the maximum reflectivity which can be expected is 50%, corresponding to reflection of all the neutrons of one spin state and none of the 'wrong' spin state.

If you do not wish to polarise and monochromate at the same time, two types of wide-bandwidth polarisers are currently in common use: polarising supermirrors and polarised ^3He gas.

Polarising supermirrors consist of alternating layers of magnetic and nonmagnetic materials. The scattering-length density of a ferromagnetic material depends on the spin state of the neutron and can be written as $N(b \pm p)$ for the two neutron spin states: parallel and antiparallel to the magnetisation of the layers. N is the atomic number density and b and p are the nuclear and magnetic scattering lengths, respectively. The combination of the two materials is chosen so that (1) they have a large contrast in scattering-length density for the 'up' spin state and (2) there is no contrast for the 'down' spin state. The result is a supermirror which reflects only one spin state and transmits the other. At very small angles, however, the 'down' spin state will also be reflected, as it encounters the interface with the substrate material underneath the supermirror. The substrate is usually made of glass or Si, which gives a critical angle for total reflection of about $m = 0.5$ for both spin states. The reflection at the substrate interface can be suppressed by adding an absorbing layer between the substrate and the supermirror. Presently, the largest critical angle which can be achieved for a polarising supermirror is about $m = 4$.

The supermirror must be kept magnetised in order to polarise effectively. This is usually done by applying a magnetisation field of a few hundred Gauss using permanent magnets. Under particular deposition conditions, a large magnetisation anisotropy can be induced in the magnetic layers, resulting in a remanent magnetisation after the external field is removed.

Supermirror polarisers can be used either in transmission or reflection geometry. In reflection geometry, the undesired spin state is transmitted through the supermirror and then absorbed in the substrate or elsewhere. In transmission geometry, the supermirror must be deposited on a nonabsorbing substrate. The undesired spin state is reflected out of the beam direction and then absorbed, while the other spin state is transmitted through both the supermirror and the substrate.

Supermirrors are a very efficient means of polarising cold-neutron beams with low divergence. For the case of small-angle scattering and reflectometry instruments, where the beam divergence is typically of the order of a few tenths of a degree and wavelengths tend to be long, supermirror polarisers can provide a transmission of more than 40% with a polarisation in excess of 95%. In these cases, a single supermirror deposited on a flat substrate and inserted in the beam at a grazing angle can do the job.

For a wide beam or when the divergence is large, the polarising supermirrors need to be assembled into packages. The most common form is the supermirror bender. These devices usually consist of thin glass plates with air gaps in between. The supermirrors are deposited on the glass plates and the neutrons are transmitted through the air channels. The assembly is curved, so as to avoid direct line of sight [Eq. (3.3)]. For a channel width of 1 mm and a length of 300 mm, we can see that a radius of curvature of 11 m is required. The transmission of the two spin states can be evaluated by considering each channel as a small curved guide. Keeping these dimensions with a critical angle of $m = 3$ for the 'up' spin state and $m = 0.5$ for the 'down' spin state, we find from Eq. (3.4) that the characteristic wavelengths for the two spin states are 2.6 Å and 15.5 Å. This is roughly the wavelength range over which such a polarising bender will work efficiently: transmitting one spin state but not the other. Supermirror packages can also be assembled from thin supermirror-coated Si wafers without air gaps, in which the neutrons are transmitted inside the Si. Supermirror benders typically have a transmission of 20–30%, depending on wavelength and divergence.

More recently, nuclear-spin-polarised ^{3}He gas has emerged as a viable alternative to supermirror-based devices. Helium-3 has a very large absorption cross section and only absorbs neutrons in the opposite spin state to the nucleus. The transmitted beam thus emerges polarised but has otherwise not been reflected, diffracted, or otherwise scattered. For this reason, ^{3}He-based polarisers and analysers are often referred to as neutron spin filters (NSFs). There are two mechanisms by which the ^{3}He nuclear polarisation can be created in a gas in a magnetic field.

In spin-exchange optical pumping (SEOP) [9], an alkali metal vapour, usually Rb, is mixed in with the ^{3}He gas. One of the ground-state electrons of the alkali metal is optically pumped with circularly polarised laser light of the appropriate wavelength, creating an electronic polarisation. During atomic collisions the polarisation of the alkali atom is transferred to a ^{3}He nucleus via the hyperfine interaction.

In metastability-exchange optical pumping [10], an RF discharge is created in order to populate a particular excited electronic state. This state is then pumped with circularly polarised laser light to preferentially populate one of the Zeeman-split energy levels, which decays into a metastable state upon emission of a photon. The electronic polarisation of the metastable state is then transferred to a ^{3}He nucleus by collisional mixing. In order to satisfy the rate balance between the depopulation of the metastable states and the atomic collisions, this mechanism is most efficient at low pressures, of the order of 1 mbar. The gas thus needs to be compressed by mechanical means in order to reach the density required to have sufficient column density to polarise a neutron beam.

Both optical pumping methods can result in gas polarisations P_{He} of up to 80% at pressures of the order of a few bars. The transmission of the two spin states is given by

$$T_{\uparrow(\downarrow)}(\lambda) = \exp(-O(\lambda)[1 \mp P_{He}]) \quad , \tag{3.11}$$

where O is the dimensionless opacity of the gas, given by

$$O = Nl\sigma_{\mathrm{a}} = 7.28 \times 10^{-2} O'(\text{bar} \times \text{cm} \times \text{Å}) \text{ at } T = 295\text{K} \quad . \tag{3.12}$$

N is the ^{3}He number density = 2.455×10^{19} cm^{-3} $\times$ pressure [bar] at $T = 295$ K, l (cm) is the length of the ^{3}He gas along the beam direction, σ_a = $2{,}966 \times 10^{-24}$ cm^2 $\times$ λ [Å] is the absorption cross section for the unwanted neutron spin state, and O' is the opacity, given by the product of the cell pressure (bar), the cell length (cm), and the neutron wavelength (Å).

The transmission of an unpolarised beam is given by

$$T_n(\lambda) = \frac{1}{2}(T_{\uparrow} + T_{\downarrow}) = \exp(-\mathrm{O}(\lambda)) \cosh(\mathrm{O}(\lambda) P_{He}) \quad , \tag{3.13}$$

and the resulting beam polarisation can be expressed as

$$P_n(\lambda) = \frac{T_{\uparrow} - T_{\downarrow}}{T_{\uparrow} + T_{\downarrow}} = \tanh(\mathrm{O}(\lambda) P_{He}) \quad . \tag{3.14}$$

The performance of a polariser or analyser is often given in terms of a figure of merit which combines the polarisation and the transmission. The most commonly used figure of merit is P^2T and is shown in Fig. 3.6 together with the transmission and the beam polarisation as a function of opacity, calculated for a ^{3}He polarisation of 75%.

The figure of merit is seen to exhibit a maximum at an opacity of about 27 bar.cm.Å. Depending on the wavelength range of interest the length or pressure of the ^{3}He polariser should be adjusted in order to always work in the vicinity of this performance maximum. The polarisation performance is seen to be very similar to that of Heusler crystals and polarising benders.

The nuclear polarisation created by the optical pumping is very far from thermal equilibrium. If the gas is polarised using SEOP, it is possible to continuously polarise the gas on the neutron beam. However, for most current applications, the gas is polarised off-line and gradually depolarises while on the

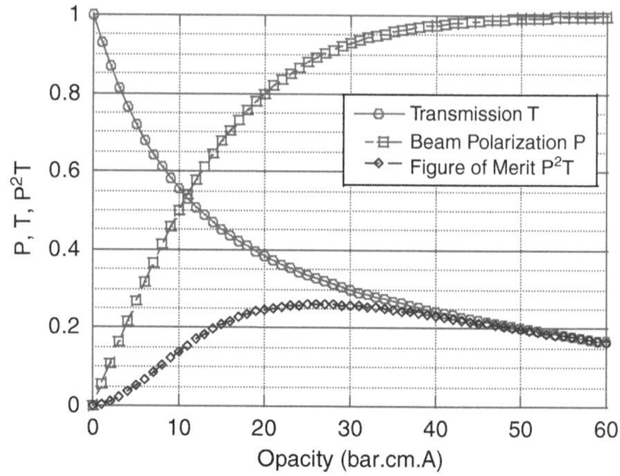

Fig. 3.6 Performance of a ^{3}He NSF cell as a function of opacity for a ^{3}He polarisation of 75%. Neutron beam polarisation (P) and transmission (T) of an unpolarised beam are shown, as is the standard figure of merit P^2T, which display a maximum at the optimum opacity

instrument with a characteristic time constant T_1. The polarisation decay is exponential with a $1/e$ time constant which is a combination of three terms:

$$\frac{1}{T_1} = \frac{1}{T_d} + \frac{1}{T_m} + \frac{1}{T_w} \quad . \tag{3.15}$$

The first term is caused by the dipole–dipole interactions between the ^{3}He nuclei. It scales inversely with pressure and is approximately 800 h/bar. The second term is caused by magnetic field gradients, and the third term is due to collisions with the cell walls. Current best efforts allow T_1 relaxation times on the instrument of the order of 100–200 h.

To minimise the relaxation due to interactions with the cell walls, containers for polarised ^{3}He gas are usually made of glass or Si, coated with Cs or Rb metal, as displayed in Fig. 3.7. Different types of magnetic field cavities have

Fig. 3.7 Examples of glass ^{3}He polariser cells with silicon entrance and exit windows. Photograph courtesy of David Jullien (ILL)

been developed to maximise the relaxation time due to field gradients. Their geometries are optimised for the particular instrument and are usually based on solenoids, Helmholtz coils, or magnetised mu-metal. Screening from external fields can be provided by steel, mu-metal, or superconducting Meissner screens.

References

1. ILL Neutron Data Booklet, Albert-José Dianoux and Gerry Lander (ed.), *OCP Science* (July 2003).
2. V.F. Sears, Neutron Optics: *An Introduction to the Theory of Neutron Optical Phenomena and Their Applications*, Oxford University Press, (1989).
3. V.F. Sears, *Acta Cryst. A* **53**, 35 (1997).
4. A. Hiess et al., *Physica B* **385–386**, 1077 (2006).
5. S.-M. Choi et al., *J. Appl. Cryst.* **33**, 793 (2000).
6. H.M. Shimizu, *Physica B* **385–386**, 989 (2006).
7. W.G. Williams, *Polarized Neutrons*, Oxford University Press, 1988.
8. A. Freund et al., *Physica B* **120**, 86 (1983).
9. T.R. Gentile et al., *J. Res. N.I.S.T.* **110**, 299 (2005).
10. A.K. Petoukhov et al., *Physica B* **385–386**, 1146 (2006).

Chapter 4
Neutron Detectors for Imaging

L. Crow

Abstract This chapter examines neutron detection methods for imaging. The first section reviews neutron capture converters, which form the basis for thermal neutron detection, and the following sections examine various detector systems with an emphasis on imaging applications and the choices available for imaging at time-of-flight facilities.

Keywords Neutron detector · Scintillation neutron detector · Neutron converter · Gas neutron detector · Solid state neutron detector · Imaging neutron detector

4.1 Overview of Neutron Converters

As neutral particles, neutrons reveal themselves through interactions with atomic nuclei. Neutrons with sufficient kinetic energy can produce signals when nuclei, recoiling from elastic collisions, ionize the surrounding material. Proton recoil in particular is a standard detection mechanism for fast neutrons [1, 2] and enabled initial discovery of the neutron [3]. Neutrons with kinetic energy less than a few electron volts cannot, however, produce ionization directly; detection requires neutron capture reactions, which release large amounts of energy in the form of charged particles and gamma rays. These are the epithermal, thermal, and cold neutrons of greatest interest in neutron imaging applications. In practice, a neutron converter, an isotope chosen for its neutron capture reaction, forms the basis for a neutron detector. Desirable converter characteristics include large neutron absorption cross section, minimal gamma-ray cross section, and energetic charged particle products. Limited or absent gamma-ray production is also desirable. Table 4.1 shows some converters currently used in neutron scattering or imaging instrumentation.

L. Crow (✉)
Neutron Facilities Development Division, Oak Ridge National Laboratory,
Oak Ridge, TN 37831, USA
e-mail: crownljr@ornl.gov

I.S. Anderson et al. (eds.), *Neutron Imaging and Applications*,
Neutron Scattering Applications and Techniques, DOI 10.1007/978-0-387-78693-3_4,

Table 4.1 Thermal neutron cross sections and charged particle production for some materials used as neutron converters [4–6]

Isotope	Reaction	Neutron absorption cross section (barns) (at 2,200 m/s)	Charged particles and energies (keV)	Gamma-ray production?
^{3}He	^{3}He(n,p)^{3}H	5333	p: 573, ^{3}H: 191	No
^{6}Li	^{6}Li(n,α)^{3}H	940	^{3}H: 2727, α: 2055	No
^{10}B	^{10}B(n, α)^{7}Li	3835	α: 1472, ^{7}Li: 480	Yes
^{nat}Gd	^{nat}Gd(n,γ)	49700	Conversion electron 29–191	Yes
^{157}Gd	^{157}Gd(n,γ) ^{158}Gd	259000	Conversion electron 29–182	Yes
^{235}U	^{235}U(n,f)	681	Various fission products	Yes

Both the nuclear properties and the material characteristics affect the choice of converter for a particular application. Helium-3 and ^{6}Li have relatively large cross sections, but the most important feature is that these capture reactions release all of their energy in charged particles [6]. This makes these converters the best choices for low-noise detectors which need to operate at high rates in accessible areas. Helium-3 gas proportional counters have modest position resolution, but very low gamma-ray sensitivity, and are the most widely used detectors for neutron scattering. Lithium-6 is a very efficient scintillation converter and appears in the most widely used thermal neutron scintillators, ^{6}Li glass and ^{6}LiF:ZnS [7]. Boron-10 has a larger cross section than ^{6}Li, and unlike ^{3}He can be in solid materials with very high atom density. The ^{10}B capture reaction has the disadvantage of producing a gamma-ray and heavier, shorter-ranged charged particle products (α and ^{7}Li). The compound $^{10}BF_3$ has long been used in proportional counters, and ^{10}B is used in surface conversion gas detectors. Boron-10 is also used in semiconductor detectors, microchannel plate detectors, and a few scintillators.

Gadolinium has a number of disadvantages as a converter but is of interest in imaging because of the enormous resonance at thermal neutron energies of ^{157}Gd (^{155}Gd also has a very large cross section). The ^{157}Gd capture results in a spectrum of high energy gamma rays, which are difficult to count or shield, along with conversion electrons. Gadolinium-based detectors rely on the internal conversion electrons to transfer the energy to a phosphor, gas, or lattice. The cross section does not have the inverse velocity dependence of most converters. This feature makes comparison of energy-dependent neutron response difficult, but with white-beam imaging, the preferential response to thermal neutrons helps suppress fast and epithermal neutron background.

Neutron capture reactions where the primary products are gamma rays may be used for applications such as prompt or decay gamma activation analysis. The characteristic gamma rays identify the isotopes which have absorbed the neutrons. This method may be used for chemical analysis or for neutron integrated flux measurements. Isotopes with higher energy absorption resonances have applications in energy-specific neutron detection [8]. A significant effort for chemical spectroscopy and imaging is discussed in Chapter 13 of this volume. Isotopes which capture neutrons and produce short-lived radioactive products maybe used in indirect imaging or autoradiography, using the decay beta particles.

4.2 Data Acquisition for Continuous and Time-Dependent Applications

Some neutron detection systems operate in *current* or *integrating* mode, where the individual neutron events are not distinguished. This can work well when the time dependence of the signal is not an important parameter within the measurement period. With intense sources and fast gating, integrating detectors can be used at relatively fast frame rates, but cannot provide the time resolution (within a few microseconds) required for time-of-flight methods.

Counting rather than integration allows use of statistical methods at low data rates, especially when signal and background are low. The vast majority of neutron scattering measurements use counting detection schemes. In addition, counting detectors facilitate time stamping or binning for time-of-flight measurements using event mode readout. Neutron time-of-flight with a pulsed source (or a chopped continuous beam) offers a method of measuring neutron interactions as a function of neutron kinetic energy. In addition to applications in diffraction and inelastic neutron scattering, time-of-flight offers opportunities in imaging for energy-dependent or element-dependent contrast variation.

4.3 Photographic Film Detectors

Photographic film provides an analogue readout, always in signal integrating mode. Film has the advantages of compact detector assembly and immediate image generation. Most film methods can also provide position resolution on the order of tens of microns, which was difficult until recently with electronic readout. The frame rate is governed by the relatively long exposure times, due to the low detection efficiency, and the requirement to physically remove and replace the film. Film continues in wide use at continuous source installations [9] but is not adaptable to time-of-flight techniques.

A direct capture camera uses prompt particle emission from neutron capture to illuminate the film (Fig. 4.1). Gadolinium foil/X-ray film detectors have served as the standard tool during much of the development of neutron radiography [9], and continue in use today [10]. The resolution of such detectors,

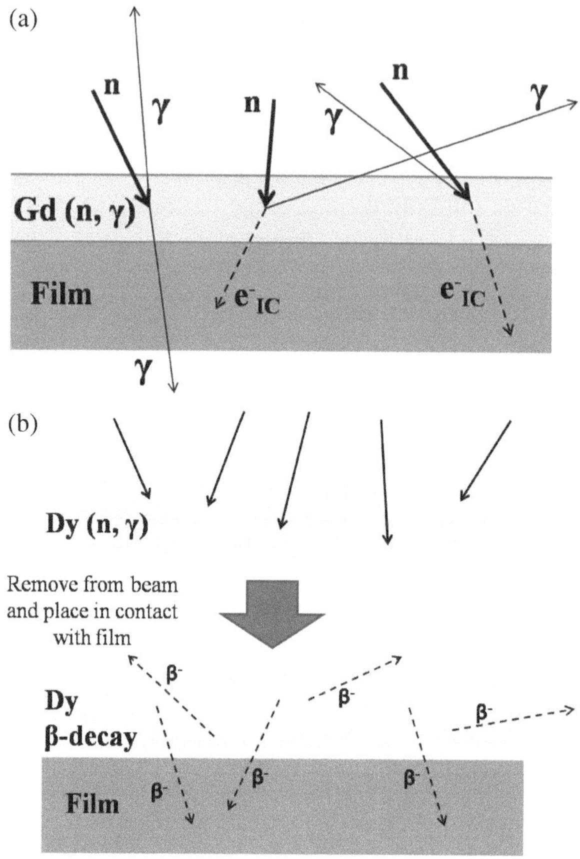

Fig. 4.1 The figure illustrates the dominant processes in direct and indirect screen capture for two common screen choices. (**a**) A direct Gadolinium screen and X-ray film neutron imaging process is shown schematically. The neutrons are captured by the Gd nuclei [^{157}Gd (n,γ) ^{158}Gd]. The high energy prompt gamma rays escape, but some of the internal conversion electrons deposit energy in the film and contribute to the photographic image. (**b**) In an indirect process, a dysprosium screen captures neutrons during irradiation [^{164}Dy (n, γ) ^{165}Dy]. The screen is removed from the neutron beam and placed in contact with film. An image is produced by β^- particles from dysprosium decay [^{165}Dy (β^-, ^{165}Ho), $t_{1/2} = 2.33$ h]

usually a few tens of microns, is determined by the grain sizes in the converter screen and the film, and the efficiency is dependent on the gadolinium conversion electron capture in the film. Newer methods are more efficient, but the position resolution of these screens remains competitive [11]. This detector system requires a mechanism to open and close the incident beam to define the exposure time and allow removal and replacement of the film and converter screen and a facility to develop the film. This works well at continuous sources with relatively long exposure times but is difficult to adapt for fast sample changes.

The direct screen and film method requires suppression of gamma-ray, X-ray, and charged particle radiation from both source and sample. If the sample produces significant prompt emissions or is already radioactive, a screen consisting of a short-lived activation material, typically a β-emitter such as In, Dy, or Au [9, 12], can serve as an indirect imaging device. The screen activates when the sample and screen are exposed to the neutron beam. The screen is then moved from the beam and placed in contact with the film. The decay from neutron activation produces the image, and incident gamma effects are absent. The indirect method needs long exposure times and image recording time, but the gamma insensitivity allows high-resolution imaging of difficult samples such as spent nuclear fuel elements (Chapter 5).

A specialized technique known as neutron autoradiography uses activation products within the sample to produce images. This technique works in principle with any object with variations in composition. Direct autoradiography couples the sample, usually containing boron, to an alpha sensitive film and records the prompt alpha emissions [13, 14]. It has been used to monitor boron compound uptake for boron neutron capture therapy [15, 16]. Indirect autoradiography works with activation of components within a sample. Autoradiography can provide details of paint pigment distribution in historical paintings [17]. The activation gamma-ray energies reveal pigment composition, which indicates the age and authenticity of the painting. The resultant images illustrate the artist's composition technique and may show earlier drafts of the painting or even earlier paintings on the same canvas.

The track-etch method uses charged particle products of neutron capture to record tracks directly in an emulsion film. The track density, viewed through a microscope, forms the image. For ^{6}Li or ^{10}B, the typical particle ranges are between 5 and 30 μm, which effectively sets the resolution limit. The most established track-etch film is boron-loaded nitrocellulose [18]. The track-etch method requires long exposure times.

Neutron scintillators may be used with film but are more commonly used with faster electronic light detection systems.

4.4 Scintillator and Storage Phosphor Detectors

Neutron Scintillators: A scintillator is a material that absorbs ionizing radiation and emits low-energy photons. A neutron scintillator may operate by recoil [19], but a thermal neutron scintillator requires a converter that emits charged particles, either nuclear products from the capture reaction or internal conversion electrons.

For a number of years, the most widely used thermal neutron scintillators have been ^{6}Li glass and ^{6}LiF/ZnS [20]. Cerium-activated ^{6}Li glass is an intrinsic scintillator (i.e., the converter atom and phosphor are part of the same material). Lithium-6 glass, used in some diffraction detectors [21–23], is not often used in imaging because a thickness of a millimeter or more is required for good efficiency, gamma-ray sensitivity is higher than desired, and the 8,000 photons/neutron light output is relatively small. The binary scintillator ^{6}LiF/ZnS is created by mixing the converter particles with phosphor particles, where both materials have sizes of a few microns (consistent with charged particle ranges). The original formulation of ^{6}LiF/ZnS [20] provides about 160,000 photons/neutron light output. Continuing efforts to increase efficiency for scattering detectors [24] by increasing thickness and lithium density have not proven advantageous to imaging, but recent work for imaging detectors [11] improves position resolution by reducing thickness to about 50 μm at the cost of lower neutron capture efficiency.

Several additional ^{6}Li and ^{10}B scintillators have been proposed and tested [25–29]. Gadolinium-based scintillators, for example Gd_2O_2S (GOS), Gd_2SiO_5 (GSO), and a microcolumnar gadolinium-cesium iodide hybrid [30–33], have limited light yield because only the conversion electron energy is available, but the very large cross section means that high-capture efficiency can be achieved with only a few microns of material. For this reason, gadolinium scintillators have potential applications in high-resolution (tens of microns) neutron imaging.

Photographic Scintillation Detectors: The optical part of a scintillation detector collects the scintillation light, records it, and uses the light collection pattern to locate the neutron. Photographic film offers a very small package in the beam location, but has far lower light collection efficiency and requires a film change for readout. Faster readout can be achieved using a television camera viewing the scintillator; sensitivity can be increased using an image intensifier [34].

The advent of high-performance charge-coupled device (CCD) cameras has transformed this technology into a competitive high resolution imaging system [11] that can collect and read out images far faster than film. A typical system consists of a scintillator screen, a 45° front surface mirror, a focusing lens, and a CCD camera [35–39]. Lehmann et al. have performed an optimization of these components for the ICON facility at SINQ [11] (Fig. 4.2). Thick scintillator screens optimized for efficiency degrade the spatial resolution; standard 250 μm or more thick ^{6}LiF/ZnS can be replaced with 50 μm or less ^{6}LiZnS layers, or with 10 μm GOS. A front surface mirror is used to eliminate multiple reflection effects. This high resolution system uses a cooled CCD camera with a custom lens system and achieves a resolution of 50 μm.

Most photographic readouts, whether film or CCD, effectively integrate over a large number of neutron events. Such readouts give good imaging results in steady-state source instruments but are less well suited to time-of-flight measurements. Fast gated cameras are a possible method for using the time structure from pulsed sources and for imaging dynamic processes [40]. The use of camera systems with pulsed sources will be simplified as more fast cameras become available.

A mirror keeps the CCD camera outside the neutron path to protect it from radiation damage, but the increased distance decreases the light collection efficiency. A fiber optic taper [41, 33] greatly increases the light collection efficiency. The glass in the taper also provides partial shielding for the camera sensor; however, it is not yet known whether a CCD can operate for long

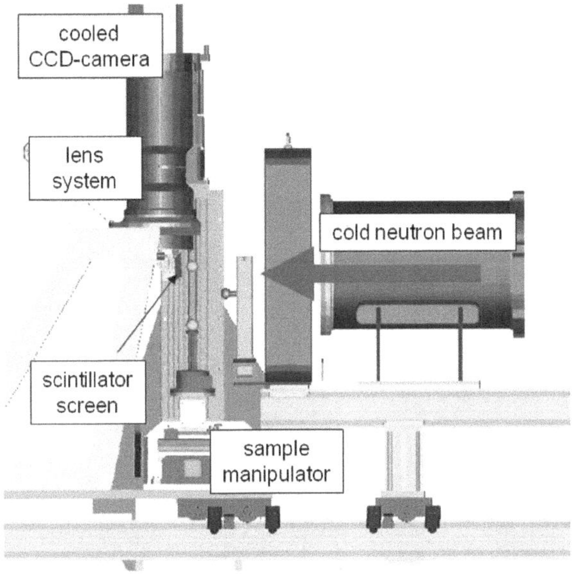

Fig. 4.2 The figure shows a scale drawing and a perspective view of the microtomography scintillator and CCD detector system at the Paul Scherrer Institut, Villigen, Switzerland. The scintillator screen, a 45° mirror, the lens system, and the cooled CCD camera form the neutron detector system. (This image was provided by Eberhard Lehmann of PSI by private communication)

periods in this geometry. Amorphous silicon flat panel photodiode arrays provide a fast readout system which could address dynamic measurement and time-of-flight requirements while providing a large solid angle for light collection. Preliminary tests at operating neutron beam lines [42–44] have successfully demonstrated the concept with good readout performance, and the test results for tomography are promising. Further development requires improved dynamic range.

Counting Scintillation Detectors: For neutron counting with scintillators, photomultipliers (PMTs) provide the necessary fast photon counting capability. Avalanche photodiodes are a possible alternative with higher quantum efficiency, but at present they have higher dark counts at higher cost. Although a scintillator with a PMT can be operated as a single pixel detector, most scattering applications use multiple PMT elements to build position-sensitive detectors. Anger cameras with neutron scintillators [21–23] can now achieve area resolution of 1.3 mm or smaller while counting (and time stamping) neutrons at tens of kilohertz, and possibly faster (Fig. 4.3). Fiber light collection using end-on fibers [45, 46] or wavelength-shifting fibers [24] gives light collection over large areas with a relatively small investment in light detectors, appropriate for diffraction instruments which require pixel sizes of a few millimeters. Fiber collection can, like the

Fig. 4.3 (**a**) The figure shows an optical detector assembly of an Anger camera module for the SNAP high pressure diffractometer at the Spallation Neutron Source (SNS) at Oak Ridge National Laboratory. Nine 8×8 channel position-sensitive photomultiplier tubes are assembled to form a single Anger module. (**b**) A SNAP module is shown with the bezel, which contains a 15 cm square ^{6}Li glass scintillator. This detector has a two-dimensional position resolution of 1.3 mm full width at half maximum (FWHM). (**c**) A crossed wavelength-shifting fiber assembly for a detector module to be mounted on the Vulcan engineering diffractometer at SNS is shown. The wavelength-shifting fibers collect the blue scintillation light from a ^{6}LiF/ZnS(Ag) scintillator screen and guide the shifted green light to a coded photomultiplier array. (**d**) A completed Vulcan module operates for initial calibration. This 80 cm × 40 cm detector module has 5 mm wide by 37.5 mm tall pixels, with resolution matched to the diffractometer

Anger camera, achieve resolution approaching 1 mm [47], but it is not likely that these technologies can achieve the submillimeter resolution needed to compete in radiography or tomography.

Storage phosphors and image plate readers: Storage phosphors are functionally similar to scintillators, except that instead of a prompt light emission, energy is stored in the lattice. The phosphor is read by photostimulation (typically by scanning with a laser), inducing light emissions. Neutron image plates using a gadolinium converter with a storage phosphor (Fig. 4.4), developed in the 1990s

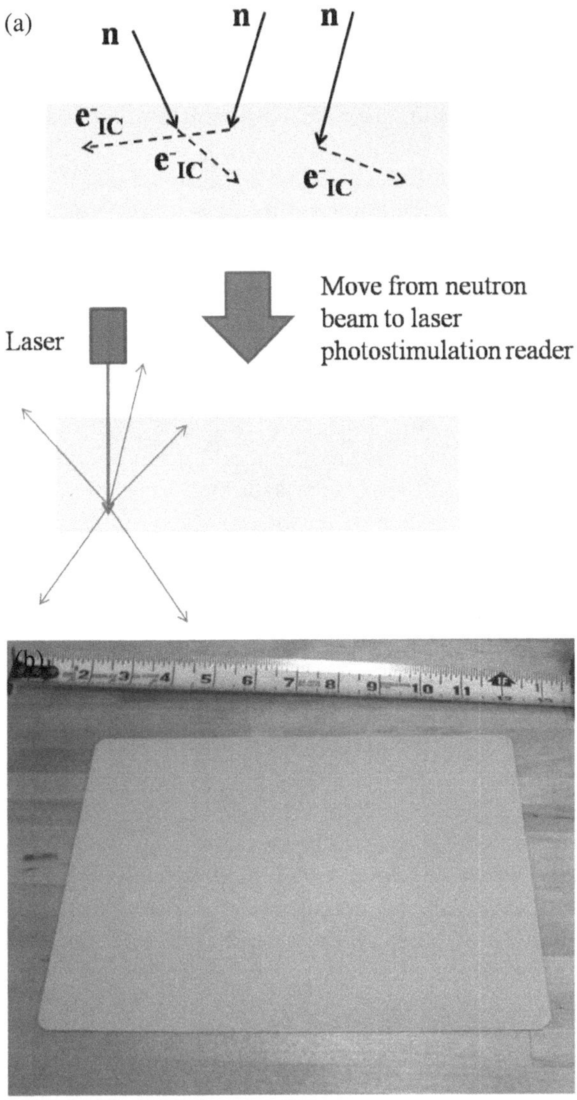

Fig. 4.4. (**a**) A ^{157}Gd-based image plate accumulates energy from internal conversion electrons produced in neutron capture in its storage phosphor. The image plate is transferred to a reading system. A laser scans the image plate and photostimulates the release of blue photons from the storage phosphor in the image plate. (**b**) A 25 cm square neutron image plate is shown

[48, 49], have become commercially available [50] in the last decade and are now in regular use for diffraction [51] and radiography [52] at steady-state sources. Image plates are much more efficient than any of the photographic film

techniques. Spatial resolution is a function of image plate thickness but is on the order of 25–100 μm [11]. In addition, the automated readout is much faster than conventional film processing (and in electronic form), and the image plates themselves are reusable. Still, image plates are integrating detectors and cannot be gated for use at time-of-flight sources. While some facilities use image plate readers developed for X-ray image plates [52], development of readers optimized for neutron applications is continuing [53].

4.5 Gas Detectors

In a gas detector, ions produced within the gas accelerate due to a charge bias applied within the gas volume. If the acceleration is large enough, the signal is amplified as the ions collide with gas atoms and produce additional ionization. In this way, an ionization event, such as a neutron capture, will produce an electron "charge cloud," which drifts to the biased anode. Depending on the bias and the gas, the charge may be equivalent to the original energy deposition (*ionization* mode), amplified proportional to the original energy (*proportional* mode), or a nonproportional avalanche (*Geiger-Müller* mode). Neutron detectors with gas converters such as $^{10}BF_3$ [54] and ^3He [55, 56] usually operate in the proportional range to take advantage of the excellent discrimination allowed by the energetic charged particles.

Helium-3 proportional counters are the most widely used detector systems for neutron scattering, where resolution can be matched to the requirements of the instruments. The development of position-sensitive proportional counters has led to imaging detectors for neutron-scattering applications such as small-angle neutron scattering (SANS), neutron reflectometry, and diffraction. Readout systems for area detectors include delay line [57], charge division [58], and the now more widely used multiwire centroid coincidence method [59–61]. Linear position-sensitive detectors using charge division [62] may be arranged to form area detectors for SANS [63, 64] and a wide variety of neutron-scattering applications [65, 66] (Fig. 4.5). The counting readout systems are readily adaptable to time-of-flight work. Gas converter detectors cannot compete easily for submillimeter position resolution because of the thick gas volume and the long capture product ranges (typically on the order of a millimeter or longer). The best wire chamber detector resolution to date is about 400 μm [67]. Gas detector development continues with efforts to overcome the present count rate limitations of wire chambers, typically between 40,000 and 100,000 counts per second per chamber due to drift time. A new high-performance multiwire gas chamber detector known as MILAND[1] [68] is expected to achieve 1 mm resolution at a count rate of up

[1] MILAND (for Millimeter Resolution Large Area Neutron Detector) is a joint research activity of the European Union supported under the Fifth Framework Program (FP5).

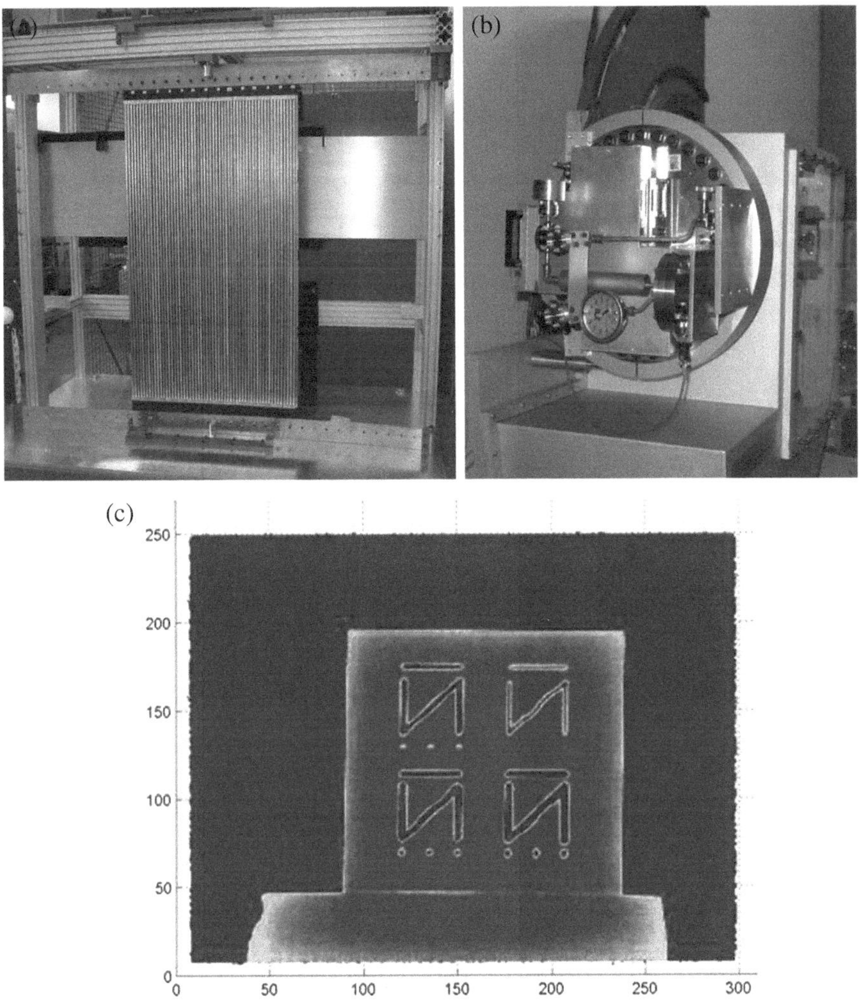

Fig. 4.5 (**a**) An array of 8 mm diameter × 1 m long linear position-sensitive ^{3}He proportional tubes forms an area detector for small angle neutron scattering, with resolution of about 8 mm. The modules shown here will be installed on EQSANS at SNS. (**b**) A 20 cm × 17 cm multiwire proportional counter manufactured by Brookhaven National Laboratory operates in the SNS liquid reflectometer. (**c**) This multiwire detector, with 6.5 bar of ^{3}He, has a resolution of 1.3 mm FWHM, as shown in this calibration mask image

to 1,000,000 counts per second over the entire detector area using readout with separate amplifiers on each wire.

Strategies for higher performance gas detectors use micropattern electrode structures to effectively reduce drift distances and times and in the process divide the readout of a detector chamber into smaller segments.

These detectors may use gas converters or solid converters within the counting gas volume. In microstrip gas counters (MSGC) [69–72], the anode and cathode strips are tens of microns apart on a glass substrate, reducing the drift times to a few hundred nanoseconds. Gas scintillation proportional counters are variants of the MSGC that use the light from the charge amplification to localize the neutron events [73–75]. Gas electron multipliers [76–78] use biased stacks of perforated kapton foils to achieve high gas amplification in a very small distance. Another in this same subclass of gas detectors, the Micromegas (for MicroMEsh Gaseous Structure), has a converter layer and a very small amplification gap of about 100 μm [79–80] between two parallel glass or epoxy plates.

4.6 Solid-State Detectors

Semiconductor detectors: If the charged particles from a neutron capture enter a biased semiconductor device, the resultant ionized region will produce a pulse. If the pulse is large enough, the device becomes a neutron detector that can be coupled to fast readout counting electronics. Potential advantages of semiconductor detectors include high data rate, small size, and very low power requirements. With very short charged particle paths and close integration to digital readout systems, these detectors should be capable of very high data rates and high resolution. A neutron converter can be intrinsic to the semiconductor (part of the semiconductor structure) or extrinsic (adjacent to the semiconductor region). Intrinsic semiconductors proposed or tested include boron phosphide [81] and boron nitride [82], but no intrinsic semiconductor has been successfully developed as a neutron detector. McGregor [83] has reviewed semiconductor neutron detectors in detail and demonstrated the efficiency limits for flat layers of ^{6}Li and ^{10}B compounds on semiconductor surfaces. Successful detection requires that one of the neutron capture product charged particle tracks end within the semiconductor. The track lengths of 30 μm or less place severe restrictions on the usable converter thickness and restrict the detection efficiency to less than about 12% even for cold neutrons. Flat layers on charged particle detector application-specific integrated circuits (ASICs) may increase readout performance [84] but will not address the efficiency issue. Modifications to the converter geometry such as placing the converter in wells or channels in perforated semiconductor structures (Fig. 4.6) have produced higher efficiency [85].

A recent MgB_2 superconducting detector prototype makes use of meander traces of MgB_2 a few microns wide on an insulating substrate. With the MgB_2 at superconducting temperature, single neutron capture events can produce normal conducting regions for very short times [86] and produce signal pulses. The resulting detector can operate at high counting rates.

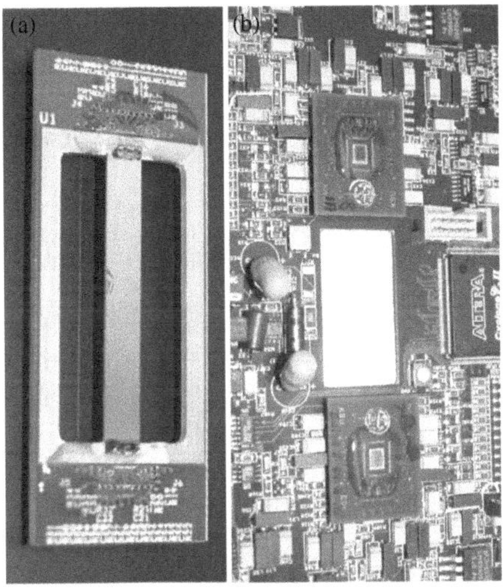

Fig. 4.6 (**a**) A silicon wafer detector element has been perforated with grooves at 100 μm spacing. Each groove has been filled with ^{6}LiF. The pixel size is 100 μm by 4 cm (the detector height). (**b**) The detector is mounted at the center of the circuit board; the application-specific integrated circuits to operate the detector are shown above and below the detector window

Resolution of tens of microns is achievable in principle with such a detector, but multiple layers would be needed to reach good efficiency for thermal neutrons.

Glass microchannel plates (MCPs), long used in optical image intensifiers, amplify electronic signals by accelerating electrons through evacuated microscopic channels, typically with diameters of 5–10 μm and spacing 20 μm or less. Microchannel plates become neutron detectors if the glass material contains a converter, typically ^{10}B, in sufficient concentrations [87]. The charged particle capture product ranges are a few microns in glass so that the resolution is potentially comparable to the channel spacing. Nova Scientific has carried out a development program to optimize the microscopic geometry and composition of MCP neutron detectors and is continuing development of readout electronics. Amorphous silicon readout [88] provides good position resolution but is an integrating system. The newer delay line readout [89] gives the detector system counting capability with 25 ns time resolution and possible rates up to 1 MHz, with spatial resolution better than 40 μm FWHM, the best yet achieved with a counting neutron detector (Fig. 4.7). Another development, also based on Nova MCPs, uses the Medipix multipurpose ASIC readout which can operate as a counting system or as a higher resolution centroiding system [90] that can approach 15 μm position resolution with gating that should be fast enough for time-of-flight applications.

Fig. 4.7 A 40 mm diameter microchannel plate detector built by NOVA Scientific, similar to the detector described in [91], is shown. The microchannel plate assembly is visible through the sapphire neutron window. The delay line readout enables counting operation at high rates with 40 μm resolution

4.7 Conclusion

The current state of neutron imaging detectors presents a range of options for continuous sources and some promising paths for achieving the timing capabilities needed for pulsed sources. The neutron storage phosphor image plates yield excellent resolution at modest operating costs. The CCD camera systems are fast and flexible and are currently the primary detection technique for most major imaging user facilities. The potential for improvement in scintillators or cameras puts pulsed source operation within reach, but gating and camera readout at high rates are big challenges. The microchannel plate detectors operate at very high count rates, are operable as counting detectors now, and may have the shortest path to operation as timing detectors.

References

1. J.A. Harvey and N.W. Hill, *Nucl. Instrum. Meth.* **162**, 507 (1979).
2. G.F. Knoll, *Radiation Detection and Measurement, 3rd ed.* Wiley, New York (2000).

3. J. Chadwick, *Proc. Roy. Soc.* **A136**, 703 (1932).
4. V.F. Sears, *Neutron News* **3**, 26 (1992).
5. C.M. Lederer and V.S. Shirley (eds.); E. Browne, R.E. Doebler, and J.M. Dairiki (principal authors); A. Shihab-Eldin, L.J. Jardine, J.K. Tuli, and A.B. Buyrn, *Table of Isotopes, 7th edition.* Wiley, New York (1978).
6. B. Gebauer, Comprehensive summary session 4, *Proceedings of the International workshop on Position-Sensitive Neutron Detectors—Status and Perspectives*, (PSND2001), http://www.hmi.de/bensc/psnd2001.
7. A.R. Spowart, *Brit. J. Nondestr. Test.* **11**, 2 (1969).
8. G. Gorini, E. Perelli-Cippo, M. Tardocchi, C. Andreani, A. D'Angelo, A. Pietropaolo, R. Senesi, S. Imberti, A. Bracco, E. Previtali, G. Pessina, N.J. Rhodes, E.M. Schooneveld. *Nucl. Instr. Meth.* **A 529**, *293* (2004).
9. H. Berger, *Neutron Radiography – Methods, Capabilities and Applications.* Elsevier, Amsterdam (1965).
10. R. Ilic and M. Najer, *Nucl. Instrum. and Meth.* **180**, 635 (1981); R. Pugliesi and M. L.G. Andrade, *Appl. Radiat. Isot.* **48**, 339 (1997).
11. E.H. Lehmann, G. Frei, G. Kühne, P. Boillat, *Nucl. Instrum. Meth.* **A 576**, 389 (2007).
12. M. Mullner and H. Jex, *Nucl. Instrum. Meth.* **103**, 229 (1972).
13. Z.B. Alfassi, and T.U. Probst, *Nucl. Instrum. Meth.* **A 428**, 502, (1999).
14. C.P. Bean, R.L. Fleischer, P.S. Swartz and H.R. Hart. *J. Appl. Phys.* **37,** 2218 (1966).
15. M. Takagaki, and Mishima, Yu., *Nucl. Tracks Radiat. Meas.* **17**, 531–535, (1990).
16. J.E. Woollard, T.E. Blue, J.F. Curran, T.F. Mengers, and R.F. Barth, *Nucl. Instr. Meth.* **A 299**, 600 (1990).
17. C. Laurenze-Landsberg, C. Schmidt, B. Schroder-Smeibidl, and L.A. Mertens, *Notiziario Neutroni e Luce di Sincrotone* **11**, 24 (2006).
18. A.A. Harms and M.S. Moniz, *Nucl. Instrum. Meth.* **122**, 567 (1974).
19. R. Batchelor, W.B. Gilboy, J.B. Parker and J.H. Towle, *Nucl. Instrum. Meth.* **13**, 70 (1961).
20. A.R. Spowart, *Nucl. Instrum. Meth.* **75**, 35 (1969).
21. M. Heiderich, R. Reinartz, R. Kurz and J. Schelten, *Nucl. Instrum. Meth.* **A 305**, 423 (1991).
22. M.G. Strauss, R. Brenner, H.P. Chou, A.J. Schultz and C.T. Roche, *Position-Sensitive Detection of Thermal Neutrons.* Academic Press, London, p. 175 (1983).
23. R. G. Cooper, J. D. Richards, C. Donahue, T. Visscher, *IEEE Nucl. Sci. Symp.* (2007).
24. M.L. Crow, J.P. Hodges, and R.G. Cooper, *Nucl. Instrum. Meth.* **A 529**, 287 (2004).
25. J.P. Chaminade, O. Viraphong, F. Guillen, C. Fouassier and B. Czirr, *IEEE Trans. Nucl. Sci.* **48**, 1158 (2001).
26. J.P. Hodges, M.L. Crow, and R.G. Cooper, patent pending (2007).
27. T. Kojima, M. Katagiri, N. Tsutsui, K. Imai, M. Matsubayashi, and K. Sakasai, *Nucl. Instrum. Meth.* **A 529**, 325 (2004).
28. M.J. Knitel, *New Inorganic Scintillators and Storage Phosphors for Detection of Thermal Neutrons*, Ph.D. Thesis, TU Delft, (1998).
29. C.W.E. van Eijk, A. Bessière and P. Dorenbos, *Nucl. Instrum. Meth.* **A 529**, 260 (2004).
30. K. Takagi, T. Fukazawa, *Appl. Phys. Lett.* **42**,43 (1983).
31. C.L. Melcher, J.S. Schweitzer, R.A. Manente, and C.A. Peterson, *IEEE Trans. Nucl. Sci.* **38**, 506 (1991).
32. J.S. Brenizer, H. Berger, C.T. Stebbings, and G.T. Gillies, *Rev. Sci. Instrum.* **68**, 3371 (1997).
33. I. Shestakova, V. Gaysinskiy, J. Antal, L. Bobek, and V.V. Nagarkar, *Nucl. Instrum. Meth.* **B 263**, 234 (2007).
34. Harold Berger, Wilfrid F. Niklas, and Adolph Schmidt, *J. Appl. Phys.* **36**, 2093 (1965).
35. T. Hibiki, K. Mishima, K. Yoneda, S. Fujine, A. Tsuruno, and M. Matsubayashi, *Nucl. Instrum. Meth.* **A**351, 423 (1994).

36. Helena Pleinert, Eberhard Lehmann, and Sonja Körner, *Nucl. Instrum. Meth.* **A399**, 382 (1997).
37. D.S. Hussey, D.L. Jacobson, M. Arif, P.R. Huffman, R.E. Williams, and J. C. Cook, *Nucl. Instrum. Meth.* **A542**, 9 (2005).
38. B. Schillinger, E. Calzada, F. Grunauer, and E. Steichele, *Appl. Radiat. Isot.* **61**, 253 (2004).
39. M. Dierick, J. Vlassenbroeck, B. Masschaele, V. Cnudde, L. Van Hoorebeke, and A. Hillenbach, *Nucl. Instrum. Meth.* **A 542**, 296 (2005).
40. B. Schillinger, H. Abele, J. Brunner, G. Frei, R. Gähler, A. Gildemeister, and A. Hillenbach, E. Lehmann, P. Vontobel, *Nucl. Instrum. Meth.* **A 542**, 142 (2005).
41. Masahito Matsubayashi and Kazuhiko Soyama, *Nucl. Instrum. Meth.* **A 529**, 384 (2004).
42. E. Lehmann and P. Vontobel, *Appl. Radiat. Isot.* **61**, 567 (2004).
43. Mirko Estermann and Joe Dubois, *IEEE Trans. Nucl. Sci.*, **52**, 356 (2005).
44. Mirko Estermann, Gabriel Frei, Eberhard Lehmann, and Peter Vontobel, *Nucl. Instrum. Meth.* **A 542**, 253 (2005).
45. N.J. Rhodes, A.G. Wardle, A.J. Boram, and M.W. Johnson, *Nucl. Instrum. Meth.* **A 392**, 315 (1997).
46. N.J. Rhodes, E.M. Schooneveld, and R.S. Eccleston, *Nucl. Instrum. Meth.* **A 529**, 243 (2004).
47. M. Katagiri, T. Nakamura, M. Ebine, A. Birumachi, S. Sato, E.M. Schooneveld, and N.J. Rhodes, *Nucl. Instrum. Meth.* **A 573**, 149, (2007).
48. Nobuo Niimura,Yuuko Karasawa, Inchiro Tanaka, Junji Miyahara, Kenji Takahashi, Hiroki Saito, Satoshi Koizumi, and Masanori Hidaka, *Nucl. Instrum. Meth.* **A 349**, 521 (1994).
49. Jens Hofmann and Christian Rausch, *Nucl. Instrum. Meth.* **A 355**, 494 (1995).
50. Fuji Film, *I & I-Imaging Inform.* (1995).
51. D.A.A. Myles, C. Bon, P. Langan, F. Cipriani, J.C. Castagna, M.S. Lehmann, and C. Wilkinson, *Physica* **B 241-243**, 1122 (1998).
52. Shigenori Fujine, Kenji Yoneda, Masahiro Kamata, and Masahiro Etoh, *Nucl. Instrum. Meth.* **A 424**, 200 (1999).
53. B. Schillinger, J. Baumann, H. Gebele, R. Schaetzing, H. Schaller, and M. Schuster, *App.l Radiat. Isot.* **61**, 451 (2004).
54. S.A. Korff and W.E. Danforth, *Phys. Rev.* **55**, 980 (1939).
55. R. Batchelor, R. Aves and T.H.R. Skyrme, *Rev. Sci. Instrum.* **26**, 1037 (1955).
56. W.R. Mills, R.L. Caldwell, and I.L. Morgan, *Rev. Sci. Instrum.* **33**, 866 (1962).
57. C.J. Borkowski and M.K. Kopp, *Rev. Sci. Instrum.* **46**, 951 (1975).
58. V. Radeka, *IEEE Trans. Nucl. Sci.* **21**, 51 (1974).
59. G. Charpak and F. Sauli, *Nucl. Instrum. Meth.* **113**, 381 (1973).
60. Veljko Radeka and Robert A. Boie, *Nucl. Instrum. Meth.* **178**, 543 (1980).
61. R. Kampmann, M. Marmotti, M. Haese-Seiller, and V. Kudryashov, *Nucl. Instrum. Meth.* **A 529**, 342 (2004).
62. R. Berliner, J.S. King and D.F.R. Mildner, *Nucl. Instrum. Meth.* **152**, 431 (1978).
63. K. Abend, W. Schmatz, J. Schelten and K.D. Müller, *Nucl. Instrum. Meth.* **83**, 111 (1970).
64. R. Berliner, D.F.R. Mildner, O.A. Pringle and J.S. King, *Nucl. Instrum. Meth.* **185**, 481 (1981).
65. R.K. Crawford, J.R. Haumann, J.E. Epperson, P. Thiyagarajan, A.J. Schultz, G.P. Felcher, D.G. Montague and R.J. Dejus, *Nucl. Instrum. Meth.* **A 299**, 17 (1990).
66. R.G. Cooper, *Nucl. Instrum. Meth.* **A 529**, 394 (2004).
67. V. Radeka, N.A. Schaknowski, G.C. Smith, and B. Yu, *Nucl. Instrum. Meth.* **A 419**, 642 (1998).
68. B. Guerard, http://jra2.neutron-eu.net/jra2/ (2008).
69. A. Oed, *Nucl. Instrum. Meth.* **A 263** 351(1988).
70. J.F. Clergeau, P. Convert, D. Feltin, H.E. Fischer, B. Guerard T. Hansen, G. Manzin, A. Oed, and P. Palleau, *Nucl. Instrum. Meth.* **A 471**, 60 (2001).

71. Hiroyuki Takahashi, Prasit Siritiprussamee, Kaoru Fujita, Hisako Niko, Masaharu Nakazawa, Shunji Kishimoto, Takashi Ino, Michihiro Furusaka, and Hidenori Toyokawa, *Nucl. Instrum. Meth.* **A 573**, 175 (2007).
72. B. Gebauer, S.S. Alimov, A.Yu. Klimov, F.V. Levchanovski, E.I. Litvinenko, A.S. Nikiforov, V.I. Prikhodko, G. Richter, V.V. Rogov, Ch. Schulz, V.I. Shashkin, M. Wilhelm, and Th. Wilpert, *Nucl. Instrum. Meth.* **A 529**, 358 (2004).
73. Masayo Suzuki, Tan Takahashi, Hidekazu Kumagai, Masayasu Ishihara, and Hisao Kobayashi, *Nucl. Instrum. Meth.* **A 333**, 484 (1993).
74. G. Manzin, B. Guerard, F.A.F. Fraga, and L.M.S. Margato, *Nucl. Instrum. Meth.* **A 535**, 102 (2004).
75. J.C. Buffet, J.F. Clergeau, R.G. Cooper, J. Darpentigny, A. De Laulany, C. Fermon, S. Fetal, F. Fraga, B. Guerard, R. Kampmann, A. Kastenmueller, G.J. Mc Intyre, G. Manzin, F. Meilleur, F. Millier, N. Rhodes, L. Rosta, E. Schooneveld, G.C. Smith, H. Takahashi, P. Van Esch, T.L. Van Vuure, and K. Zeitelhack, *Nucl. Instrum. Meth.* **A 554**, 392 (2005).
76. M. Klein, H. Abele, D. Fiolka, and Chr. J. Schmidt, *AIP Conf. Proc.* **596**, 193 (2001).
77. D.D. DiJulio, A.I. Hawari, and R. Berliner, *Nucl. Instrum. Meth.* **A 579**, 71 (2007).
78. F.A.F. Fraga, L.M.S. Margato, S.T.G. Fetal, M.M.F.R. Fraga, R. Ferreira Marques, A. J.P.L Policarpo, B. Guerard, A. Oed, G. Manzini, and T. van Vuure, *Nucl. Instrum. Meth.* **A 478**, 357 (2002).
79. S. Andriamonje, D. Cano-Ott, A. Delbart, J. Derre, S. Diez, I. Giomataris, E.M. Gonzalez-Romero, F. Jeanneau, D. Karamanis, A. Lepretre, I. Papadopoulos, P. Pavlopoulos, and D. Villamarin, *Nucl. Instrum. Meth.* **A 481**, 120 (2002).
80. Charles L. Britton, Jr, William L. Bryan, Alan L. Wintenberg, Robert J. Warmack, Timothy E. McKnight, Shane S. Frank, Ronald G. Cooper, Nancy J. Dudney, Gabriel M. Veith, and Andrew C. Stephan, *IEEE Trans. Nucl. Sci.* **51**, 1016 (2004).
81. Y. Kumashiro, and Y. Okada, *Appl. Phys. Lett.* **47**, 64 (1985).
82. D.S. McGregor, T.C. Unruh, and W.J. McNeil, *Nucl. Instrum. Meth.* **A 591**, 530 (2008).
83. D.S. McGregor, M.D. Hammig, Y.-H. Yang, H.K. Gersch, and R.T. Klann, *Nucl. Instrum. Meth.* **A 500**, 272 (2003).
84. J. Uher, T. Holy, J. Jakubek, E. Lehmann, S. Pospisil, and J. Vacik, *Nucl. Instrum. Meth.* **A 542**, 283 (2005).
85. W.J. McNeil, S.L. Bellinger, B.J. Blalock, C.L. Britton, Jr., J.L. Britton, S.C. Bunch, S.A. Cowley, C.M. Henderson, T.J. Sobering, R.D. Taylor, and D.S. McGregor, *Nuclear Science Symposium Conference Record, 2007. NSS '07. IEEE* **3**, 2340 (2007).
86. Takekazu Ishida, Masatoshi Nishikawa, Shigehito Miki, Hisashi Shimakage, Zhen Wang, Kazuo Satoh, Tsutomu Yotsuya, Masahiko Machida, and Masaru Kato, *Physica* **C 460–462**, 618 (2007).
87. G.W. Fraser and J.F. Pearson, *Nucl. Instrum. Meth.* **A 293**, 569 (1990).
88. R.M. Ambrosi, G.W. Fraser, R.A. Street, J.I.W. Watterson, R.C. Lanza, J. Dowson, A. F. Abbey, B. Feller, G. Downing, P. White, and T. Stevenson, *Nucl. Instrum. Meth.* **A 542**, 271 (2005).
89. Oswald H.W. Siegmund, John V. Vallerga, Adrian Martin, Bruce Feller, Muhammad Arif, Daniel S. Hussey, and David L. Jacobson, *Nucl. Instrum. Meth.* **A 579**, 188 (2007).
90. A.S. Tremsin, J.V. Vallerga, J.B. McPhate, O.H.W. Siegmund, W.B. Feller, L. Crow, and R.G. Cooper, *Nucl. Instrum. Meth.* **A 592**, 374 (2008).

Section B

Chapter 5
Neutron Radiography

A.K. Heller and J.S. Brenizer

Abstract Neutron radiography and its related two-dimensional (2D) neutron imaging techniques have been established as invaluable nondestructive inspection methods and quantitative measurement tools. They have been used in a wide variety of applications ranging from inspection of aircraft engine turbine blades to study of two-phase fluid flow in operating proton exchange membrane fuel cells. Neutron radiography is similar to X-ray radiography in that the method produces a 2D attenuation map of neutron radiation that has penetrated the object being examined. However, the images produced differ and are often complementary due to the differences between X-ray and neutron interaction mechanisms. The uses and types of 2D neutron imaging have expanded over the past 15 years as a result of advances in imaging technology and improvements in neutron generators/sources and computers. Still, high-intensity sources such as those from reactors and spallation neutron sources, together with conventional film radiography, remain the mainstay of high-resolution, large field-of-view neutron imaging. This chapter presents a summary of the history, methods, and related variations of neutron radiography techniques.

Keywords Neutron imaging · Neutron radiography · Neutron converter · Direct radiographic method · Neutron imaging standards · Neutron computed radiology

5.1 Introduction

Neutron radiography is a powerful nondestructive imaging technique that produces a two-dimensional (2D) attenuation map of neutrons that have penetrated an object being examined. The method was initially developed after X-ray radiography, and the techniques share many similarities in setup and practice. X-ray

A.K. Heller (✉)
Department of Mechanical and Nuclear Engineering,
The Pennsylvania State University, University Park, PA 16802, USA
e-mail: axh174@psu.edu; jsb18@psu.edu

I.S. Anderson et al. (eds.), *Neutron Imaging and Applications*,
Neutron Scattering Applications and Techniques, DOI 10.1007/978-0-387-78693-3_5,

and neutron radiography are often complementary techniques, especially when low-energy neutrons are used. X-rays interact with orbital electrons and are strongly tied to the physical density of the examined object. Neutrons interact with an object's nucleus rather than its orbital electrons, so there is usually no tie to the object's electron density, but rather its elemental composition. Because the technique is based on attenuation from a well-collimated beam, either scattering or absorption will result in intensity variations to create an image. Low-Z materials such as hydrogen are easily imaged due to scattering, while boron and cadmium are readily imaged due to their strong absorption.

New methods similar to radiography, but using nonfilm image detectors, have been developed. To remove the confusion originally created by also referring to nonfilm methods as neutron radiography, new terms such as radioscopy and computed radiology have been adopted. These other methods are dealt with in more detail in Chapter 6 of this volume.

5.2 A History of Neutron Radiography

Although the discovery of the neutron by English physicist James Chadwick [1] occurred in 1932, the history of neutron radiography begins in 1935 when, in Germany, Kallmann and Kuhn [2] performed the first experiments specifically concerned with generating images using neutrons. The neutron radiographs produced by these experiments were not of high quality and took 4 h due to the small accelerator neutron source available during the study, which yielded a low-intensity thermal neutron beam. Regardless, these early experiments did provide insight into some of the possible uses of neutron radiography and the detection methods used to generate neutron radiographs.

Around the same time, another experimenter in Germany, Peter [3], was conducting a similar set of experiments using a much more intense neutron source. Because of this, the neutron radiographs produced by Peter were of fair quality and could be obtained in 1–3 min. Because of the Second World War, further development of neutron radiography did not occur until the mid-1950s. Indeed, Peter had to wait until 1946 to publish his results and Kallmann and Kuhn until 1948.

Despite the War, nuclear reactor technology development continued, with advances during and after the War that would increase the intensity of neutron fluxes available to researchers by many orders of magnitude. The first instance of a neutron radiograph being produced by a beam of thermal neutrons from a reactor occurred in 1956. Thewlis and Derbyshire [4] used the 6 MW BEPO[1] graphite reactor at Harwell in England for their neutron source, producing radiographs of better quality than those made previously by Peter and Kallmann. This work helped to illustrate a number of possible applications for the technique, specifically the inspection of neutron shielding materials and the study of organic specimens.

[1] British Experimental Pile '0,' the UK's second reactor.

Beginning in 1960, several independent studies to further the technique were undertaken at various laboratories by individuals such as Berger [5, 6], Watts [7], Barton [8], Criscuolo and Polansky [9], and Schultz and Leavitt [10]. It was during this time period that neutron radiography established itself as a viable nondestructive inspection method.

In the 1970s, the majority of research reactors throughout Europe and America had facilities in place capable of taking quality neutron radiographs. It soon became evident, however, that standardization was needed if further development of neutron radiographic techniques was to be achieved. Initial steps in this direction were taken in 1973 at Birmingham, England [11], and at Gaithersburg, Maryland (USA), in 1975 [12], but the pivotal moment occurred in 1979 when, through The Commission of European Communities, a Neutron Radiography Working Group was formed. This group initiated an extensive research program aimed at determining the best techniques for producing neutron radiographs and the accuracy of the dimensional measurements associated with them. This European effort, coupled with American and Canadian interest in the same area, led to the first of many *World Conferences on Neutron Radiography* [13] in 1981, the same year the group published its *Neutron Radiography Handbook* [14].

In the early 1980s, electronic, real-time neutron radiography emerged [15], which made use of neutron image intensifier tubes to increase image production rates and videocassettes as the storage medium. The dynamic nature of this technique found a wide variety of applications from investigations of internal combustion engines [16] to analysis of two-phase flow [17]. By the mid-1980s, images were being digitized and stored on computers [18], allowing the application of advanced imaging processing techniques for quantitative analysis [19].

From the 1990s to the present day, the imaging systems used to perform neutron radiography and the techniques used to analyze the images have continued to advance. The enhancement of existing detectors with, for example, thinner scintillation screens, and the development of new detectors like the microchannel plate, have yielded continual increases in resolution. Digital cameras with faster readouts and higher dynamic range are now used, and the analysis of uncertainty associated with measured optical density has provided more accurate results. These advancements have allowed neutron radiography to evolve from a nondestructive inspection method into a measuring method using neutrons as microscopic probes [20].

5.3 Basic Principles

5.3.1 Sources

Neutron radiography requires, of course, a source of neutrons. There are three general types of neutron sources: accelerator, radioisotope, and nuclear reactor. Accelerator-based neutron sources are ones that accelerate and direct a beam of charged particles such as protons, deuterons, and alphas onto a target, which then results in the emission of a neutron. There are a variety of combinations of incident particle and target material that can be used, a list of which may be found in von der Hardt [14]. Accelerators produce thermal neutrons in the range from 10^7 to 10^{10} n/cm^2/s and offer the benefit of intermittent operation and portability. Sealed-neutron generator tubes are easy to use but generally produce low fluxes compared with high current accelerators, which are moderately complex to operate and have a target output that may deteriorate with use. Large spallation sources can achieve fluxes of more than 10^{14} n/cm^2/s, but such sources are very complex and not portable.

A radioisotope-based neutron source makes use of radioactive isotope decay to generate its neutron flux in much the same way as accelerators: by allowing the gamma rays or alpha particles emitted by a radioactive isotope to bombard a neutron emitting target. A number of (γ,n) and (α,n) radioisotopic neutron sources exist, tables of which may be found in von der Hardt [14] and Berger [8]. Radioisotope-based neutron sources have low thermal neutron fluxes, within the range of 10^5–10^9 n/cm^2/s, and also suffer from decreasing output, from both deterioration of the target and decay of the source. The inability to turn off the radiation is often considered a disadvantage. The major advantages these systems possess are their simple operation and ease of portability.

Nuclear reactor-based neutron sources provide intense neutron beams and thus high-quality neutron radiographs. Thermal neutron fluxes obtainable in such facilities range from 10^{10} to 10^{15}n/cm^2/s or even higher. It should be noted that large spallation neutron facilities may produce beams with intensities equal to or higher than those produced by nuclear reactors. However, reactors typically provide a lower cost per neutron than accelerators. Their major disadvantages are the high costs associated with construction; licensing and regulatory requirements; complexity to operate; and lack of portability.

5.3.2 Moderation

The neutrons born in each of the sources discussed above possess high energies with a continuous spectrum of energies peaking from 0.85 MeV from fission (in reactors) up to 14 MeV (in accelerators). Conventional neutron radiography, however, requires neutrons in the thermal/epithermal energy range of 0.025 eV–10 keV. Thus, some form of moderator with low neutron absorption cross section (to maximize flux) and high scattering cross section (to maximize energy loss) is required to slow down the neutrons to this energy range. The often-used moderator materials of water, heavy water, graphite, beryllium, and polyethylene meet these criteria. In this, the nuclear reactor has an inherent advantage: the moderation of its core already produces a low-energy spectrum resulting in fewer neutrons lost in the moderation process.

5.3.3 Collimation

Once low-energy neutrons are produced, they must be formed into a usable beam. Neutrons are emitted and then scattered randomly in the moderator and, because of their neutral charge, they cannot be focused like electrons. Those neutrons traveling in the desired solid angle can, however, be selected by the introduction of a tube into or adjacent to the moderator. This has the effect of allowing neutrons to stream down the tube axis toward the object being radiographed. The walls of the collimator tube are lined with a neutron opaque material having a high absorption cross section (such as boron, gadolinium, and cadmium), which prevents stray neutrons from entering and also reduces low-angle scattering within the collimator. The most common collimator design is a divergent collimator (Fig. 5.1) with a small entrance aperture and a larger exit. This maximizes the neutron flux and permits a larger field at the imaging plane. The angular spread of the emerging beam is dependent upon the ratio of the collimator tube length (L) to its aperture diameter (D), referred to as the L/D ratio. A higher L/D results in a narrower beam spread at the expense of a lower neutron flux. This ratio is a characteristic parameter of each collimator. Extensive discussions regarding collimator design may be found in von der Hart [14] and Domanus [21].

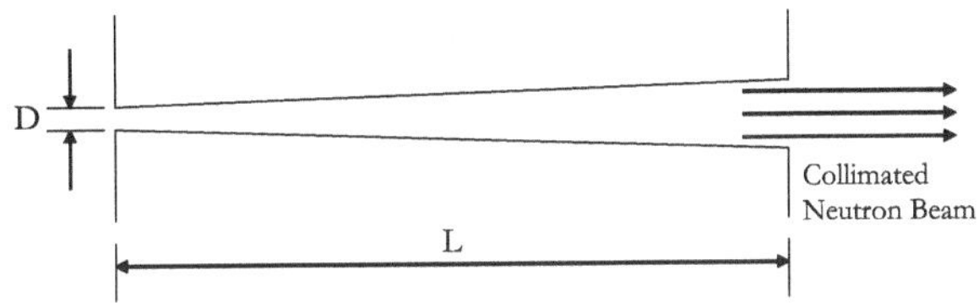

Fig. 5.1 Divergent collimator illustrating the ratio of the collimator tube length (L) to its aperture diameter (D) (the L/D ratio)

5.3.4 Detectors

After a neutron's birth in the source, its moderation to thermal energies, and its escape along the collimator tube, it will encounter the object to be radiographed. Any neutrons that successfully penetrate the object must then be detected to produce the radiograph. In neutron radiography, a detector collectively refers to both an intermediate medium, called a converter (which emits an alpha, beta, gamma, or light when neutrons are absorbed) and the sensor used to detect this emitted radiation, called the image recorder. The converter material is used because it emits a much more readily detectable radiation.

When the image recorder is film, one possible converter material is a gadolinium foil which emits an electron with every absorbed neutron. The converter foil is placed in direct contact with the film's emulsion and the emitted electrons expose the emulsion, producing an image. A typical spatial resolution using a single-coated fine-grain radiographic film, a vapor deposited gadolinium converter, and a vacuum cassette is 10 μm.

Another possible converter is a scintillation screen, which will expose the film's emulsion with light. A scintillation screen is 30–100 times faster at producing an image on radiographic film than a gadolinium foil. However, due to light spread within the scintillator, the spatial resolution is reduced. A typical spatial resolution for a scintillator and film is 100 μm. Both of these converters continuously emit radiation for the duration of neutron exposure and can therefore be used in low neutron flux environments with long integration exposures.

An electronic form of imaging where a scintillation converter is optically coupled with an analog or digital camera image recorder can also be used. This allows the rapid capture of successive neutron radiographic images that can be viewed directly and stored on videocassette or digital media, preserving dynamic information. This electronic form of producing radiographic images is referred to as radioscopy.

Other types of analog and electronic detectors exist and vary depending upon the imaging technique used. They are discussed in later sections of this chapter.

To summarize, a neutron radiography system consists of a neutron source, a moderator to thermalize the neutrons, an aperture and a collimator to organize the neutrons into a beam, and a detector to visualize the image (Fig. 5.2).

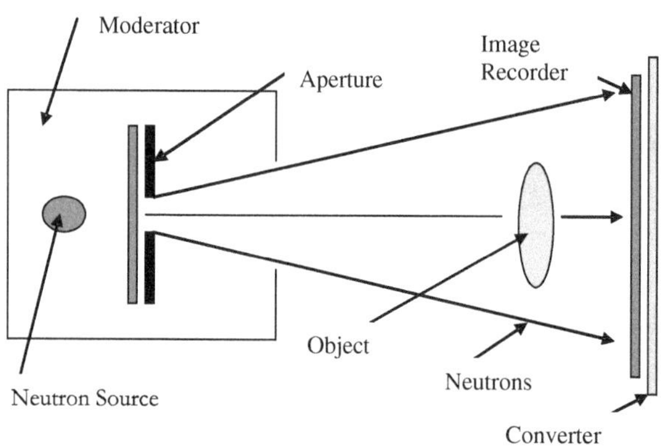

Fig. 5.2 Illustration of a typical neutron imaging system

5.4 Image Analysis

Any analysis of a radiographic image, be it film or electronic, begins with an understanding of how the image is formed. The relationship between the incident neutron intensity upon an object to be radiographed and the transmitted neutron intensity (ignoring scattering) is the simple exponential attenuation law

$$\phi = \phi_o e^{-\Sigma_i t} \quad . \tag{5.1}$$

The transmitted neutron intensity, ϕ, is a function of the incident neutron intensity, ϕ_0, and the product of the total macroscopic cross section and thickness of the object, $\Sigma_t t$ [14].

In the case of film, the degree of film darkening (photographic density, D_e) is related to the neutron exposure by the film's characteristic response curve. D_e will have a logarithmic nature as described by

$$D_e = G^*(\log E) \quad , \tag{5.2}$$

where E is the exposure of the film (transmitted neutron intensity multiplied by time, ϕT) and G is the slope in the linear portion of the characteristic response curve for the film being used; it is a parameter describing the manner in which a particular film responds to an exposure. This is the manner in which images are *formed* on film. One must bear in mind that when the image is being *viewed*, the processed film's photographic density is described by

$$D_e = \mathrm{In}\left(\frac{I_o}{I}\right) \quad , \tag{5.3}$$

where I_o is the incident light (such as from a light box) and I is the transmitted light through the film.

In nearly all forms of digital imaging, the resulting grey level value of any pixel making up the image may be described by

$$G = C^*\phi + G_{offset} \quad , \tag{5.4}$$

where G is the numerical grey level value of the pixel within an image, C is the electronic gain of the camera or imaging system (a constant), ϕ is the transmitted neutron intensity and G_{offset} is the dark current, an additive offset due to electronic noise.

These equations form the basis of all radiographic image analysis. With them, one may manipulate images to isolate terms and perform quantitative analyses or provide the basis for qualitative comparisons. Extensive descriptions and applications of these equations may be found in [14, 20–22].

5.5 Direct Radiographic Method

The direct radiographic method refers to the technique by which a radiographic image is generated when the converter is in direct contact with the image recorder. Traditionally, this has referred to detectors using film. The converter screen used more often than not is a gadolinium foil, which absorbs neutrons and emits gamma rays that are internally converted to low-energy electrons. These electrons expose the film's emulsion, after which the film must be developed. Converter screens of this type are usually 25-μm thick and are either gadolinium foils laminated to aluminum or vapor-deposited gadolinium on aluminum for ease of handling. Vacuum cassettes are used to ensure good contact between the converter and image recorder, which is vital in reducing image blur. The resolution of this converter–imager recorder combination is 10–20 μm. This resolution is quite impressive, considering it is obtainable over very large areas using films up to 35.5 cm by 43 cm.

In a similar fashion, a light-emitting scintillation screen may be used as the converter. In these screens, a neutron-absorbing material that yields a charged particle is mixed with a phosphor material that produces light. An example is a mixture of lithium-6 and ZnS(Ag). Lithium-6 emits an alpha particle when absorbing a neutron, and it is the kinetic energy of the alpha particle that causes the ZnS(Ag) to emit light. This light then exposes the film emulsion. Scintillation screens are thicker than gadolinium foils and, because of the light spread within them, the obtainable resolution (75–100 μm with the best screens) is less than that of gadolinium foils.

The direct imaging method has a major disadvantage when it involves nuclear applications. A radioactive object emitting gamma rays as it decays or a neutron beam contaminated with gamma rays will directly produce an image on the film. Since the gamma rays are from a different source than the neutrons, the images will be different and the film will be blurred or what is known as "gamma fogged." Another disadvantage of the method is the time associated with the film development, which inherently prohibits the investigation of dynamic processes.

There are, however, a wide variety of applications for which this method is well suited. In the area of turbine manufacturing, the direct radiographic method can easily perform a quality control check on the presence of residual materials used in the blade's production (Fig. 5.3). In fact, the technique has been used for quality control inspections for moisture, corrosion, adhesive defects, proper lubrication, and quality of seals in the aerospace and automotive industries. It is also used in the health monitoring and maintenance of in-service components such as aircraft flight control surfaces [23]. The direct method also

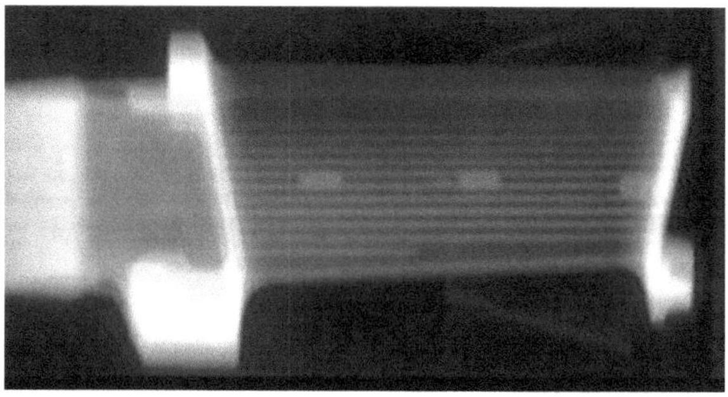

Fig. 5.3 Neutron radiograph of a turbine blade using a gadolinium contrast agent to accentuate blocked channels

plays a major role in the field of research, where it has been used to investigate two-phase flow behavior in heat pipes, water distribution in fuel cells, and water permeability in concrete, to list a few examples.

5.6 Indirect Radiographic Method

The indirect radiographic method applies when the converter screen is not in direct contact with the image recorder. This is also referred to as the transfer technique because of the manner in which the image is produced. A metal foil converter is placed in the beam independent of the image recorder. The foil builds up a radioactivity through neutron absorption, producing what is called an activation image. After being removed from the beam, the activated foil is placed in contact with the image recorder and the decay radiation emitted (low-energy electrons) "transfers" the activation image to the image recorder. In this method, the image recorder is film, just as with the direct radiographic method, but the converter screen used is indium, dysprosium, or gold because of the need for the foil to rapidly activate for image transfer and rapidly decay for reuse. The image quality produced by these techniques is the same as that of the direct techniques and provides spatial resolutions of 10–20 μm.

The technique is slower than the direct method, but one major advantage is that the converter foil used is insensitive to gamma radiation from the object or beam, making it an ideal candidate for use in nuclear applications such as the investigation of spent fuel rods. Gamma rays produced by radioactive decay in these applications will not fog up the radiographic image. The fact that this technique is slower than the direct method does not, however, preclude it from being applied to the same fields as the direct radiographic method. Indeed, the indirect radiographic method is equally applicable to the same industries, research, and quality control as the direct radiographic method.

5.7 Track-Etch Method

Technically a form of the direct radiographic method because of the direct contact between the converter and image recorder, the track-etch method is discussed independently here because of its uniqueness. The image recorder is nitrocellulose film, a dielectric material capable of detecting charged particles by the radiation damage caused within it, and it is placed between two alpha-particle emitting converter screens, such as boron or lithium. Unlike the low-energy electrons emitted by the metal foil converter screens used in the direct and indirect radiographic methods, the alpha particles take short and relatively straight paths through the nitrocellulose film, giving good resolution. The radiation damage within the film is made visible by etching it in a hot base solution, such as sodium hydroxide. A vacuum cassette is used to ensure tight contact between the film and the converter screens.

The track-etch method is insensitive to gamma rays and also to visible light, allowing the etching to be performed in daylight. However, the exposure time needed for this technique is longer than the direct and indirect methods, and the contrast is also weaker. A unique advantage is the ability to stop the etching process at intermediate stages, which can then be continued after evaluating the resultant neutron radiograph. In the end, several radiographs of varying densities and contrasts can be had. Because of the gamma-ray insensitivity, radiography of radioactive objects, such as spent nuclear fuel, is the primary application for this method.

5.8 Electronic Imaging Methods

Film is not the only form of image recorder possible. There are techniques that use electronic means, both analog and digital, for the image recorder. The production of radiographic images via electronic image recorders, initially referred to as real-time neutron radiography, is now known as radioscopy [24]. Electronic image recorders depend upon the capture of light to record an image and, because they have the ability to rapidly acquire these images, they are universally coupled with a converter that has an equally rapid (if not faster) light response: a scintillation screen. Thus, these methods are frequently used to view and store dynamic radioscopic images.

In electronic imaging systems, the image recorder can be a video camera if an analog system, or a charge-coupled device (CCD) camera if a digital one. In the case of an analog system, the camera is connected to a television and VCR for the real-time viewing and recording of the dynamic radioscopic images. In a digital system, the CCD camera is connected to a computer that displays and stores the radioscopic images in a digital format. A combination of these systems may also exist wherein the analog television camera can be connected to a computer via an analog-to-digital converter card which will digitize the camera's analog signal and store it.

The camera, be it analog or digital, is coupled to the scintillation screen via a light-tight box to prevent polluting the radioscopic image with outside light. The intensity of light produced by the scintillation screen is linear with the intensity of incident neutron flux, and there are a number of reactor facilities where the neutron flux intensity is sufficiently large for the light produced by the scintillation screen to be seen on a TV monitor. However, there are many facilities that do not possess a neutron flux of such intensity, and in these circumstances an image intensifier is placed within the light-tight box between the scintillation screen and the camera. The image intensifier boosts the light from the scintillation screen by about 10^4, still providing good image definition and allowing the radiographic image to be viewed. In addition, a mirror is often placed between the camera and image intensifier (or scintillation screen if an intensifier is not needed), forming a right angle and allowing the camera to be positioned outside the beam path to reduce exposure.

In certain circumstances, an electronic image recorder may be used, but not with the intent of dynamic imaging. If very high signal-to-noise ratios are needed, such as in computed tomography applications, the integration time can be increased to obtain the required ratio. The use of a digital camera makes the image acquisition easier by eliminating the need to develop a film and, because the image is digital, provides ease of image manipulation.

5.9 Nonfilm Imaging Methods

Recently new sensors called photostimulable luminescence (PSL) imaging plates have been developed that can be used for X-ray and neutron imaging. Photostimulated luminescence is a phenomenon where a phosphor is first exposed to light or charged particle radiation [25]. After removal from the exposing radiation, the PSL plate retains a stored image which can be read out by later exciting the previously exposed phosphor to longer wavelength light. The 2D flexible storage phosphor that can store a latent image from radiation is called an imaging plate. The overall imaging process—where an imaging plate is first exposed to gamma-ray, X-ray, or neutron beams in a manner similar to traditional radiography and is subsequently read by means of photo stimulation to obtain a radiographic image—is called computed radiology (CR) [24].

Imaging plates can be used as a direct replacement for radiographic film. There are several advantages of using CR instead of traditional film. Imaging plates can be erased and reused for potentially thousands of exposures. Physical damage from handling is the common limiting factor. Readout is accomplished in a scanner-type device that reads out the latent image, stores a digital image, and erases the imaging plate for the next use. Imaging plates have been shown to have a much greater sensitivity range than photographic film, allowing a wide linear dynamic range of eight orders of magnitude. This is very advantageous for neutron imaging, where neutron intensity at the imaging plane is usually lower than desired. CR has the additional advantage of eliminating film processing, thereby making the imaging process faster, independent of processing chemicals, and more environmental friendly than film processing. While not inexpensive compared with simple radiography with film and hand processing, CR systems offer significant cost savings over film when large volumes of radiographs are needed.

The major disadvantage of CR is the spatial resolution, typically $\sim$100 μm, compared with the $\sim$15 μm obtainable with film [26, 27, 28].

5.10 Standards

As early as 1969 it was recognized that some standardization in neutron imaging was needed. Haskins presented two reviews of neutron radiography standards in the United States [29, 30], and a later paper by Newacheck and Tsukimura updated these earlier papers [31]. The American Society for Testing and Materials International Committee E07 on Nondestructive Testing has developed a suite of standards for neutron radiography. There are currently no standards for nonfilm neutron imaging.

E748, *Standard Practices for Thermal Neutron Radiography of Materials*, provides a good introduction to neutron radiography [32]. The document is tutorial in nature, describing common practices, facilities, and necessary equipment. This standard also provides example applications and some basic guidance for determining the practicality of the method.

E545, *Standard Test Method for Determining Image Quality in Direct Thermal Neutron Radiographic Examination*, has become the world standard for determining the relative overall quality of neutron radiographs [33]. It is not intended to be used for controlling the acceptability or quality of materials or components. Radiographic quality is based upon the evaluation of images obtained from two different indicators, the beam purity indicator and the sensitivity indicator. The information obtained from radiographic images using these devices is used to determine a facility's neutron radiographic category.

E803, *Standard Test Method for Determining the L/D Ratio of Neutron Radiography Beams*, is a method widely used by radiographers to characterize neutron beams [34]. Knowledge of the L/D ratio is important for understanding the geometric "unsharpness" of imaged objects as a function offset from the imaging detector.

A fourth standard, E1496, *Standard Test Method for Neutron Radiographic Dimensional Measurements*, presents a method that can be used to obtain quantitative length dimensions reproducibly from a radiographic image [35].

5.11 Conclusions

Neutron radiography and its related 2D neutron imaging techniques have established themselves as invaluable nondestructive inspection methods and quantitative measurement tools. They have been used in a wide variety of applications ranging from inspection of aircraft engine turbine blades to studying two-phase fluid flow in operating proton exchange membrane fuel cells.

Advances in digital cameras, CR using imaging plates, and the development of more intense and robust neutron generator tubes will expand the use of neutron imaging. These advances will also increase portability, making it possible to take imaging systems to the field rather than bringing objects to a fixed imaging facility. Improvements in digital image processing coupled with the increased computational power of modern processors provide the ability to collect and analyze images rapidly. This allows imaging of dynamic events at spatial and temporal resolutions sufficient to provide the qualitative and quantitative information needed for many new applications. However, high-intensity sources, such as those from reactors and spallation neutron sources, with conventional film radiography will remain the mainstay of high-resolution, large field-of-view neutron imaging into the foreseeable future.

References

1. J. Chadwick, Possible Existence of a Neutron, *Nature* **129**, 312 (1932).
2. H. Kallmann, Neutron Radiography *Research* 1, 254 (1948).
3. O.Z. Peter, Neutronen-Durchleuchtung, *Naturforsch.* **1**, 557 (1946).
4. J. Thewlis and R.T.P. Derbyshire, *Report AERE M/TN 37,* U. K. Atomic Energy Research Establishment, Harwell England (1956).
5. H. Berger, *Proceedings Symposium on Physics and Nondestructive Testing, Argonne, Illinois ANL-6346*, Argonne National Laboratory, Argonne Illinois (1960).
6. H. Berger, *Neutron Radiography: Methods, Capabilities and Applications*, Elsevier Publishing Company, New York (1965).
7. H.V. Watts, *Report ARF 1164-6,* Armour Research Foundation, Chicago, Illinois (1960).
8. J.P. Barton, Introductory Remarks *Neutron Radiography: Proceedings at the Second World Conference on Neutron Radiography*, Paris, France (June 16–20, 1986), J. Barton, J. Farny, J. Person, H. Röttger, D. Reidel, eds. Pub. Co., Dordrecht, Holland, p. 13 (1987).
9. E.L. Criscuolo and D. Polansky, *Proceedings Missiles and Rockets Symposium, Concorde, California, 1961,* U. S. Naval Ammunition Depot, Concord, California, p. 112 (1961).
10. A.W. Schultz and W.Z. Leavitt, *Report WALTR 142.67,*Watertown Arsenal Laboratories, Watertown Massachusetts (1961).
11. M.R. Hawkesworth (technical editor) Radiography with Neutrons, Conference held 10–12 September, 1973, at the University of Birmingham, London, British Nuclear Energy Society (1975).
12. H. Berger (editor) *Practical Applications Of Neutron Radiography And Gauging, ASTM STP 586* (1976).
13. J.P. Barton and P. von der Hardt (editors) *Neutron Radiography Proceedings of the First World Conference*, San Diego, California, U.S.A. , December 7–10, 1981. D. Reidel Publishing Company, Dordrecht: Holland/Boston: U.S.A./London, England (1983).
14. P. von der Hardt and H. Röttger (editors), *Neutron Radiography Handbook,* D. Reidel Publishing Company, Dordrecht: Holland/Boston: U.S.A./London, England (1981).
15. J.S. Brenizer and M.F. Sulcoski Real-Time Neutron Radiography at the University of Virginia in *Use and Development of Low and Medium Flux Research Reactors* Harling, O.K., Clark, L., and von der Hardt, P., eds., Karl Theiemig Graphische, W. Germany, p. 958 (1984).
16. J.D. Jones, J.T. Lindsay, C.W. Kaufmann, A.T. Vulpeti and B. Peters, Real Time Neutron Imaging Applied to Internal Combustion Engine Behavior SAE *Technical Paper Series No. 850560*, International Congress and Exposition (1985).
17. M. Tamaki, K. Ohkubo, Y. Ikeda, and G. Matsumoto Analysis of Two Phase Flow in Heat Pipe by Neutron Radiography *Proceedings of the Second World Conference on Neutron Radiography*, Paris, France, June 16–20, 1986 Barton, G. Farny, J. Person, H. Röttger, D. Reidel, eds. Pub. Co., Dordrecht, Holland, pp. 609–616 (1987).

18. J.S. Brenizer, B. Hosticka, R.W. Jenkins Jr and D.D. McRae An Advanced Video System for Real-Time Neutron Radiography *Proceedings of the Second World Conference on Neutron Radiography*, Paris, France, June 16–20, 1986, Barton, G. Farny, J. Person, H. Röttger, D. Reidel, eds. Pub. Co., Dordrecht, Holland, pp. 571–578 (1987).
19. R. Polichar and D. Shreve Processing of Real-Time Images for Quantitative Neutron Radiography *Proceedings of the Second World Conference on Neutron Radiography*, Paris, France, June 16–20, 1986, J. Barton, G. Farny, J. Person, H. Röttger, D. Reidel, eds. Pub. Co., Dordrecht, Holland, pp. 587–593 (1987).
20. T. Hibiki and T. Mishima Prediction of Measurement Error Due to Low Gray Scale and Spatial Resolution of an Imaging System on Quantification of Neutron Radiographic Image *Nuclear Instruments and Methods in Physics Research Section A*, vol. 338, pp. 204–211 (1997).
21. J.C. Domanus (editor) *Practical Neutron Radiography*, Kluwer Academic Publishers Dordrecht: Holland/Boston: U.S.A./London: England (1992).
22. R.A. Quinn and C.C. Sigl (editors) (1980) *Radiography in Modern Industry*, 4th edition, Eastman Kodak Company, Rochester NY.
23. T.R. Chalovich, L.G.I. Bennett, W.J. Lewis, and J. Brenizer, Jr., Development of Neutron Radioscopy for the Inspection of CF188 Flight Control Surfaces, *Applied Radiation and Isotopes*, vol. 61, pp. 693–700 (2004).
24. E07 on Nondestructive Testing ASTM E1316 Standard Terminology for Nondestructive Examinations, *Annual Book of ASTM Standards* vol 03.03 (2007).
25. E07 on Nondestructive Testing E2007 Standard Guide for Computed Radiology (Photostimulable Luminescence (PSL) Method) *Annual Book of ASTM Standards*, vol. 03.03 (2007).
26. M. Thoms, D. Myles and C. Wilkinson Neutron Detection with Imaging Plates Part I. Image Storage and Readout, *Nuclear Instruments and Methods in Physics Research A*, vol. 424, pp. 26–33 (1999).
27. M. Thoms Neutron Detection with Imaging Plates Part II. Detector Characteristics *Nuclear Instruments and Methods in Physics Research A*, vol. 424, pp. 34–39 (1999).
28. H. Kolbe, E Lehmann, W. Gunia and S. Körner Applications and Characteristics of Imaging Plates as Detector in Neutron Radiography at SINQ *Nuclear Instruments and Methods in Physics Research A*, vol. 424, pp. 40–47 (1999).
29. J.J. Haskins ASTM Activities in Neutron Radiography *Practical Applications of Neutron Radiography and Gauging, ASTM STP 586*, H. Berger (editor) American Society for Testing and Materials, Philadelphia, pp. 106–113 (1976).
30. J.J. Haskins Neutron Radiography Standards in the United States of America *Neutron Radiography Proceedings of the First World Conference*, San Diego, California, U.S.A. , December 7–10, 1981, Barton JP and von der Hardt P (editors) D. Reidel Publishing Company. Dordrecht: Holland/Boston: U.S.A./London: England, pp. 985–991 (1983).
31. R.L. Newacheck and R.R. Tsukimura Current Status of the ASTM E545 Image Quality Indicator System *Neutron Radiography (3) Proceedings of the Third World Conference on Neutron Radiography* Osaka, Japan (May 14–18, 1989), S. Fujine, K. Kanda, G. Matsumoto, J. Barton, eds. Pub. Co. Kluwer Academic, Dordrecht, The Netherlands, pp. 875–883 (1990).
32. E07 on Nondestructive Testing E748 Standard Practices for Thermal Neutron Radiography of Materials *Annual Book of ASTM Standards*, vol. 03.03 (2007).
33. E07 on Nondestructive Testing E545 Standard Test Method for Determining Image Quality in Direct Thermal Neutron Radiographic Examination *Annual Book of ASTM Standards*, vol. 03.03 (2007).
34. E07 on Nondestructive Testing E803 Standard Test Method for Determining the L/D Ratio of Neutron Radiography Beams *Annual Book of ASTM Standards*, vol. 03.03 (2007).
35. E07 on Nondestructive Testing E1496 Standard Test Method for Neutron Radiographic Dimensional Measurements *Annual Book of ASTM Standards*, vol. 03.03 (2007).

Chapter 6
Neutron Tomography

W. Treimer

Abstract This chapter provides an introduction to neutron tomography. The basic factors influencing instrument design and capability are discussed and the mathematical methods used for image reconstruction are explained. Some of the more promising new techniques, which are likely to have an increasing range of applications, are described in more detail. These include energy (wavelength)-dispersive radiography and tomography, real-time radiography, phase contrast, refraction, and small-angle tomography. In the latter methods, there is an interesting overlap/complementarity between the real-space imaging aspects and the Fourier space imaging aspects, particularly as this gives rise to the possibility of imaging to a resolution of 10 μm or better.

Keywords Foundations of CT · Geometry · Point spread function · Reconstruction · Phase contrast · Refraction and USANS tomography · Neutron cross section · Scintillator · Neutron flux · Neutron intensity · Spatial resolution · Time resolution · Detector · Point spread function · Fourier transform · Fourier space · Modulation transfer function · Attenuation · Absorption · Scattering · Image reconstruction · Projection · Filtered-back projection · Grey value · Energy-dispersive radiography · Energy-dispersive tomography · Bragg-edge radiography · Bragg-edge tomography · Wavelength · Wedge transmission · Real-time radiography · Phase contrast · Refraction · Small-angle tomography

W. Treimer (✉)
Department of Mathematics, Physics and Chemistry, University of Applied Sciences (TFH) Berlin, Luxemburgerstr. 10, D-13353 Berlin; Helmholtz Center Berlin for Materials and Energy, Department SF1, Glienicker Str. 100, D-14106, Berlin
e-mail: treimer@tfh-berlin.de, treimer@helmholtz-berlin.de

I.S. Anderson et al. (eds.), *Neutron Imaging and Applications*,
Neutron Scattering Applications and Techniques, DOI 10.1007/978-0-387-78693-3_6,

6.1 Introduction

Tomography and radiography are well known from their uses in medicine, and many people have already come into contact with one or more of these methods. They are closely related because tomography is based on radiography, which is a two-dimensional (2D) attenuation coefficient distribution of a ray-path-integrated projection from a three-dimensional (3D) object (Chapter 5). In the past, X-ray film-converter systems were used for neutron imaging; but since the mid-1990s digital imaging systems, e.g., image plates or charge-coupled device (CCD) cameras with scintillator screens, have become increasingly available (Chapter 4). The quality of digital images is now comparable to that obtained using film. The advantages of CCD camera-based imaging systems are easy image processing, high bit depth (optical density), and real-time imaging and information handling (delivery, storage, post-processing). Thus radiographic and tomographic techniques for neutron imaging have become easier to handle and to apply with higher performance since the advent of PC-based electronic media. It is only a short step from a radiograph to a tomogram, so some fundamental parameters for both techniques are briefly summarized here.

For neutron radiography and tomography, four parameters characterizing the neutron beam are of principal interest: the incident flux intensity Φ, the wavelength spectrum spread $\Delta\lambda/\lambda$, the divergence ϕ, and the given spin state of the neutron beam in the case of polarized neutrons (Chapter 10). The geometry required for a radiography setup, for example the distance between sample and detector, is influenced by the size of the sample and by the pixel size (smallest scanning part) of the detector. The flux and the wavelength spread are directly connected because the smaller the $\Delta\lambda/\lambda$, the lower the neutron flux. This also holds true in principle for the achievable spatial resolution—the smaller the pixel size (i.e., the higher the spatial resolution), the smaller the integrated flux at the pixel. Therefore, for conventional radiography and tomography (i.e., attenuation-based neutron imaging techniques), the full spectrum (white beam) is used to get acceptable results in short times. For various other cases of tomography, the use of monochromatic neutrons is necessary, so a certain $\Delta\lambda/\lambda$, typically 1–5%, is required to determine structures in samples that are not revealed by conventional tomography.

The basic properties of the neutron, for example, the relationships between wavelength, energy, and velocity, and the scattering and absorption cross sections, have already been introduced in Chapter 1. Scattering processes are unwanted effects in a classical radiograph or tomogram because they cause a decrease of sharpness in the image. However, it must be noted that scattering processes can also be used as imaging signals using different imaging techniques (Section 6.4). X-ray scattering cross sections generally have an atomic number–dependent relationship, while neutron cross sections typically have a "statistical" behavior, as shown in Fig. 6.1 (see also Fig. 2.2).

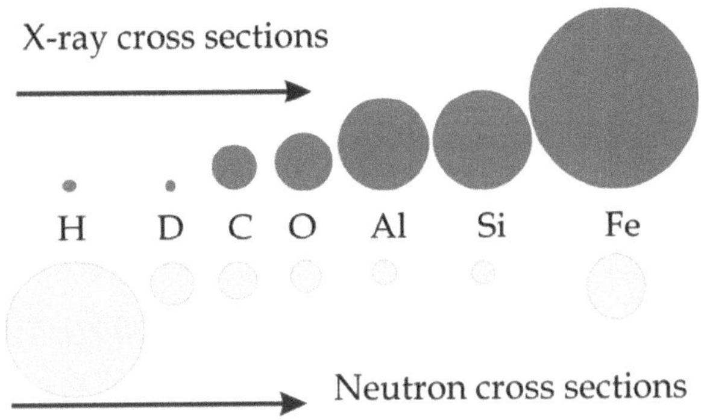

Fig. 6.1 X-ray and neutron cross sections for different elements. The sizes of the circles symbolize the strengths of the scattering cross sections

6.2 Geometric Considerations and Spatial Resolution

For tomography (and radiography), neutrons are collimated by slits, apertures, or collimator systems to reduce the range of directions in which the radiation propagates and to obtain a defined point-to-point image. For a (virtual) source of a given size, a point P in the object (sample) plane is enlarged to have a diameter d as a result of the incident divergence $\phi \sim L/D$, where L is the distance of the point from the source and D the source dimension, and of the distance l_d from the detector to the object (Fig. 6.2).

However, in order to obtain an image in a reasonable exposure time, the available neutron flux is also important. The exposure time that is reasonable depends on the problem to be solved and can range from fractions of seconds to several minutes. In every case, the exposure time must be optimized to the available neutron flux.

The neutron intensity incident on a sample to be imaged, I_0, is determined by the neutron flux I_{source} coming from the source decreased by $\sim l/L^2$. The horizontal collimation ϕ_h and vertical collimation ϕ_v define a solid angle $\Delta\Omega = \phi_h \times \phi_v$ so that the intensity I_0 is proportional to the area of the surface of the sphere $\Delta A = ds_h \times ds_v$. The elements $ds_{h,v}$ can be determined as $\phi_{h,v}$ R. So for a given divergence ϕ (which

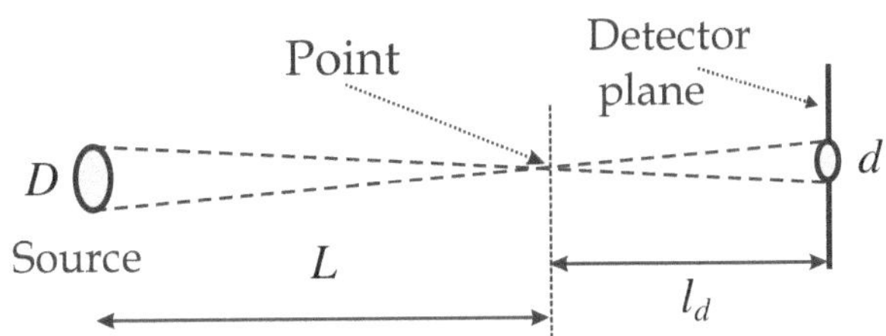

Fig. 6.2 Geometric point broadening: A point P in the object plane is enlarged (blurred) up to a size d as a result of a given L/D

can be measured), the incident intensity I_0 in front of a sample can be calculated from a given source flux I_{source}. To realize a point-to-point image, the size of the source and/or the beam divergence must be kept small. This means that a large L/D ratio corresponds to a small horizontal and vertical divergence. Furthermore, one must consider that a point in an object is also enlarged and blurred to an area in the image by the object-detector (e.g., film, CCD camera) distance, and by scattering effects.

Every point in a sample is enlarged to an area with diameter d on the detector as $d = l/(L/D)$ if L/D defines the (inverse) divergence $\phi_{h,v}$ (Fig. 6.2). One can calculate that even for $L/D = 1000$ (divergence $\phi_{h,v} \sim 0.057° = 0.001$ rad), a point 50 mm from the detector is enlarged to 50 μm. Additional broadening of an object point is also caused by the process of converting neutrons into visible light; for example, the thickness of the scintillator material and the optical system focusing the point onto the CCD chip might inherently decrease the sharpness of the image. However, it must be noted that in the case of refraction, or of phase radiography and tomography, the general rule that a shorter distance between object and detector means a sharper image does not hold.

The imaging system, i.e., the imaging of a given point in a sample, can be described by the so-called point spread function (PSF) because it involves not only the interaction with the sample but also the detector resolution (pixel size), the spread of radiation due to divergence and scintillation (detection) processes, and optical systems (CCD cameras). The PSF is also known as the impulse response (Chapter 7). The spatial 2D resolution of a system is best measured by determining its modulation transfer function (MTF; Fig. 6.3). The MTF can be derived from the Fourier transform of the PSF, i.e., $|\mathrm{MTF}| = \mathrm{FT}\{\mathrm{PSF}(x,y)\}$, the PSF being calculated as [2]

$$\mathrm{PSF}(x,y) = \frac{\delta^2/2\pi}{\left[1+\delta^2\{x^2+y^2\}\right]^{\frac{3}{2}}}, \tag{6.1}$$

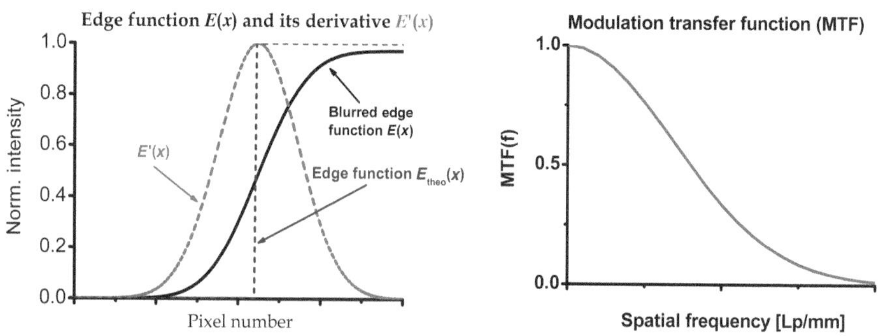

Fig. 6.3 The point spread function of an imaging system can be derived from the first derivative of a "real" (blurred) edge (*left*). The MTF is derived from the Fourier transform of the PSF as $|\mathrm{MTF}| = \mathrm{FT}\{\mathrm{PSF}\}$ (*right*)

where $\delta = 1/d_{\text{conv}}$, d_{conv} being the distance of the converter foil to the image plane. Usually d_{conv} can be taken as the thickness of the converter if the film is attached to it or the CCD camera is focused on it.

So the measured image $I(x,y)$ is the convolution of the 2D PSF with the 2D projection of the object function $O_P(x,y)$ (the 2D Radon transform of the 3D object), plus background (bg), normalized to the pixel-efficiency $\varepsilon_{\text{pixel}}\,(x,y)$ of the detector:

$$I(x,y) = \{\text{PSF}(x,y) \otimes O_P(x,y) + bg(x,y)\} \cdot \varepsilon_{\text{pixel}}(x,y)\ . \qquad (6.2)$$

Because of different efficiencies of the detector pixel and background, these two parameters must be determined very accurately to achieve definite images or reconstructions. They are derived from flat field and background images, the first obtained without the sample in the beam and the second when the beam shutter is closed.

The physics underlying the detection of neutrons using CCD cameras in conjunction with converter screens is quite sophisticated. Neutrons must be detected and finally converted into visible light (which can also be done by imaging plates, converter foils, or semiconductor devices). This process is based on the absorption of a neutron by a Li^6 atom, which then decays into H^3 and H^4 atoms. These react with ZnS-Ag particles and the reaction is accompanied by the emission of visible light. This light must be focused onto the chip of the CCD camera by a mirror system in the neutron beam (to avoid radiation damage — Fig. 6.4) and processed for data acquisition.

The CCD is connected to a conventional PC that steers the experiment and collects the data to be reconstructed (Section 6.3). The 2D image can be processed and stacked into a 3D image so that details can be extracted and visualized as 2D or 3D images. The use of modern CCD cameras as 2D neutron

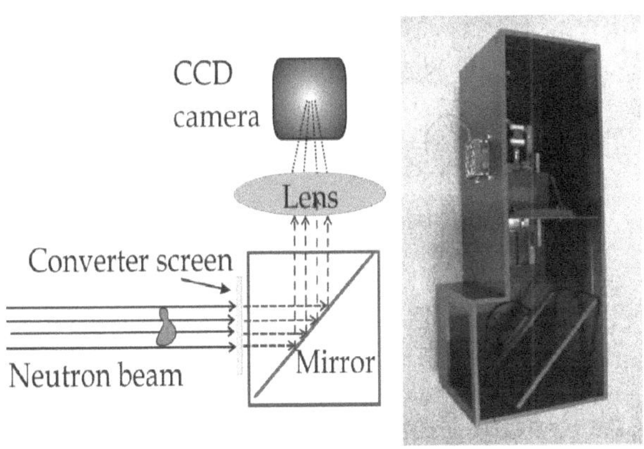

Fig. 6.4 *Left*: Neutrons hitting the ZnS screen are converted into visible light that is collected on the CCD of a camera. *Right*: Light-proofed box containing mirror and CCD optics and camera. For details see, for example, Lehmann et al. [3]

detectors has nearly superseded the film-converter technique, because all the image-processing techniques known from medical imaging can also be applied to neutron radiography and tomography [3].

The number of applications of neutron tomography is quite large, and there are several papers and books on this topic. However, some new developments are significant. The experimental conditions have improved, and there is an increased need for applications of neutron tomography related to industry and cultural heritage. The new developments are energy-dependent radiography, (rapid) real-time radiography, refraction or phase contrast radiography and tomography, small-angle scattering radiography and tomography, and radiography and tomography using polarized neutrons (the latter is covered in Chapter 10). However, these are currently not routine methods, and their potential must be further evaluated.

6.3 Mathematical Foundations of (Neutron) Tomography

6.3.1 Scanning

The methodological foundations underlying the reconstruction of 2D and 3D images of objects have been described in detail by Kak and Slaney [4] and Rosenfeld and Kak [5]. Other important text books, for example by G. Herman [6], describe different methods such as the algebraic reconstruction technique (ART). This chapter follows the method of Kak and Slaney and explains the essential steps for tomography from the measurement to the reconstruction of a single slice. Stacking of such slices yields a 3D image. By means of image processing, parts of the 3D reconstruction can then be extracted and displayed separately from the rest of the data set.

An arbitrary object $O(x,y,z)$ is considered to comprise n slices, of equal thickness Δz, that all lie in planes parallel to the (x,y) plane and perpendicular to the z-axis (Fig. 6.5). Each slice represents a cut through the object O, which

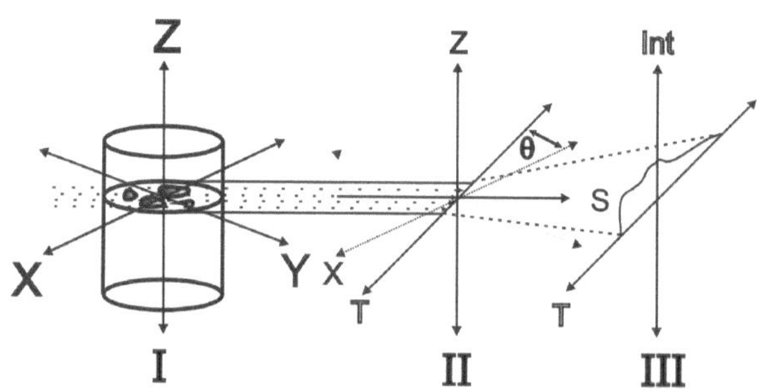

Fig. 6.5 Scanning geometry of an object (I) in the $\{x\text{-}y\text{-}z\}$ coordinate system. A slice in the $(x\text{–}y)$ plane is scanned through angle θ (II) and the transmitted intensity is registered in the t-s rotated coordinate system (II, III). For details, see text

has to be reconstructed. It is considered as a 2D function $f_n(x,y)$ that describes, for example, the position-dependent attenuation coefficient $\mu_n(x,y)$, or any other 2D function that can be measured and the signal described as a line integral. Any function $f(x,y)$ can be taken into account for tomography as long as it is bounded and finite in a given region and zero outside this region. (Note that the object can be larger than the scanning region, but the reconstruction of the outer part is not unique). The condition of finite size is easy to realize for solid or liquid samples or gas components in a box, but it is not so easy to handle tomography of electric and magnetic fields. The problem in generalizing this function for tomographic measurements is in determining the interaction that a (pencil) beam (e.g., neutron, X-ray) experiences as it passes through the object. The goal of tomography is to reconstruct this 2D function (i.e., slice) from projections measured in a unique way.

The slice is scanned through angles θ with $0 < \theta < 180°$ (360° in the case of refraction tomography), and the transmitted intensity (e.g., of neutrons) is registered as a function of the translation, i.e., the position parameter t (Fig. 6.6). The transmitted intensity is given by Lambert's law as

$$I(x,y) = I_0 \exp\left(-\int_{\text{path}} \mu(x,y) \cdot \mathrm{d}s\right), \tag{6.3}$$

where I_0 is the intensity of the incident neutron beam and $f = \mu(x,y)$ is the 2D function to be reconstructed. A new rectangular coordinate system (t,s) is

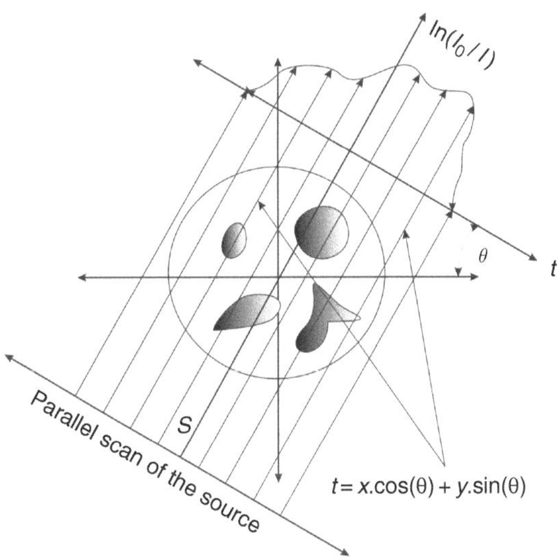

Fig. 6.6 Scanning of a slice in the x–y plane (compare with Fig. 6.5). Note that z is the axis of rotation, which coincides with the z-axis of the $\{s\text{-}t\text{-}z\}$ system

defined that expresses the rotated detector system (t,s) with respect to the fixed sample system (x,y) (or vice versa, if the sample is rotated as is done in neutron tomography). Transforming the (x,y) system into the (t,s) system, t can be expressed as

$$t = x \cdot \cos(\theta) + y \cdot \sin(\theta) \ . \tag{6.4}$$

The path across the sample, given in terms of t and θ, by eliminating s, is $\delta(t - x\cos\theta + y\sin\theta)$; i.e., the δ function guarantees that only points fulfilling Eq. (6.4) that lie on the pencil beam will contribute to the projection $P_\theta(t)$.

One can define a so-called projection $P_\theta(t)$ as

$$P_\theta(t) = \ln\left(\frac{I_0}{I}\right) = \int\limits_{\text{path}} \mu(x,y) \cdot \mathrm{d}s \ . \tag{6.5}$$

The set of all projections $P_\theta(t)$ of $\mu(x,y)$ is called the Radon transform of $\mu(x,y)$. From these projections, a 2D image can be reconstructed using the "Fourier slice theorem." This states that the Fourier transform of a projection $P_\theta(t)$ of $\mu(x,y)$ is a subset of a function $M(u,v)$, which is the Fourier transform of $\mu(x,y)$, i.e.,

$$\mathrm{FT}\{P_\theta(t)\} = S_\theta(w) \subset \mathrm{FT}\{\mu(x,y)\} = M(u,v) \ , \tag{6.6}$$

where $u = w\cos(\theta)$ and $v = w\sin(\theta)$. To demonstrate this relation, we again follow the method given in [4] and [5]. Assume $\mu(x,y)$ to be bounded and finite, so that $\mu(x,y)$ has a Fourier transform $M(u,v)$:

$$M(u,v) = \int\limits_{-\infty}^{\infty}\int\limits_{-\infty}^{\infty} \mu(x,y) \cdot \mathrm{e}^{-2\pi \cdot i \cdot (ux+yv)} \cdot \mathrm{d}x\mathrm{d}y. \tag{6.7}$$

To prove the slice theorem, consider the simple case of $\theta = 0$. In this case, the two coordinate systems (x,y) and (u,v) coincide, so the projection $P_\theta(t)$ is simply the integral (Eq. [6.10]):

$$P_{\theta=0}(t) = \int\limits_{-\infty}^{\infty} \mu(x,y) \cdot \mathrm{d}y \ . \tag{6.8}$$

The Fourier transform of the object becomes (as a result of $\nu = 0$)

$$\begin{aligned} M(u,0) &= \int_{-\infty}^{\infty}\int_{-\infty}^{\infty} \mu(x,y)\cdot \mathrm{e}^{-2\pi\cdot i\cdot(ux)}\cdot \mathrm{d}x\mathrm{d}y \\ &= \int_{-\infty}^{\infty}\left\{\int_{-\infty}^{\infty} \mu(x,y)\cdot \mathrm{d}y\right\}\mathrm{e}^{-2\pi\cdot i\cdot(ux)}\cdot \mathrm{d}x\ . \end{aligned} \tag{6.9}$$

Now, one can substitute into Eq. (6.7) and get

$$M(u,0) = \int_{-\infty}^{\infty} P_{\theta=0}(x)\cdot \mathrm{e}^{-2\pi\cdot i\cdot ux}\cdot \mathrm{d}x\ . \tag{6.10}$$

If the Fourier transform of $P_\theta(t)$ is given by

$$S_\theta(w) = \int_{-\infty}^{\infty} P_\theta(t)\cdot \mathrm{e}^{-2\pi\cdot i\cdot w\cdot t}\cdot \mathrm{d}t\ , \tag{6.11}$$

then $S_{\theta=0}(w)$ is equal to $M(u,0)$. Because the orientation of the object in the (u,v) system is arbitrary with respect to the (x,y) system, this relation allows it to be used for all angles θ, keeping in mind that the Fourier transform conserves the rotation made in real space. The Fourier transform of $P_\theta(t)$ yields the values of the Fourier transform of $\mu(x,y)$. A dense set of values in the Fourier space (given in polar coordinates because of the polar scanning of the object) can be approximated to rectangular coordinates and back-transformed into real space. This may sometimes lead to diffuse reconstructions as a result of incomplete fitting for high frequencies in the Fourier domain (Fig. 6.7).

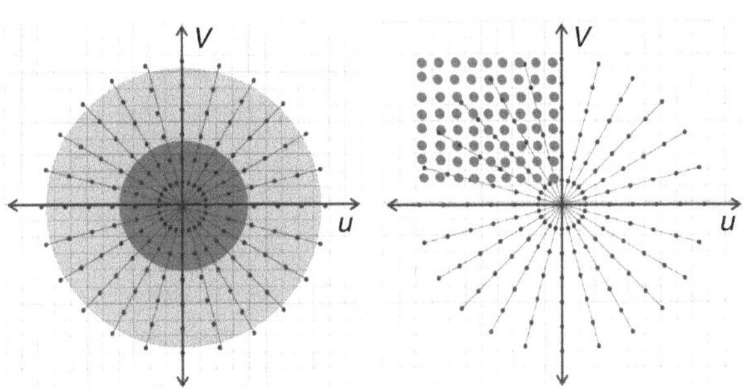

Fig. 6.7 Plot of Fourier-transformed projection P (*left*) and interpolation of the data onto a rectangular coordinate system (*right*). The interpolation for higher frequencies (large (u,v) values) becomes imprecise

How many projections yield the best result? This is answered by the Shannon theorem, which states that a unique reconstruction from a sampled object is obtained if the object is sampled with a frequency twice as high as the highest frequency in the Fourier-transformed image. For parallel mode scanning, using N steps for one projection, it can easily be derived that M projections (angles θ) are required to fulfill the Shannon theorem for tomography. If D is the diameter of the object to be scanned, and Δx is the difference between two scanning points, then the number of scanning points per projection is

$$N = \frac{D}{\Delta x} , \tag{6.12}$$

fulfilling the Shannon condition. The rotation of the sample around 180° means a path length y for a point sitting on the circle with the radius $D/2$. We require that $\Delta x = \Delta y$, so that the number of projections M must be

$$M = \frac{\frac{D}{2}\pi}{\Delta y} = \frac{D \cdot \pi}{2 \cdot \Delta y} . \tag{6.13}$$

Using $\Delta x = \Delta y$ results in a relation between the number of scanning points of a projection and the number of angles M

$$M \geq \frac{\pi}{2} \cdot N . \tag{6.14}$$

N and M determine the quality of the reconstruction. If M does not fulfill the inequality, i.e., if fewer than M projections are registered, then the reduced $D = D^*$ that fulfills the Shannon condition can be calculated for a smaller number of projections.

6.3.2 Image Reconstruction

As mentioned in the previous section, the back-transform (inverse Fourier transform) of the Fourier transform $S_\theta(w)$ can yield a 2D reconstruction of the slice $\mu = \mu(x,y)$. However, a more efficient (and elegant) method is the filtered back-projection (FBP), which is described in detail in [4] and [5]. The basic idea is embedded in the scanning mode: the sample (slice) is scanned from 0° to 180°, which involves $P_\theta(t) = P_{\theta+180}(-t)$ and suggests the use of a polar coordinate system rather than a rectangular one. The projections are registered in polar coordinates, so that the Fourier-transformed projections of $\mu(x,y)$, the $S_\theta(w)$ data, are discrete polar function values. So it seems to be appropriate to write $\mu(x,y)$ as a Fourier transform representation in polar coordinates. The function $\mu(x,y)$ written as the Fourier representation with $M(u,v)$ in rectangular coordinates is

$$\mu(x,y)\cdot = \int_{-\infty}^{\infty}\int_{-\infty}^{\infty} M(u,v)\cdot e^{2\pi\cdot i\cdot(ux+vy)}\,\mathrm{d}x\mathrm{d}y\ . \tag{6.15}$$

The same representation in polar coordinates is

$$\mu(x,y) = \int_{0}^{2\pi}\int_{-\infty}^{\infty} M(w,\theta)\cdot \mathrm{e}^{+2\pi\cdot i\cdot w\cdot(x\cos(\theta)+y\sin(\theta))}w\cdot \mathrm{d}w\cdot \mathrm{d}\theta\ . \tag{6.16}$$

The transformation of the (x,y) system into the (w,θ) system uses the relations

$$\begin{aligned} u &= w\cos(\theta) \\ v &= w\sin(\theta) \\ \mathrm{d}u\cdot \mathrm{d}v &= w\cdot \mathrm{d}w\cdot \mathrm{d}\theta\ . \end{aligned} \tag{6.17}$$

Substituting these formulae in the $\mu(x,y)$ Fourier representation yields (with the slice theorem)

$$\begin{aligned} \mu(x,y) &= \int_{0}^{\pi}\int_{-\infty}^{\infty}\left[M(w,\theta)\cdot \mathrm{e}^{+2\pi\cdot i\cdot w\cdot t}|w|\cdot \mathrm{d}w\right]\mathrm{d}\theta \\ &= \int_{0}^{\pi}\left[\int_{-\infty}^{\infty} S_{\theta}(w)\cdot \mathrm{e}^{+2\pi\cdot i\cdot w\cdot t}|w|\cdot \mathrm{d}w\right]\mathrm{d}\theta\ . \end{aligned} \tag{6.18}$$

The integral given in the square brackets can be considered as the Fourier transform of $P_{\theta}(t)$. However, $S_{\theta}(w)$ is multiplied by the $|w|$ function, which is a special filtering function of the frequencies in the frequency space. We define

$$Q_{\theta}(t) = \int_{-\infty}^{\infty} S_{\theta}(w)\cdot |w|\cdot \mathrm{e}^{+2\pi\cdot i\cdot w\cdot t}\cdot \mathrm{d}w\ . \tag{6.19}$$

With $t = x.\cos(\theta) + y.\sin(\theta)$, $\mu(x,y)$ then becomes simply

$$\mu(x,y) = \int_{0}^{\pi} Q_{\theta}(x\cos(\theta) + y\sin(\theta))\,\mathrm{d}\theta\ . \tag{6.20}$$

A product in Fourier space, when inverse Fourier-transformed, corresponds to the convolution of the inverse Fourier-transformed functions, i.e., we use the relation

$$\mathrm{FT}^{-1}\{\mathrm{S}_\theta(\mathrm{w}) \cdot |\mathrm{w}|\} = \mathrm{P}_\theta(\mathrm{t}) \otimes \mathrm{FT}^{-1}\{|\mathrm{w}|\} \ . \tag{6.21}$$

The convolution operation is denoted by the $\otimes$ symbol. The function $|w|$ is not a square integrable function, so it has no inverse Fourier transform. However, $\mathrm{FT}^{-1}\{|w|\}$ can be approximated by several filter response functions (convolution kernels, filter functions). The reconstruction of $\mu(x,y)$ from projections $P_\theta(t)$ now works in the following way. Each $P_\theta(t)$ is convoluted with a proper filter function (e.g., Shepp-Logan, $\left|\frac{\omega}{2\pi}\right| \sin c(\omega)$) and the resulting values are "smeared" over the (x,y) planes as demonstrated in Fig. 6.8.

To control the tomographic measurement, the recorded projections are arranged in a "sinogram," as shown in Fig. 6.9. Each vertical line represents one projection at θ and is plotted in gray levels as a function of the pixel number.

Figure 6.10 shows four different reconstructions. The projections $P_\theta(t)$ are converted into gray levels, convoluted with a filter kernel, and back-projected over the total (x,y) plane. Adding up all measured projections yields the desired 2D reconstruction of the slice. The more the projections overlap, the more the star-artifacts are suppressed. With such reconstructions, 3D images can be made by stacking up all the reconstructed slices and 3D volume data details (e.g., having different attenuation values) can be extracted from the rest of the body (Fig. 6.11).

A number of striking applications of neutron absorption tomography can be found in [3] and [7].

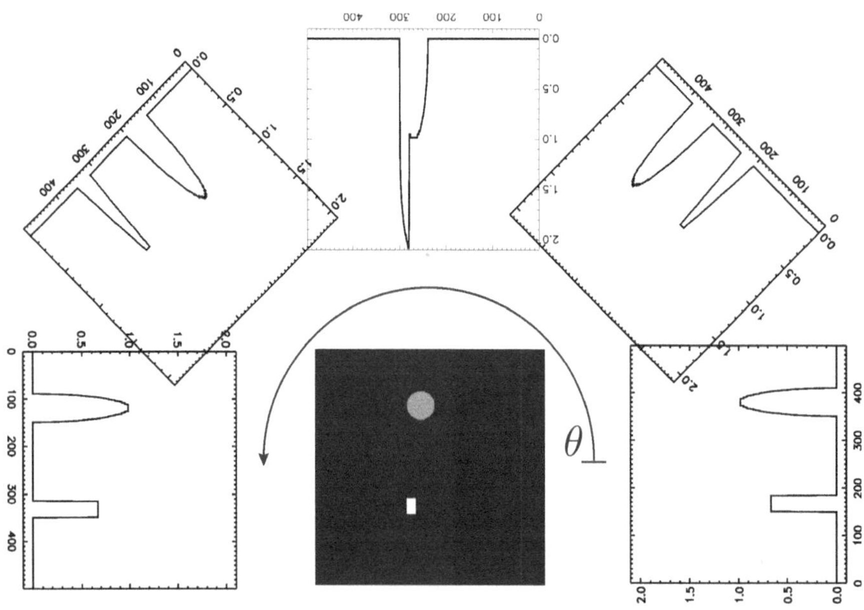

Fig. 6.8 Scanning of a simple 2D object (*square* and *circle*) and corresponding projections (figure courtesy of T. Donath, GKSS)

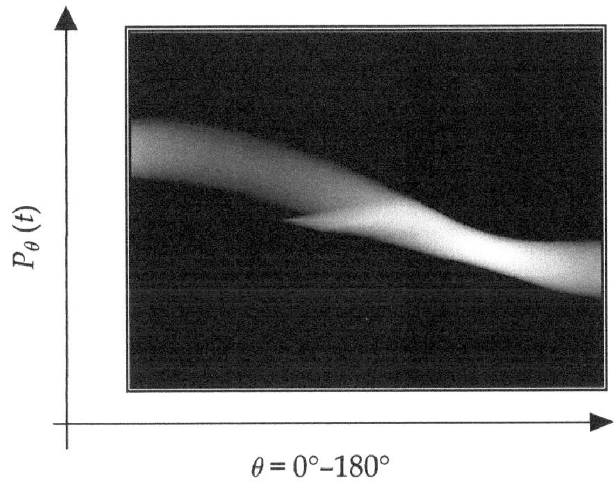

Fig. 6.9 The sinogram of Fig. 6.8. The *y*-axis shows the projection as *gray* values of Fig. 6.8

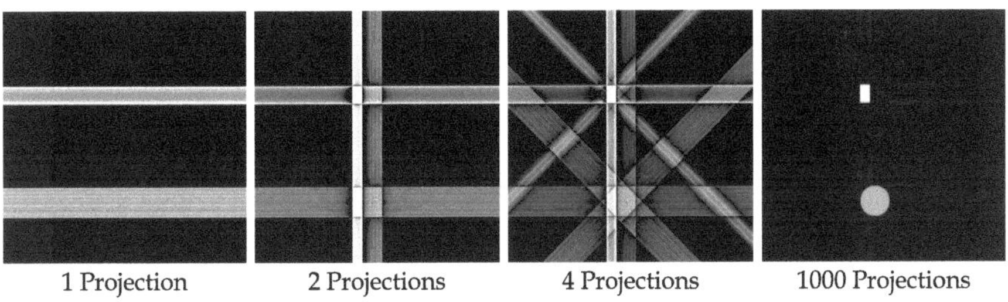

Fig. 6.10 Four 2D projections to illustrate the filtered back-projection procedure

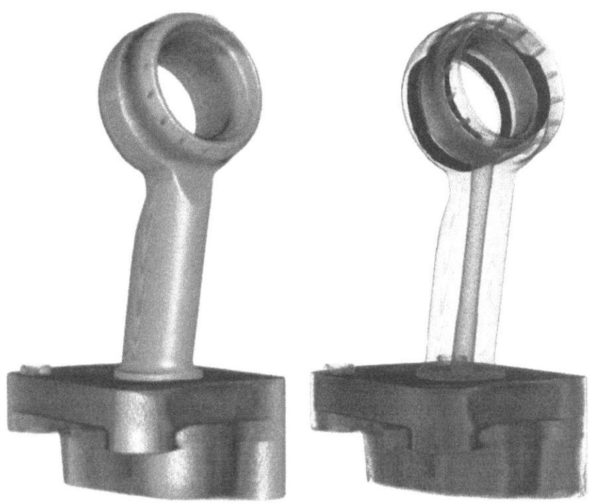

Fig. 6.11 Three-dimensional reconstructions of parts of a fuel injector device (height 115 mm, base 45 by 60 mm, ring outer diameter 35 mm). Metallic components (here not specified) can be extracted or suppressed by image processing

6.4 Experimental Techniques and Results

This section summarizes some techniques that are becoming increasingly relevant in neutron tomography and that will be used for a range of applications in the future.

6.4.1 Energy-Dispersive and Bragg-Edge Radiography and Tomography

The transmission (attenuation) of neutrons through matter depends on the composition of the material itself, the particular absorption (and scattering) cross sections, the wavelength (energy) of the neutrons used, temperature, isotopic composition and, also quite important, the crystalline structure and its orientation in the material. Materials consisting of polycrystalline components attenuate neutrons in a unique way because the thermal and cold neutron wavelength spectrum (λ = 0.05 nm to 1 nm) spans the typical crystal lattice constants. Neutrons having wavelengths of the order of the lattice constants d can fulfill the well-known Bragg condition and be scattered away from the incident direction.

If certain regions or parts of a material contain aluminum, for example, then the dependence of the attenuation can be observed from the wavelength that affects a radiograph, as shown in Fig. 6.12a and b (see also Fig. 1.4). This behavior is caused by the superposition of scattering interactions with the sample crystallites with a well-defined lattice spacing, but in random orientation. The integrated reflecting power increases with the Bragg angle, so that $\theta = 90^\circ$ and $\lambda = 2d$ result in sharp jumps in the attenuation spectrum. This

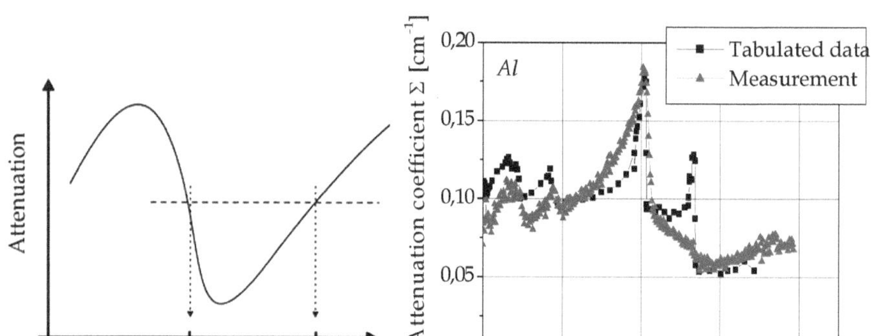

Fig. 6.12 (**a**) Attenuation of an element (polycrystalline) as a function of neutron wavelength (**b**) attenuation coefficient for aluminum in the range 2–6 Å [8]

behavior can be enhanced if well-oriented crystalline structures (textures) are present. The spectrum in Fig. 6.12 was measured with the CONRAD setup at the Helmholtz Center Berlin (HZB) (formerly the Hahn-Meitner Institut [HMI]), using a graphite double monochromator [8].

A similar behavior of the attenuation cross section is found for Fe, Cu, brass, and other crystals having lattice spacings of the order of the neutron wavelength. In the close neighborhood of a so-called Bragg edge, the attenuation (scattering away from the incident beam direction) rises as a result of the increasing integrated reflectivity [9] and then falls very quickly because the Bragg condition for a certain set of atomic planes is no longer fulfilled. This can be used to detect regions of strong deformations, as shown in Fig. 6.13. Here a steel plate was deformed and radiographs were taken at wavelengths in the range of 2.2 to 6.4 Å in steps of 0.1 Å. A 2D mapping of the Bragg edge for each pixel in the deformed plate is shown in the bottom image. This map can be related to the residual stress distribution in the plate.

This dependence of the attenuation upon the neutron wavelength enables sophisticated radiographies. If a material has the same linear attenuation coefficient Σ for two different wavelengths, it can be made transparent in a normalized image as follows. If $I_1(\lambda_1)$ and $I_2(\lambda_2)$ are the transmitted intensities at λ_1 and λ_2

$$I_1 = I_{01} \cdot \mathrm{e}^{-\Sigma(\lambda_1)\cdot d}, \quad I_2 = I_{02} \cdot \mathrm{e}^{-\Sigma(\lambda_2)\cdot d}, \tag{6.22}$$

then the ratio of the normalized intensities is

$$\frac{\frac{I_1}{I_{01}}}{\frac{I_2}{I_{02}}} = \frac{I_1 \cdot I_{02}}{I_2 \cdot I_{01}} = \frac{\mathrm{e}^{-\Sigma(\lambda_1)d}}{\mathrm{e}^{-\Sigma(\lambda_2)d}} = \mathrm{e}^{[\Sigma(\lambda_2)-\Sigma(\lambda_1)]d}. \tag{6.23}$$

In the case of $\Sigma(\lambda_1) \sim \Sigma(\lambda_2)$, for example, as shown in Fig. 6.12 (a), the exponential function becomes nearly unity and there will be no contribution of this material to the normalized image; i.e., this material becomes transparent in this image.

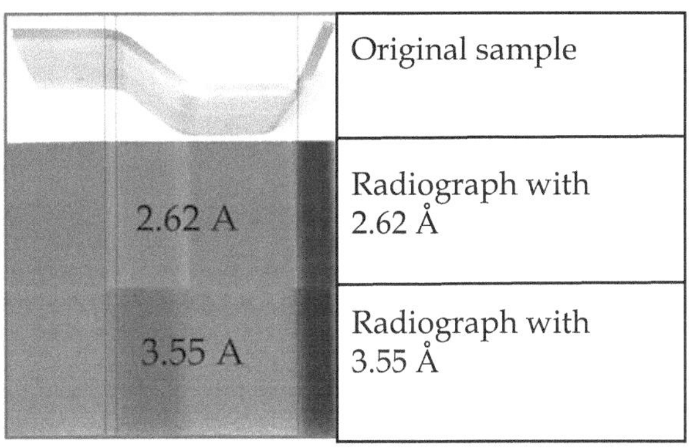

Fig. 6.13 Radiographs of a bent metal plate (50 × 10 × 5 mm, top picture) performed at two different wavelengths (2.62 and 3.55 Å). For details see [8]

Such measurements have been performed, for example, using CONRAD, a radiography and tomography instrument for cold neutrons at the BER II reactor (HZB). To be able to make radiographs at different wavelengths, a double monochromator system of two graphite crystals was used so that the wavelength could be tuned within the range of 2.8 to 6.2 Å [8]. It must be taken into account that the incident intensities I_{01} and I_{02}, corresponding to different wave lengths λ_1 and λ_2, are different owing to the wavelength spectrum. A set of step wedges of known composition and thickness (5, 10, 15, and 20 mm) was imaged at different neutron energies and the radiographs properly normalized to each other. The different behaviors of the attenuation coefficients are plotted for 3.14 and 6.40 Å and the transmitted intensities compared with each other as shown in Fig. 6.14.

This wavelength dependence of the neutron attenuation enables other sophisticated radiographies, similar to the examples shown in Fig. 6.15, to enhance or

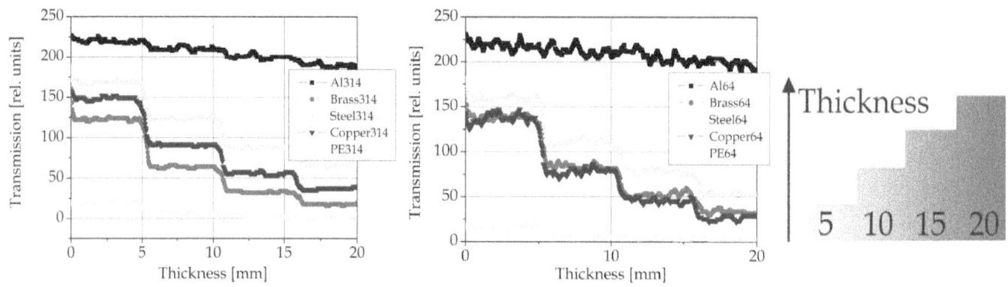

(a) (b) (c)

Fig. 6.14 Different wedge transmissions for two wavelengths, (**a**) 3.14 Å and (**b**) 6.40 Å, for aluminum brass, steel, copper, and polyethylene (PE). (**c**) Step wedges: 5–20 mm thickness. To visualize the effect of normalization, the transmitted intensities have been converted into gray values (compare with Fig. 6.15)

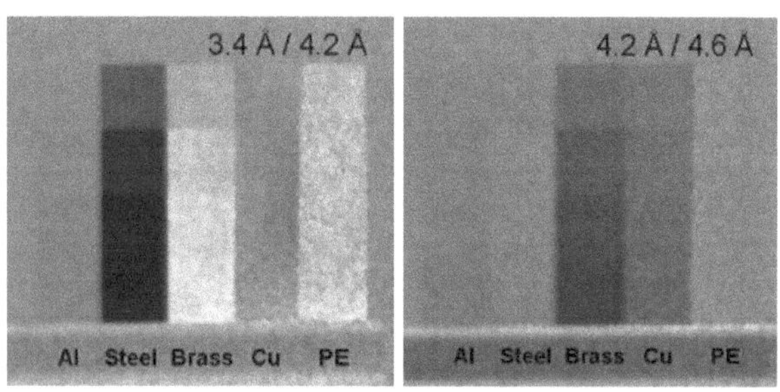

(a) (b)

Fig. 6.15 Ratios of properly normalized radiographs of polyethylene and metal wedges for different wavelengths. Some materials become transparent for particular wavelength ratios, for example, copper in (**a**) and steel in (**b**) [8, 9] suppress components in a sample under investigation. This technique of normalized intensities yields an exponential expression, which can significantly increase or decrease parts in an image of a sample and so improve the interpretation of images.

Some other examples of energy-dispersive radiography are shown in Chapter 12.

6.4.2 Real-Time Radiography

A number of technical processes must or should be directly controlled on-line. Engineers and scientists want to see, for example, how oil flows in machines and how water flows in fuel cells (Chapter 11) or is taken up in plants (Chapter 17) and so on, because from these observations they can modify their theories or designs in order to optimize machines, to save energy, and to understand natural processes. The process to be visualized determines the method of collecting data, because one must distinguish between repetitive and simple emerging nonrepetitive processes. In the case of nonrepetitive processes, each frame must image (by absorption) an instantaneous state of the process within a certain time interval Δt. The process within a time interval Δt itself can be visualized by registering as many frames as possible. In the case of an oscillating piston of a model aircraft engine, the motion was triggered so that different positions of the piston could be imaged at different times [8, 9].

The sharpness of the images depends on the number of neutrons that can be collected within one pixel (Δx) within the time interval Δt (i.e., per frame) and how well defined the triggering is. It can easily be understood that the exposure time Δt must be small compared with the velocity of the process for a given Δx; i.e., any detail of the moving object being imaged should move less than half a pixel within Δt [11, 12]. See Chapter 12 for examples of what is currently achievable.

6.4.3 Phase Contrast, Refraction, and Small-Angle Tomography

In considering neutrons not as particles but as de Broglie waves, all properties that are connected with the nature of these waves must be taken into account. The most important feature of this representation is the coherence property of the waves. One can adapt from optics the description of Fresnel and calculate the coherent area of a source emitting coherent (neutron) waves. This area is given by the first Fresnel zone and its size A_{FZ} is

$$A_{\mathrm{FZ}} = \frac{R \cdot r}{R + r} \cdot \pi \cdot \lambda \cdot \frac{1}{\sin(\theta)} \,. \tag{6.24}$$

Here R is the distance from the point source to the object, r is the distance between object and detector, and θ is the angle between the ray propagation vector and the plane of observation (usually $\theta = 90°$). (Compare with drawings in classical optic books). The size A_{FZ} determines the coherent source for any interference effects to be observed. For standard geometries in neutron tomography (and radiography), $R \sim 500$ cm, $r \sim 1$ cm, and $\lambda \sim 0.5$ nm, so A_{FZ} is $\sim 1.6 \times 10^{-7}$ cm^2 and the diameter of the coherent source can roughly be estimated as $\sim 4\ \mu$m. To use the coherent properties of such a beam, a point-like source must be created by a small pinhole, which determines—together with the divergence of the beam—the coherent area of the real neutron beam. The diameter d_{coh} characterizing the coherent size of a source can be estimated by simple geometric considerations as given by [13].

$$d_{coh} = \lambda \cdot \frac{1}{2}\frac{L}{D} \cdot \qquad (6.25)$$

To achieve a reasonable d_{coh}, the source (i.e., D) must be kept small and L must be large (meters).

To understand the phase contrast, consider a (plane) wave traveling through an object that could be smaller or larger than the coherent wave front of the wave (Fig. 6.16).

Phase contrast originates from a coherent superposition of diffracted waves; the resulting intensity variation can be observed only in the coherent part of the wave that propagates through the object and experiences position-dependent phase shifts caused by different indices of refraction in the sample. This modulates the wave front, and a change of propagation direction will be observed. The strongest phase contrast is observed in the near field region, $z < d_{coh}^2/\lambda$, where the modulation of the amplitude is best; it broadens with increasing z [13]. (z is the distance from object to detector and $d_{coh} \sim$ dimension of the coherence width.) Fringes can be registered that are caused, for example, by an edge that would be washed out if the beam divergence and/or

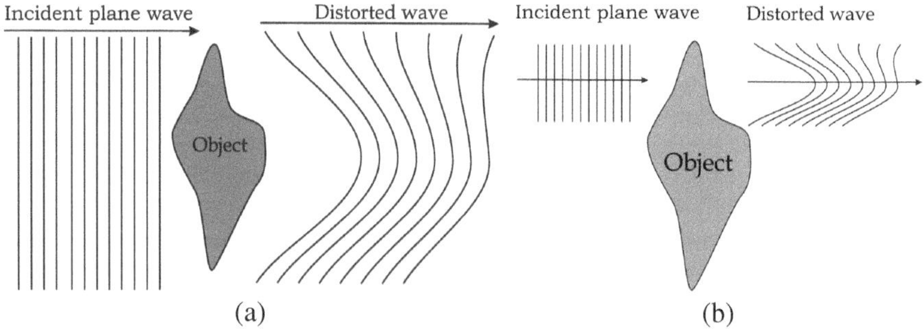

Fig. 6.16 An object is illuminated by a coherent wave front, which can be (**a**) larger or (**b**) smaller than the object

source size were too large. It can be calculated that the optimal contrast occurs at $z = d_{coh}^2/2\lambda$ [13]. One approach for realizing phase contrast was performed by neutron radiography some years ago [14]. The first experiments based on phase effects (refraction) were performed in 1989 using a double crystal diffractometer [15]. The first neutron radiography based on phase contrast was performed in 2000 at the National Institute for Standards and Technology (USA) using a pinhole and a wavelength of 4.43 Å and with the sample placed 1.8 m from the source [16, 17]. However, phase contrast imaging is much more often used with synchrotron radiation because of the higher brilliance of the source.

One can calculate the phase contrast, determining the Fourier-transform of the structure or of the phase boundary of two structures. This can be done by ray tracing, assuming that a certain width of the edge is coherently illuminated. Here we assume that an incident plane wave hits a phase step (boundary of two different materials with a difference of 10 μm in thicknesses). However, in general, one can also use ray-tracing, adding up all amplitudes emerging from the structure that cause the phase gradient that contributes to a measuring point showing an intensity distribution as plotted in Fig. 6.17, top.

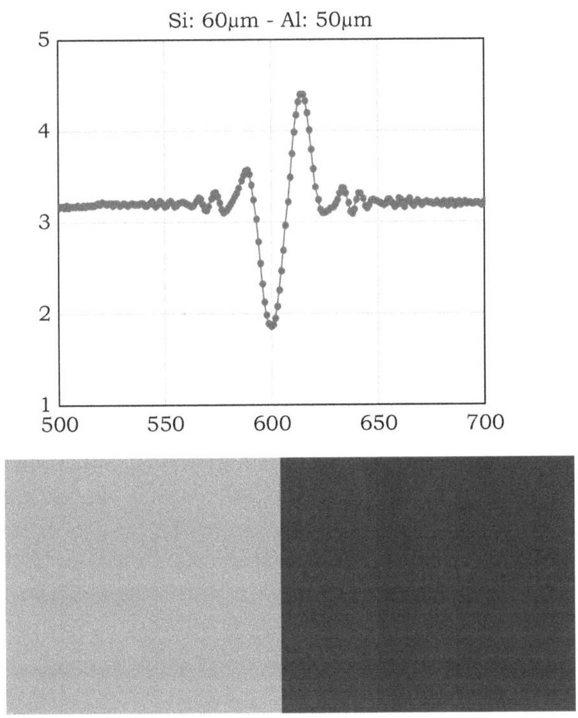

Fig. 6.17 Calculated phase contrast (*top*; *x*-scale in μm, *y*-scale in arbitrary units) for a phase step consisting of silicon and aluminum (*bottom*) [12]

The phase development quite close behind the sample yields strong intensity oscillations that can be measured at different distances from the object [18]. Very often, another concept is used to describe this phase contrast correctly via geometric considerations of refraction by edges [19]. The latest developments use a phase grating-based shearing interferometer; expressed simply, this senses the phase-modulated (refracted) beam behind the sample using an analyzing grating. The measured intensity depends on the difference in phase shift introduced by two adjacent regions in the sample that are coherently illuminated. Several images must be measured with varying phase grating positions and the phase change due to the object then calculated. Additional details concerning this technique can be found in Chapter 8 and in [20, 21].

6.4.4 Refraction and Small-Angle Scattering Tomography

As was shown earlier, the size of the coherence area of the neutron (X-ray) wave with respect to the object determines the kind of interaction—refraction or small-angle scattering. If the coherent width (lateral coherence length) is smaller than the object, then refraction occurs (Fig. 6.16). If the object is smaller than the coherent width, small-angle scattering occurs. To study this behavior, we consider a function $n(x,y)$ that describes the position-dependent index of refraction $n(x,y) \neq 1$ inside the sample and $n(x,y) = 1$ outside it. The Snellius law states that $n_1\sin(\varepsilon_1) = n_2\sin(\varepsilon_2)$, ε_1 being the angle of incidence and ε_2 the angle of refraction; n_1 and n_2 are the indices of refraction of medium 1 and medium 2.

From geometrical optics it is also well known that the larger the ε_1 is, then the larger ε_2, up to the case of total reflection. Neutrons leaving the sample will have a final angle of deflection Θ, which is the sum of all individual deflections δ_ι ($\delta_\iota = \varepsilon_\iota - \varepsilon_{\iota+1}$) along the path through the sample (Fig. 6.18):

$$\Theta = \sum_{i=1}^{n} \delta_i \, . \tag{6.26}$$

A deflection δ_ι from the preceding direction is present for adjacent volumes with different indices of refraction and if $\vec{k} \cdot \nabla n(x,y)$ ($\vec{k}$ is the neutron wave vector). One can determine $\{\nabla n(x,y)\}$, and from that $n(x,y)$, by measuring the Θ of each path (line integral) and varying the orientation θ of the sample from 0 to 360°. If $\vec{k} = (k_{\|}, k_{\perp})$ is represented in terms of components parallel and perpendicular to$\nabla n(x,y)$, and $\nabla n(x,y)$is the component of $\nabla n(x,y)$ perpendicular to $k_{\perp}$, then a point $P_\theta(t)$ of a projection is given by [22]

$$P_\theta(t) = P(t,\theta) = \int_{\text{path}} \nabla n(x,y) \cdot \bar{k}_{\perp} \cdot \mathrm{d}s \, . \tag{6.27}$$

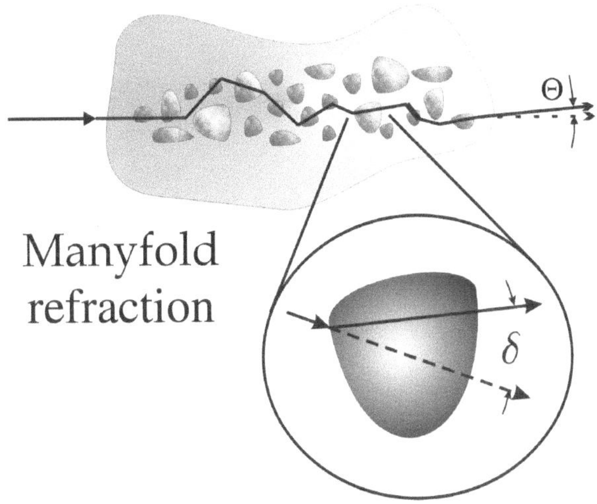

Fig. 6.18 Neutrons are refracted several times in the sample (e.g., by inhomogeneities) and exit with a deviation angle Θ

The change of the phase $\Delta\varphi(x,y)$ (here Δ is not the Laplacian operator) can be expressed as the sum of all deviations that the ray experiences:

$$\Delta\varphi(x,y) = -k \cdot \int_{\text{path}} \delta(x,y) \cdot \mathrm{d}s \,. \tag{6.28}$$

The data corresponding to the phase changes are measured as refraction angles $\Theta = \Theta(t,\theta)$, which represent the orientation of the object in the (x,y) system. Θ is large (i.e., can be measured) if the incident ray is nearly parallel to the interface that separates two regions with different indices of refraction; so Θ can be written as

$$\Theta(\theta,t) = \frac{1}{k}\nabla_{\perp}(\Delta\varphi(x,y)) \,. \tag{6.29}$$

So $P_\theta(t)$ becomes

$$\begin{aligned} P_\theta(t) &= \int_0^t \Theta(x, y(x,\theta,t')) \cdot \mathrm{d}t' \\ &= \int_{-\infty}^{\infty}\int_{-\infty}^{\infty} \delta_{\text{Dirac}}(x\cos(\theta) + y\sin(\theta) - t) \cdot \mathrm{d}x \cdot \mathrm{d}y \,. \end{aligned} \tag{6.30}$$

To measure the (resulting) refraction angle Θ, the incident direction must be known; Θ and $P_\theta(t)$ can be determined from the experiment. The total set of functions $P_\theta(t)$ is then another form of the Radon transform of $n(x,y)$, so the reconstruction of $n(x,y)$ from projections can follow well-known mathematical procedures. An estimation of the total angle of deflection is given by the calculation of the index of refraction n, which is expressed for neutrons by

$$n = 1 - \frac{\lambda^2 N b_c}{2\pi} \;. \tag{6.31}$$

In this expression, $\lambda \sim 5 \times 10^{-10}$ [m], N = number of unit cells $\mathrm{m}^{-3} \sim 5 \times 10^{28}$, and b_c = coherent scattering length $\sim 10^{-15}$ [m] of the sample; i.e., $1-n \sim 10^{-5}$ to $10^{-6} \sim \mu$rad gives the size of the angle of refraction that must be measured. This can be done only with a special double crystal instrument, as described in [22].

In a case in which the extent of the object illuminated by the neutron wave is smaller than the lateral coherence width, the whole object acts as a phase shifter and small-angle scattering is observed (compare Figs. 6.16 and 6.19).

In the case of (ultra) small-angle neutron scattering (USANS) tomography, $\mu(x,y)$ must be replaced by a suitable function that represents the USANS effects in the sample. One feature of small-angle scattering—single scattering events assumed—is the individual scattering function of the particles in the sample. From this scattering function, the size and shape of the particle can be deduced. The measured scattering function is the convolution of the particle shape function with the instrumental resolution function, which is the rocking curve (RC) or 2D scattering pattern without the sample. The broadening $B_\theta(t)$ [nm^{-1}] of a scattering function due to (ultra) small-angle scattering (t again means the scanning parameter defined in Eq. [6.4]) can be (roughly) approximated by a Gaussian distribution and, involving multiple scattering, written as [24, 25]

$$B_\theta(t) = \sqrt{\int\limits_{\mathrm{path}} \frac{\sigma(x,y) \cdot N(x,y)}{R(x,y)^2} \cdot \mathrm{d}s} \;, \tag{6.32}$$

where σ is the scattering cross section, $N(x,y)$ is the particle density distribution, and R is a parameter with the dimension of a length specifying an average size or correlation length in the scattering object (corresponding to the Gaussian approximation). Now one can use B or B^2 as an imaging signal to use (ultra)

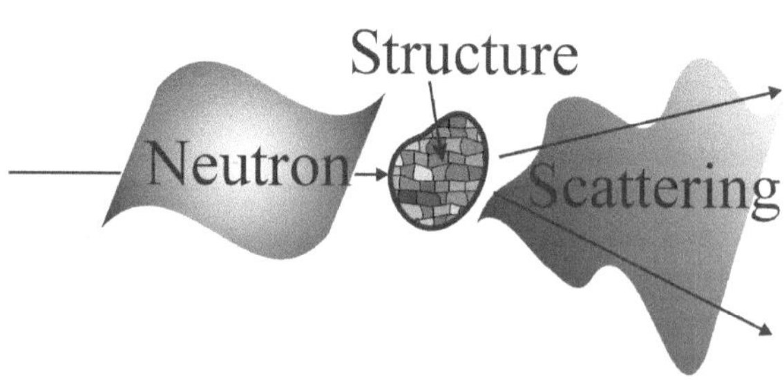

Fig. 6.19 The coherent incident wave front is larger than the object. The scattered wave contains all phase changes due to the object and therefore yields its 2D Fourier transform

small-angle scattering for image reconstruction. A function $f_B(x,y)$ can be defined as

$$f_B(x,y) = \frac{\sigma(x,y) \cdot N(x,y)}{R^2(x,y)} \,, \tag{6.33}$$

so that a projection $P_\theta(t)$ can be rewritten as $P_\theta(t) = B^2(\theta, t)$

$$P_\theta(t) = B^2(\theta, t) = \int\limits_{\text{path}} f_B(x, y(x, \theta, t)) \cdot \mathrm{d}s \,. \tag{6.34}$$

With this proper description of $P_\theta(t) = B^2(\theta,t)$, the standard methods of reconstruction such as the FBP can be used. Note that, in both cases, the wave properties of a neutron are responsible for the phase change due to the inner sample structure; i.e., the individual phase shifts in the sample add coherently to, and result in a final phase change of, the wave behind the sample. So, in both cases, we can call this phase contrast because the fundamental interaction is the change of the phase front due to structures in the sample. In the case of refraction, the wave is refracted many times before it leaves the object and has a different direction from the incident direction in front of the sample. The change of the wave front (mean direction) of the beam can be measured for each path of the neutron through the sample, forming (for a given θ) a projection $P_\theta(t)$ that indicates for each t a certain angular shift of the intensity distribution.

The same occurs in the case of small-angle scattering. Here the neutron wave is scattered behind the sample as a result of phase shifts in the sample. This causes a deformation of the wave front behind the sample (Figs. 6.16 and 6.19) and is measured as a broadening $B_\theta(t)$ of the intensity distribution $I_\theta(x,y)$.

Usually one must deal with three different effects—absorption, refraction, and small-angle scattering—because all of them will occur simultaneously in a sample. The signals can be represented as follows:

1. Refraction shifts the total RC without a decrease in the total (integrated) intensity and leaves the shape unchanged (Fig. 6.20a).
2. Small-angle scattering also has no influence on the total scattered intensity; however, the shape of the RC is changed, the peak intensity is decreased, and the wings of the curve are enhanced (Fig. 6.20b).

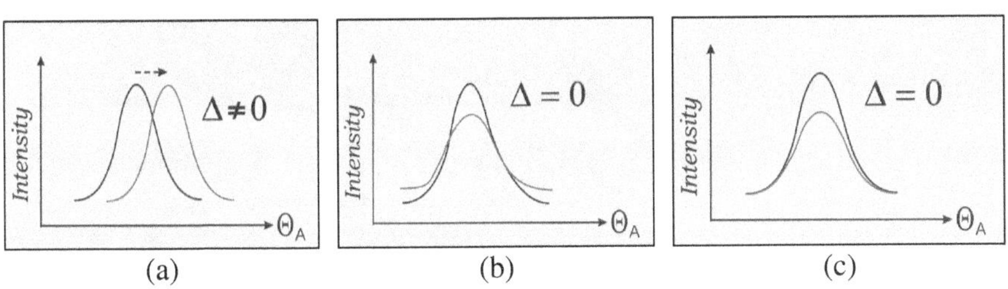

Fig. 6.20 Rocking curves due to refraction (**a**), small-angle scattering (**b**), absorption (**c**). Θ_A is the rotation angle of the analyzer crystal

3. As a result of absorption caused by the sample, the total intensity is decreased, but the shape of the RC remains nearly unchanged (Fig. 6.20c).

Note that the angular shift due to refraction is on the order of μrad, the broadening of the order of 10^{-3} to 10^{-5} nm^{-1}, so both signals are rather difficult to measure with standard tomography setups. Moreover, the fundamental problem with such imaging signals (refraction and small-angle scattering) is to separate absorption, refraction, and small-angle scattering clearly from one another. In a real experiment, all three interactions may happen simultaneously, so a single measurement should register all of them at the same time. In principle, one can assess them separately; however, the best way is to determine all three signals simultaneously as is done with a special double crystal diffractometer, shown in Fig. 6.21.

The sample under investigation is placed between the silicon monochromator and the silicon analyzer crystal, as shown in Fig. 6.21. Both are perfect crystals; the analyzer is bent so that the reflecting net planes continuously change their orientation with respect to the incident beam, which enters perpendicular to the front face of the analyzer crystal (Fig. 6.21). Neutrons traveling parallel to the surface are reflected by those net planes that fulfill the Bragg condition, i.e., entering within the so-called Darwin (angular) width ($\sim$ 10–60 μrad). The strong angular correlation between both crystals guarantees an angular sensitivity of order μrad or, in terms of momentum transfer q units, some 10^{-4} nm^{-1}. Any deviation of the incident beam direction is measured with a 1D or 2D position-sensitive detector. Each detector channel corresponds to a certain deviation angle or momentum transfer $q = \left(\frac{4\pi}{k}\right) \cdot \sin(\theta)$, where $k = 2\pi/\lambda$ and $\theta =$ half scattering angle, defining the peak position of the RC at $q = 0$ and as well for $\Theta = 0$ (cp Fig. 6.20).

The sample is scanned by steps perpendicular to the neutron beam. For each step, the total RC, and so the shift (center of gravity of the curve), is measured. The worst cases yield no absorption contrast, and only refraction contrast yields a good reconstruction, as demonstrated in Fig. 6.22.

With this technique a 3D-refraction tomography could also be realized, stacking reconstructed slices upon each other (Fig. 6.23) [26].

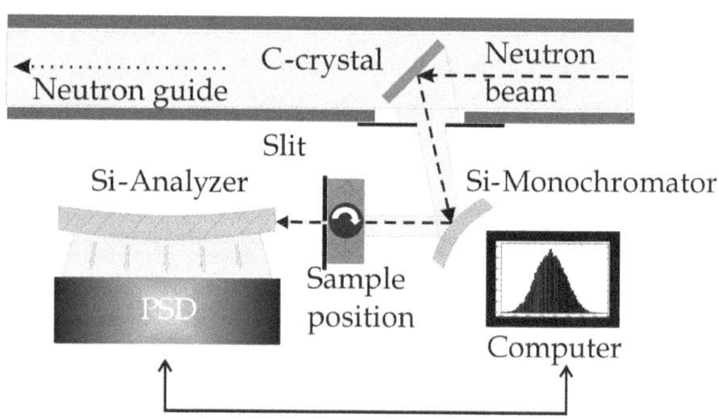

Fig. 6.21 The double crystal diffractometer V12a at the HZB

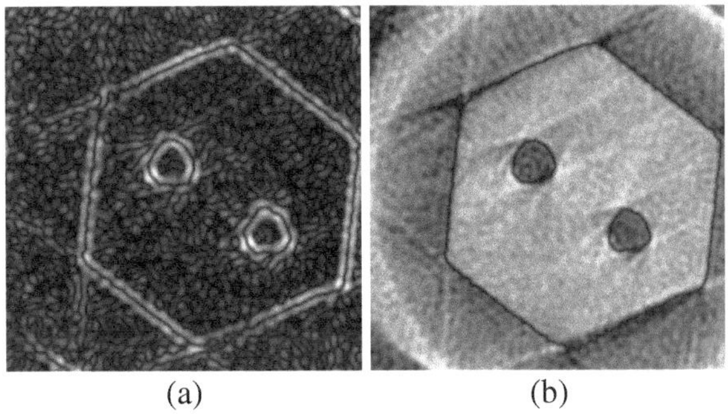

Fig. 6.22 Reconstructions of (**a**) the gradient and (**b**) the integrated gradient of the index of refraction (sample aluminum, diameter 6 mm, holes 2 mm, 60 projections)

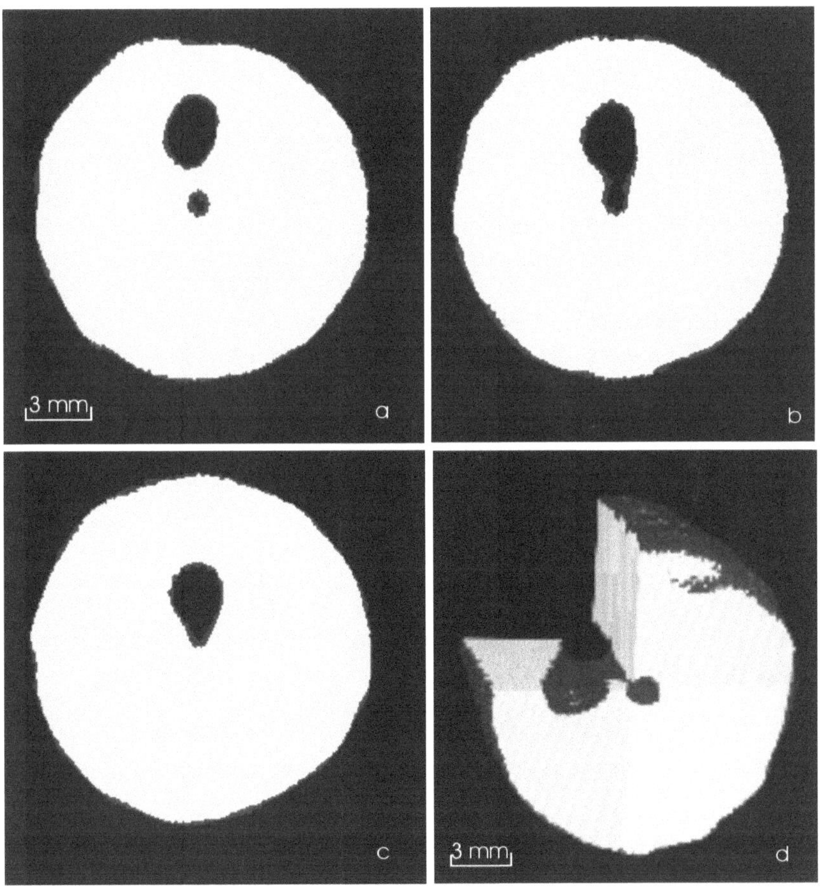

Fig. 6.23 Three-dimensional reconstruction based on pure phase (refraction) contrast (aluminum sample, diameter 15 mm, central hole 1 mm, oblique hole 3 mm)

The most astonishing result was the detection of clusters of particles (ß-carotene) having a size of approximately 150 nm, dissolved in D_2O (~ 5–12%). In this experiment all effects—refraction, USANS, and absorption—were measured simultaneously with the double crystal diffractometer. The shift of the RC yields the final deviation and determines the refraction, the broadening of the small-angle scattering, and the decrease of the integral intensity of the absorption of the neutron beam. All these data are derived from a single curve. Combining USANS and refraction data yields a complete reconstruction. With pure attenuation, and only incomplete refraction or USANS data, imperfect images are reconstructed [24, 27].

Chapter 16 describes the use of modeling to reconstruct the shapes of biological macromolecules with a typical size of <100 nm based on small-angle scattering data. In combination with the methods described here, this offers the possibility of covering the full range of length scales from macroscopic to nanoscopic with neutron imaging methods.

6.5 Outlook

There are several other fields of application of neutron tomography that should be mentioned here. In stress and strain investigations (Fig. 6.13), the investigation time can be drastically reduced by 3D scanning of a sample using Bragg-edge and energy-resolved techniques (Chapter 12). There seems to be a great interest in investigating quasi-in-situ water distribution in operating fuel cell stacks by neutron tomography (Chapter 11 and e.g., [30]), and this also holds true for applications in the fields of moisture detection and geological sampling (Chapter 17) and noninvasive investigation of cultural heritage objects (Chapter 13 and, e.g., [3]). Another interesting topic will be the use of polarized neutrons (Chapter 10). The imaging signal is the change of the spin direction with respect to the orientation in front of the sample; i.e., the 2D-depolarization image of a neutron beam will be measured. The neutron depolarization technique is well established; the problem is to calculate a unique reconstruction, for example, of two-dimensionally (three-dimensionally) randomly arranged small magnetic fields in samples, such as complicated magnetic flux distributions in solids, magnetic domains, etc. Several attempts to solve this problem have been published [31]; however, a reconstruction technique applied to refraction or small-angle tomography similar to that discussed in [22] or iterative techniques seem to be most challenging and promising.

References

1. S.G. Glasstone, M.C. Edlund, *Kernreaktortheorie*, Wien, Springer-Verlag, p. 30 (1961).
2. A.A. Harms, D.R. Wyman, *Mathematics and Physics of Neutron Radiography*, D. Reidel Publishing Company, Dordrecht, Holland (1986).

3. E. Lehmann, P. Vontobel, R. Hassanein, Neutron Tomography as Tool for Applied Research and Technical Inspection, *Adv. Solid St. Phys.* **45**, 389–405 (2005).
4. A.C. Kak, M. Slaney, *Principles of Computerized Tomographic Imaging*, IEEE Ed., N.Y. (1999).
5. A. Rosenfeld and A.C. Kak, *Digital Picture Processing, Computer Science and Applied Mathematics*, Academic Press Inc. (1982).
6. G.T. Herman, *Image Reconstruction from Projections: The Fundamentals of Computerized Tomography*, Academic Press, New York (1980).
7. E.H. Lehmann, P. Vontobel, G. Frei, The non-destructive study of museums objects by means of neutrons imaging methods and results of investigations, *Il Nuovo Cimento*, 30 C, N. 1, 93–104 (2007).
8. W. Treimer, A. Hilger, M. Strobl, N. Kardjilov, I. Manke, a wavelength tunable device for neutron radiography and tomography, *Appl. Phys. Lett.* **89**, 203504 (2006).
9. R.W. James, The optical principles of the diffraction of X-rays, *The Crystalline State – Vol. II*, ed. Sir L. Bragg, G. Bells and Sons Ltd. 1967.
10. N. Kardjilov, A. Hilger, I. Manke, M. Strobl, W. Treimer, J. Banhart, Industrial applications at the new cold neutron radiography and tomography facility of the HMI, *Nucl. Instrum. Methods Phys. Res. A* **542**, 16–21 (2005).
11. W. Treimer, A. Hilger, N. Kardjilov, M. Strobl, Review about old and new imaging signals for neutron computerized tomography, *Nucl. Instrum. Methods Phys. Res. A* **542**, 367–375 (2005).
12. W. Treimer, N. Kardjilov, U. Feye-Treimer, A. Hilger, I. Manke, M. Strobl Absorption- and phase-based imaging signals for neutron tomography, In Bernhard Kramer (Ed.) *Adv. Solid St. Phys.* **45**, 407–420, ISSN 1438-4329 (2005).
13. X. Wu, H. Liu, Clinical implementation of x-ray phase-contrast imaging: Theoretical foundations and design considerations, *Med. Phys.* **30**, 8, 2169 (2003).
14. A. Pogany. D. Gao, S.W. Wilkins, Contrast and resolution in imaging with a microfocus x-ray source, *Rev. Sci. Instrum.* **68**, 2774 (1997).
15. K.M. Podurets, V.A. Somenkov, S.Sh. Shilshtein; Neutron radiography with refraction contrast, *Physica B*, **156**, 691–693 (1989).
16. B.E. Allman, P.J. McMahon, K.A. Nugent, D. Paganin, D.L. Jacobson, M. Arif, S.A. Werner, Phase radiography with neutrons, *Nature* **408**, 158 (2000).
17. P.J. McMahon, B.E. Allman, D.L. Jacobson, M. Arif, S.A.Werner, K.A. Nugent, Quantitative phase radiography with polychromatic neutrons, *Phys. Rev. Lett.* **91**, 14 (2003).
18. N. Kardjilov, E. Lehmann, E. Steichele, P. Vontobel, Phase contrast radiography with a polychromatic neutron beam, *Nucl. Instrum. Methods Phys. Res. A* **527**, 519–530 (2004).
19. M. Strobl, W. Treimer, N. Kardjilov, A. Hilger, S. Zabler, On neutron phase contrast imaging, *Nucl. Instrum. Methods Phys. Res. B* **266**, 181–186 (2008).
20. F. Pfeiffer, T. Weitkamp, O. Bunk, C. David, Phase radiography with neutrons, *Nat. Phys.* **2**, 258 (2006).
21. F. Pfeiffer, C. Grünzweig, O. Bunk, G. Frei, E. Lehmann, C. David, Neutron phase imaging and tomography. *Phys. Rev. Lett.* **96**, 215505 (2006).
22. W. Treimer, M. Strobl, A. Hilger, A.C. Seifert, U. Feye-Treimer, Refraction as imaging signal for computerized neutron tomography. *Appl. Phys. Lett.* **83**, 398 (2003).
23. M. Strobl, W. Treimer, A. Hilger, U. Feye-Treimer, Neutron tomography in double crystal diffractometers, *Physica B* **350**, 1–3, 155–158 15 July, (2004).
24. M. Strobl, W. Treimer, A. Hilger, Small angle scattering signals for (neutron) computerized tomography, *Appl. Phys. Lett.* **85**, 3, 488–490 (2004).
25. T.M. Sabine, W.K. Bertram, The use of multiple-scattering data to enhance small-angle neutron scattering experiments, *Acta Cryst. A* **55**, 500 (1999).
26. M. Strobl, W. Treimer, A. Hilger, First realisation of a three-dimensional refraction contrast computerised neutron tomography, *Nucl. Instrum. Methods Phys. Res. B* **222**, 653–658 (2004).

27. W. Treimer, M. Strobl, A. Hilger, H.J. Peschke, Neutron tomography using small angle scattering data, *IEEE* **52**:1, 386–288 (2005).
28. N. Kardjilov, I.Manke, M. Strobl, A, Hilger, W. Treimer, M. Meissner, T. Krist, J. Banhart, Three-dimensional imaging of magnetic fields with polarized neutrons. *Nat. Phys.* **4**, 399–403 (01 May 2008).
29. W. Treimer, BMBF project 03TR7TFH 2007–2010.
30. I. Manke, Ch. Hartnig, M. Grünerbel, J. Kaczerowski, W. Lehnert, N. Kardjilov, A. Hilger, J. Banhart, W. Treimer and M. Strobl, Quasi-in situ neutron tomography on polymer electrolyte membrane fuel cell stacks, *Appl. Phys. Lett.* **90**, 184101 (2007).
31. E. Jericha, R. Szeywerth, H. Leeb, G. Badurek, Reconstruction techniques for tensorial neutron tomography, *Physica B* **397**, 1–2, 159–161 (2007).

Chapter 7
Mathematics of Neutron Imaging

K.W. Tobin, P.R. Bingham, and J. Gregor

Abstract Imaging with neutrons at macro-world scales (e.g., $>10\,\mu m$) requires particular understanding of the non-diffracting or refracting interactions that these electrically neutral particles have with their environment. While image formation with neutrons shares some commonality with other radiation sources such as X-rays and gamma rays, neutrons provide complementary interaction mechanisms to these techniques that can uniquely probe materials and structure. Neutron sources are also historically fraught with issues regarding low source intensities, challenging beam configurations and resolution limitations that are all closely linked. In this chapter we will present a mathematical construct and methods compatible with the design and characterization of radiography systems for volumetric imaging. We begin with a review of neutron image formation, provide resolution analysis concepts and methods for both the design and the characterization of radiography systems, and conclude with a review and discussion of volumetric reconstruction techniques using analytic or iterative computed tomography algorithms.

Keywords Neutron radiography · Computed tomography · Modeling · Resolution · Contrast

7.1 Introduction

Neutron imaging has long been known to provide complementary, nondestructive imaging capabilities to X-ray and gamma-ray imaging methods [3]. Today, neutron imaging with conventional reactor-based sources enables the interrogation of complex, multicomponent systems for many applications, such as nuclear material nondestructive testing [9], characterizing flight control surfaces on aircraft [4], testing heat transfer in porous materials [30], examining

K.W. Tobin (✉)
Image Science and Machine Vision Group, Oak Ridge National Laboratory, Oak Ridge, Tennessee 37831-6075
e-mail: tobinkwjr@ornl.gov

I.S. Anderson et al. (eds.), *Neutron Imaging and Applications*,
Neutron Scattering Applications and Techniques, DOI 10.1007/978-0-387-78693-3_7,

heat exchanger systems [2, 10], development of hydrogen fuel cells (Chapter 11), inspection of cultural heritage objects (Chapter 13), and interrogating biological systems [28]. Newer, intense sources of neutrons from spallation facilities are providing the potential to interrogate time and energy-dependent phenomena as well. Therefore, a mathematical representation of the neutron image formation process can provide insight into the potential application of time- and energy-dependent, volumetric radiography to the characterization and quantitation of complex material structures.

The mathematics of neutron imaging is similar to that of other non destructive methods that use non-refracting or diffracting radiation to interrogate materials or composites of materials [3]. Figure 7.1 represents the basic elements of a generic neutron radiography system beginning with a source of neutrons, beam conditioning and shaping, neutron conversion and detection, and sampling and signal generation.

For the purposes of this discussion, neutron image formation will be treated as an encoding of the material characteristics of a three-dimensional (3D) object projected onto a two-dimensional (2D) detection array, typically comprising a converter followed by an imaging sensor as described in Chapter 5. Material characteristics of the object are represented by the total macroscopic interaction cross-section, Σ_T, which is a sum of the absorption and scattering cross-sections, $\Sigma_T = \Sigma_a + \Sigma_s$ (i.e., ignoring the potential for neutron capture and fission cross-sections in some materials). The interaction of the neutron beam, ϕ_o, with the 3D object produces a total 2D field at the converter plane, ϕ_T, that can be decomposed into the uncollided flux, ϕ_u, and the scattered flux, ϕ_s, as $\phi_T = \phi_u + \phi_s$. The uncollided flux arises from neutrons that arrive at the converter plane unperturbed by the imaged object while the scattered flux considers only neutrons that have been scattered by the object toward the converter plane.

To understand the dependencies of the scattered flux on the image formation process, we introduce a buildup factor [20, 24] defined as $B = \phi_T/\phi_u = 1 + \phi_s/\phi_u$. Rearranging to solve for ϕ_T, we can represent the total flux at the converter plane as $\phi_T = B\phi_u$, i.e., a multiplicative build up of the uncollided flux due to object scattering. This is represented using the well-known neutron

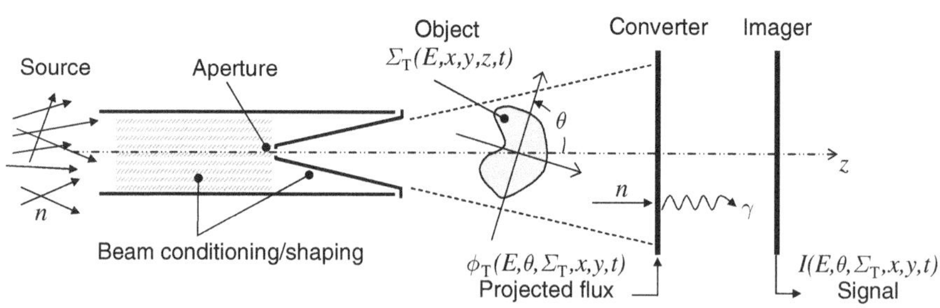

Fig. 7.1 Basic elements of a neutron radiography-imaging environment. A source of neutron is typically shaped and collimated, interacts with the material constituents of a 3D object and is projected onto a 2D imaging array

attenuation law as $\phi_T = \phi_o B\exp\{-\int \Sigma_T dz\}$, or with space-, time-, and energy-dependencies as,

$$\phi_T(E,\theta,\Sigma_T,x,y,t) = \phi_o(E,x,y,t)B(E,\theta,\Sigma_s,x,y,t)e^{-\int \Sigma_T(E,x,y,t,z)dz} \tag{7.1}$$

where x and y are spatial position in the plane perpendicular to the ray from source to detector, z represents position along ray from source to detector and E is the neutron energy. In this relationship, θ corresponds to the angular orientation of the object under test (Fig. 7.1) and is included as a prelude to later discussions regarding volumetric reconstruction from projections. Also note that the buildup factor is a value $B \geq 1$. It is functionally dependent on the material composition and geometry presented by the object under test and is therefore difficult to predict or accommodate under typical radiographic conditions. Buildup can be significant for materials containing, e.g., H, Si, Ni, Cu, and other highly scattering metals, and should be considered a source of error in quantitative analysis. For example, imaging geometries that move the object away from the detection plane can reduce the number of neutrons scattered into the detector, but this may also reduce the system resolution due to the characteristics of the beam aperture. The beam aperture and length is typically characterized by the L/D ratio, where L denotes the distance from the aperture to the object and D is the diameter of the aperture. This effect will be detailed further in Section 7.2.

Note that a primary goal of image analysis can be formulated as one of measuring the interaction cross-section of the materials in the imaged object, $\Sigma_T(E,x,y,z)$. Direct measurement of Σ_T, for example, provides indirect information regarding other material properties since $\Sigma = N\sigma = \sigma\rho N_A/M$, where σ is the microscopic cross-section, N is the number of atoms per cm^2, ρ is the material density, N_A is Avagadro's number, and M is the atomic weight [25].

The electronic signal produced by the system is further impacted beyond the geometry effects of L/D by the response of the converter, optical components, and sensor (e.g., a charge-coupled device, CCD). Making the practical assumption of a linear shift invariant (LSI) system [14], we can describe the time-, space-, and energy-dependent 2D electronic image, I, in terms of a convolution of the modulated (by the object) neutron field at the converter, ϕ_T, with a system impulse response function, h, to give,

$$\begin{aligned} I(E,\theta,\Sigma_T,x,y,t) &= \int\int \phi_T(E,\theta,\Sigma_T,\alpha,\beta,t)h(x-\alpha,y-\beta)d\alpha d\beta \\ &= \phi_T(E,\theta,\Sigma_T,x,y,t) \times h(x,y) \end{aligned} \tag{7.2}$$

In Section 7.2 we will model and analyze the impulse response of a radiography system to estimate system design resolution and characterize resolution of an existing radiography system through empirical means. In Section 7.3 we will describe considerations and methods required to perform volumetric reconstruction and estimation of Σ_T using projection data and computed tomography based on both analytic and iterative techniques.

7.2 Neutron Image Formation and Resolution Analysis

In this section, we will describe the image formation characteristics of a neutron radiography system from both a modeling and an empirical perspective. A model of resolution is important when designing a system to ensure that realistic resolution goals can be established and achieved. Once a system is operational, an empirical methodology is required to verify performance.

7.2.1 Resolution Modeling

For modeling purposes, Fig. 7.2 represents an idealized radiography system containing the basic elements described earlier in Fig. 7.1. The goal of a resolution model is to understand the impact of various system components on signal degradation. For this analysis we will neglect the impact of electron noise and counting statistics, and focus instead on the effects of the neutron aperture, D, the optical diffusion response of a scintillator, δ, the optical component, and sampling at the sensor, Δs (e.g., by a CCD camera).

The effects of these system elements are fundamentally described through the system impulse response, $h(x)$, introduced in Eq. (7.2), which will be expressed as 1D for this discussion. In an LSI system, the response, $I(x)$, is a convolution of the real-world signal, $\phi_T(x)$ in our case, with an impulse function, $h(x)$, as $I(x) = \phi_T(x)*h(x)$. The total impulse response can be decomposed into elements representing the major components of the system. For our case, based on Fig. 7.2, $h(x) = h_D(x)*h_\delta(x)*h_{CCD}(x)$, corresponding to the neutron beam aperture, scintillator, and CCD sensor, respectively. We model these components of the system using idealized functions for the physical neutron beam aperture, converter optical diffusion, and sensor sampling,

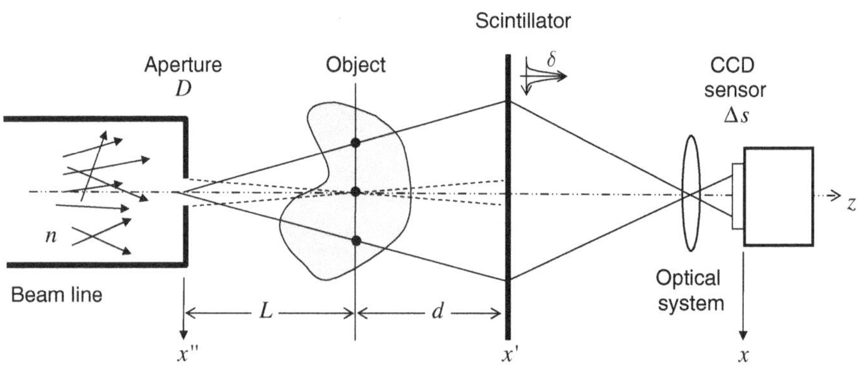

Fig. 7.2 Representation of a basic radiographic image formation system used to estimate resolution performance during the design stage. Model accounts for geometry, aperture size, scintillator resolution, and optical system resolution

$$h_D(x'') = \text{rect}\left(\frac{x''}{D}\right), h_\sigma(x') = \text{Gauss}\left(\frac{x'}{\delta}\right), h_{CCD}(x) = \text{rect}\left(\frac{x}{\Delta s}\right) \quad , \qquad (7.3)$$

where *rect(·)* is a rectangle function and *Gauss(·)* is a Gaussian function as defined in [14].

Note in Eq. (7.3) and Fig. 7.2, the introduction of different scales x'', x', and x. The aperture in plane x'' is magnified at the scintillator plane, x', which is in turn magnified (or de-magnified) in the CCD plane, x. The aperture-to-scintillator magnification is given by $M_D = d/L$, whereas the scintillator-to-CCD magnification is defined as M_{CCD}= *(object in scintillator plane/object in CCD sensor plane)*. The transformation, $x = M_{CCD}x' = M_D M_{CCD}x''$, results in the following analytical model for the impulse response, h, at the CCD imaging plane, x,

$$h(x) = \text{rect}\left(\frac{x}{M_D M_{CCD} D}\right) \times \text{Gauss}\left(\frac{x}{M_{CCD}\delta}\right) \times \text{rect}\left(\frac{x}{\Delta s}\right) \quad . \qquad (7.4)$$

The normalized modulus of the Fourier transform of the impulse response is known as the modulation transfer function [7], defined as $MTF(u) = |H(u)|/|H(0)|$, where $H(u)$ is the system transfer function calculated by taking the Fourier transform of the impulse response. The MTF describes the magnitude of the frequency response of the system and is useful for depicting and quantifying system resolution as discussed in Chapter 6.

From Eq. 7.4, we can analytically produce $MTF(u)$ through Fourier transformation and simplification to yield,

$$MTF(u) = \left| \text{sinc}\left(\frac{dM_{CCD}}{L/D} u\right) \cdot \text{Gauss}(\delta M_{CCD} u) \cdot \text{sinc}(\Delta s\, u) \right| \quad , \qquad (7.5)$$

where $sinc(x) = sin(x)/x$. Note that this expression has been put into a form explicitly containing the L/D ratio.

In Eq. (7.5), the *sinc* (·) term drives much of the response typically observed in a radiography system. For example, as $L/D \rightarrow \infty$, this term broadens out in frequency representing improved resolution. Also, as the object moves closer to the detection plane, i.e., as $d \rightarrow 0$, resolution also improves. These are both effects that are readily observed in radiography systems.

An example of the MTF is shown on the left-hand-side of Fig. 7.3 for a system with the model parameters: $L/D = 150$, $d = 10$ cm, $\delta = 50$ μm, $M_{CCD} = 0.25$. MTF is a measure of system response versus frequency content of the object being imaged. In this figure, the X axis represents frequency, with units of cycles per millimeter often referred to as line pairs per millimeter. The Y axis represents the transmission of this frequency to the image. In this example, the response to a sinusoid with 10 cyc/mm drops to approximately 55% of the sinusoid's amplitude.

Applying the Rayleigh criterion of 10% MTF [16] to this example, indicates that the highest resolvable frequency is 17 cyc/mm. System resolution is defined

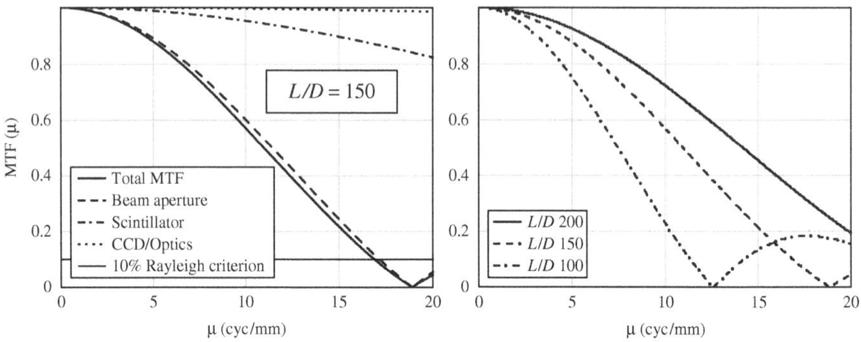

Fig. 7.3 MTF modeling results for typical radiography system parameters. The *left plot* shows the frequency response of constituent system components at a specified L/D ratio, while the *right plot* shows the system response for a variety of L/D ratios

as the period of this frequency. Therefore, resolution is 1/(17 cyc/mm) = 60 μm at the CCD imaging plane. Due to magnification in the system, the resolution at the object plane is 231 μm.

Note also for this geometry that the beam aperture (or equivalently the L/D ratio) is the primary factor impacting resolution. The scintillator screen and the optical system (defined by the CCD sample rate) produce only minor relative degradation. On the right-hand side of Fig. 7.3, the MTF is shown for a variety of L/D ratios, indicating the strong effect this factor has on the resolution of a neutron radiography system. Of course, the obvious result of this analysis is to increase the L/D ratio either by moving farther from the beam aperture or by reducing the size of the aperture. Unfortunately, both of these choices also result in a rapid reduction in the available neutron flux, requiring further signal-to-noise analysis to optimize both the geometry and the counting statistics.

7.2.2 System Performance Measurement

Measurement of system performance allows system designers to determine if the imaging system is performing as expected and to give potential users information useful in determining whether the imaging system is appropriate for their application. The key measurements of system performance are resolution and contrast discrimination. Resolution is a measure of the details that can be seen within an image while contrast discrimination measures the contrast required between an object and its background to resolve the object.

Resolution measurement methods can be categorized as direct and indirect. For direct methods, a series of test objects over a range of sizes near the resolution limits of the system are used. These test objects take many forms from line pairs to spheres to radial patterns and provide a direct visual display of the resolution on the radiographs [7].

Indirect resolution measurements use the edge response of the system to provide a more quantitative view of resolution. These measurements result in an MTF for the system. MTF can be calculated for radiography systems through measurement of the response of the system to an edge object. These edges are tilted to provide appropriate edge sampling by a pixilated imager [37, 32]. Tilted thin wires can also be used for this method [29]. The MTF is calculated from an edge image by first locating the edge within the radiograph and binning the edge response perpendicular to the line over the length of the line to get the edge response function, $E_r(x)$. Taking a derivative of the edge response function produces the edge spread function, which is then Fourier transformed to produce $F(u)$, which is used to produce the MTF, where $F(u)$ is given explicitly by,

$$F(u) = \int E_r'(x)\, e^{-iux} dx \quad , \tag{7.6}$$

where $E_r'(x) = dE(x)/dx$ represents a suitable discrete derivative process. Applying these methods to a radiograph will measure the MTF of the imaging system and verify the expected results based on the calculations from the previous section.

Neutron radiography systems are naturally extended to perform computed tomography (CT) for 3D reconstruction. Resolution of the reconstructed results will depend on the system, data collection, and reconstruction parameters as well as contrast in the object. Measurement of the empirical resolution of the 3D results after tomographic (CT) reconstruction can also be obtained from the determination of the edge response. The American Society for Testing and Materials (ASTM) has developed a standard test method for measurement of CT system performance designated E1695. In this standard, a cylindrical phantom is used to determine both spatial resolution and contrast discrimination measurements on the reconstructed data from a CT system as shown in Fig. 7.4. The shown example is a result from X-ray CT imaging; however, neutron tomography systems will produce the same type of data and the ASTM standard is directly applicable.

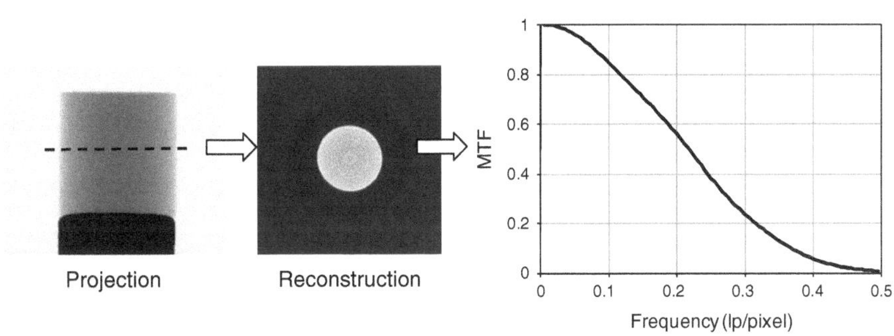

Fig. 7.4 Spatial resolution from CT reconstruction using ASTM E1695 standard

In the ASTM E1695 standard method, a cylindrical phantom is scanned and a slice normal to the central axis of the cylinder is reconstructed. A normal slice through the cylinder contains a circular cross-section of the cylinder. Just as in the tilted edge method, the MTF is calculated from the line edge response. The line edge response is found by locating the center of this circle and circularly integrating the pixels around this center point. In the tilted edge method, the tilt provides subpixel resolution across the edge. The pixel sampling on the circle edge will also provide this higher resolution edge response assuming the circle is of adequate size. Guidelines for the size selection are provided in the standard.

The MTF characterizes the spatial resolution of the system as described earlier. The second key performance parameter for CT systems is also included in ASTM E1965 and is the contrast discrimination function (CDF). The CDF measures the ability to discriminate an object from its background and is calculated using the same cylindrical phantom or spherical phantom used for the MTF calculation.

For calculation of the CDF, the region within the reconstructed slice through the phantom is subdivided into neighborhoods of size $N \times N$ as shown on the left side of Fig. 7.5. For CDF, the mean value for each neighborhood and the standard deviation of the means are calculated. This process is repeated for $N = 1,2,\ldots, M$, until there are fewer than 25 neighborhoods within the cylinder slice. Finally, for confidence in discrimination of the object from the background, the contrast needs to be three times greater than the standard deviation in the noise. Therefore, the standard deviations are multiplied by 3 and graphed as a function of neighborhood size to produce the CDF.

Since the cylinder is a uniform material, CDF can be interpreted as the average noise in the reconstructed image for various levels of pixelization. An example CDF is given on the right side of Fig. 7.5. In this example, the CDF shows that a contrast of 70% is required to discriminate a single pixel object from the background, and a contrast of 10% enables discrimination of objects larger than 3×3 pixels.

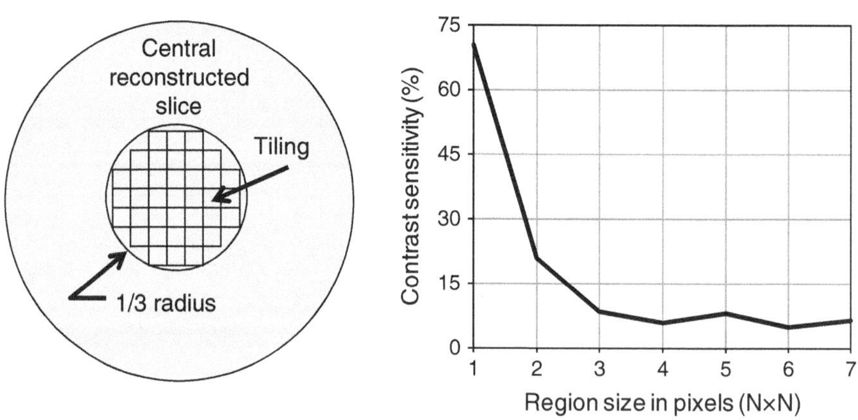

Fig. 7.5 Contrast discrimination measurement geometry (*left*) and CDF results (*right*)

The 3D MTF measurement method has been extended to use a spherical phantom [6] for two reasons. First, any tilt in the cylinder will produce an elliptical cross-section and degrade the calculated edge response, whereas any slice through a sphere will produce the appropriate circular cross-section. Second, resolution will vary over the reconstruction space for CT systems depending on the sampling and the reconstruction method used. This spherical phantom implementation locates the center of the sphere and integrates over spherical surfaces centered on this point to develop an edge response function that is an average over all directions on the spherical surface. Again using Eq. (7.6), the MTF is calculated over all directions.

With the resolution and contrast measurements presented in the graphical forms as shown in Figs. 7.4 and 7.5, a user can determine whether the imaging system is suited for a particular measurement.

7.3 Volumetric Imaging

Neutron imaging systems have followed the path of x-ray imaging systems in the introduction of rotational systems and CT reconstruction algorithms to perform volumetric imaging. An introduction to neutron tomography was provided in Chapter 6. This section builds upon the introduction by providing a mathematical description of CT and the classical methods for reconstruction of volumetric data.

Given a collection of projection images of an object, a 3D volumetric image can be reconstructed using either an analytic or an iterative algorithm. The former is based on some variant of filtered backprojection. The latter is based on optimization of a criterion function that relates the fit of the volumetric image with the observed projection data. In this section, we review the mathematics behind these conceptually different approaches and comment on advantages and disadvantages. To simplify the discussion, we apply the following model of the projection data:

$$I = I_0 \exp\left\{ - \int_K m(\alpha)\, d\alpha \right\} , \tag{7.7}$$

where I_0 and I refer to the beam intensity before and after interaction with the object, K denotes the path of the beam, and m represents the material-dependent macroscopic cross-sections. To reconstruct m, which corresponds to Σ_T, in the preceding section for neutron radiography, we apply log-normalization to isolate the cross-section (Eq. (7.1), assuming B$\sim$1),

$$\int_K m(\alpha)\, d\alpha = -\log \frac{I}{I_0} . \tag{7.8}$$

We note that in practice, the result of a dark field scan (where the source is turned off) is often subtracted from both I_0 and I to compensate for noise generated by physical processes that occur inside the imager (CCD).

Different imaging geometries require different reconstruction algorithms. We consider both parallel and divergent beam collimation of the neutron beam as shown in Fig. 7.6. We assume that the entire object is visible in each projection image and that it is placed on a turntable to allow for a circular orbit-type scan. Generalization to a helical scan is possible but beyond the scope of the present exposition.

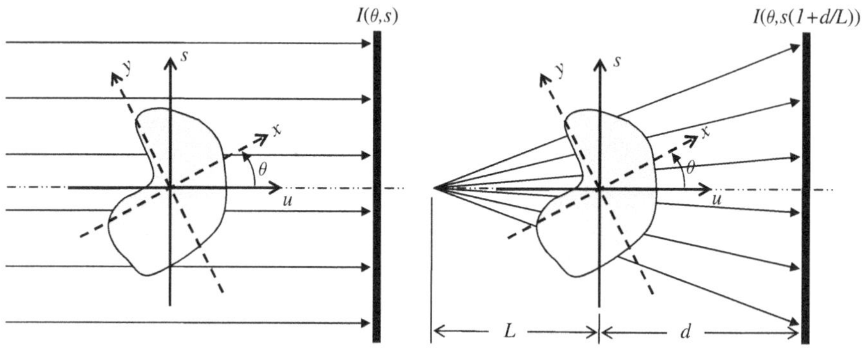

Fig. 7.6 Parallel beam (*left*) and fan beam (*right*) sampling geometries. Cone beam sampling extends the fan beam geometry axially

7.3.1 Filtered Backprojection

For 2D parallel beam data, the Fourier-slice theorem says that the Fourier transform of a 1D projection equals a comparably angled slice through the 2D Fourier space of the object being imaged [23]. Let $p\ (\theta,s)$ denote the log-normalized projection data and let $P\ (\theta,w)$ be the Fourier transform thereof. Mathematically,

$$\begin{aligned} p(\theta,s) &= \int_{-\infty}^{\infty} m(s\cos\theta - u\sin\theta, s\sin\theta + u\cos\theta)\, du \\ P(\theta,w) &= \int_{-\infty}^{\infty} p(\theta,s)\exp\{-i\,2\pi ws\}\, ds \end{aligned} \tag{7.9}$$

Merging these expressions, while introducing the variable substitutions $x = s\ cos\theta - u\ sin\theta$ and $y = s\ sin\theta + u\ cos\theta$, leads to,

$$\begin{aligned} P(\theta,w) &= \int\int_{-\infty}^{\infty} m(x,y)\exp\{-i2\pi\, w(x\cos\theta + y\sin\theta)\}dx\, dy \\ &= M(w\ \cos\,\theta, w\ \sin\,\theta) \end{aligned} \tag{7.10}$$

where M denotes the Fourier transform of m. While it would be tempting to superimpose appropriately rotated 1D Fourier transforms of a collection of projections and then obtain m through 2D Fourier inversion of the thus approximated version of M, the interpolation needed to map the radial data into the required Cartesian format would likely introduce artifacts in the resulting image.

Filtered backprojection is a better approach. The mathematical derivation can be summarized as follows [35]. When rewriting the inverse Fourier transform of M using polar coordinates, the resulting Jacobian introduces an absolute value-based frequency multiplier. That is,

$$\begin{aligned} m(x,y) &= \iint_{-\infty}^{\infty} M(u,v)\, \exp\{i2\pi\,(ux+vy)\}du\,dv \\ &= \int_0^{\pi}\left[\int_{-\infty}^{\infty} |w|\, M(w\cos\theta, w\sin\theta)\exp\{\,i2\pi\,(x\,\cos\theta + y\,\sin\theta)\}\,dw\right] d\theta \end{aligned} \tag{7.11}$$

We can apply the Fourier-slice theorem to replace $M\,(w\cos\theta,\, w\sin\theta)$ by $P\,(\theta,w)$. For simpler notation, we can furthermore express the inverse Fourier transform of $|w|\; P\,(\theta,w)$ as a convolution of $p\,(\theta,s)$ with a kernel that implements the ramp-filter corresponding to $|w|$. The result is the well-known algorithm,

$$\begin{aligned} m(x,y) &= \int_0^{\pi} \tilde{p}(\theta, x\cos\theta + y\sin\theta)\, d\theta \\ \tilde{p}(\theta,s) &= p(\theta\,,s)\,\times\, h_{\mathrm{RAMP}}\,(s) \end{aligned} \tag{7.12}$$

where the ramp-filter is often band-limited to suppress high-frequency noise and sampling based aliasing.

The extension to a 2D fan-beam geometry is not difficult. Essentially, geometric weighting is introduced to compensate for the divergent nature of the projection rays. The algorithm for equiangular sampling seen in connection with a curved ring-detector is very similar to that associated with the equidistant sampling geometry of a linear detector. The latter algorithm is given by,

$$\begin{aligned} m(x,y,z) &= \frac{1}{2}\int_0^{2\pi} g^2(\theta,x,y)\tilde{p}(\theta, g(\theta,x,y)(x\cos\theta + y\sin\theta))\, d\theta \\ g(\theta,x,y) &= L/(L + x\sin\theta - y\cos\theta) \\ \tilde{p}(\theta,s) &= L/\sqrt{L^2+s^2}\cdot p(\theta,s)\times h_{\mathrm{RAMP}}\,(s) \end{aligned} \tag{7.13}$$

where L refers to the distance from the neutron beam aperture to the isocenter of the object (c.f. Fig. 7.6). The detector is assumed to be located at the isocenter for mathematical simplicity (i.e., $d=0$). Notice that the backprojection takes place over 2π instead of just π as used for the parallel beam geometry. Also note that $L \rightarrow \infty$ leads to the parallel beam algorithm.

The extension to 3D parallel beam data is even more straightforward. In fact, the reconstruction algorithm is identical to the one above for 2D parallel beam data except for the increase in dimensionality.

The extension to a 3D cone-beam geometry based on a flat-panel detector is more involved. Tuy [40] established a sufficient condition for reconstruction that says that every image plane must contain at least one source point. While this is satisfied for helical scanning and other trajectories, it is not satisfied for a circular-orbit scan, which is considered here. This leads to an approximate algorithm of which Feldkamp's without a doubt is the most widely used.

Developed for industrial x-ray cone-beam CT, the Feldkamp algorithm [13] can be viewed as a generalization of the fan-beam algorithm. The central slice of a cone-beam volume is reconstructed in exactly the same manner as it would have been, had it been reconstructed from single-slice fan-beam data. Slices further out in the cone, however, correspond to tilted fans and are thus handled differently. While the mathematical details are too extensive for inclusion here, the end result is the following algorithm,

$$\begin{aligned} m(x,y,z) &= \frac{1}{2}\int_0^{2\pi} g^2(\theta,x,y)\tilde{p}(\theta,g(\theta,x,y)[x\cos\theta + y\sin\theta, z])\,d\theta \\ g(\theta,x,y) &= L/(L + x\sin\theta - y\cos\theta) \\ \tilde{p}(\theta,[s,t]) &= L/\sqrt{L^2+s^2+t^2}\cdot p(\theta,[s,t]) \times h_{\mathrm{RAMP}}(s) \end{aligned} \tag{7.14}$$

Notice that $z=0$ leads to the fan-beam algorithm. Also note that filtering is not applied in the axial dimension.

We mentioned that the fan-beam backprojection equation indicates that the source trajectory data must cover 2π. However, by introducing a weight function that effectively limits each projection ray to be considered just once, Parker [34] showed that it is possible to stop when $\pi+\gamma$ has been covered where γ denotes the fan angle. The idea, which is heuristic, can be extended to apply to cone-beam data as well.

Filtered backprojection algorithms, whether exact or approximate, are based on continuous mathematics. When implemented for practical use, all integrals are replaced by Riemann sums and nonintegral projection data lookup indices are handled using some form of interpolation. The filtering can take place in either the spatial domain or the frequency domain. When based on fast fourier transforms, the latter can be done very efficiently. In that case, backprojection along with interpolation dominates the computational cost. Still, given recent advances in computer technology, it is possible to reconstruct even large images in a very short amount of time.

The image quality depends on taking enough projections and having good count statistics to work with. Not taking enough projections invariably leads to streak artifacts. Indeed, they are almost always present to some degree. Poor count statistics results in a low-contrast, grainy image. Under normal imaging conditions, however, good images are produced.

7.3.2 Iterative Reconstruction

Henceforth, let vectors $\mathbf{x}$ and $\mathbf{b}$ refer to the image representing the macroscopic cross-sections and the log-normalized projection data, and let matrix $\mathbf{A}$ denote the discretization of the projection integral connecting the two. Reconstruction can then be viewed as a matter of solving $\mathbf{Ax} = \mathbf{b}$. Numerous applicable iterative algorithms exist ranging from Gauss–Seidel, Jacobi, and other stationary methods to conjugate gradient (CG) and other Krylov subspace methods.

Stationary methods can be derived using matrix splitting [36]. For example, $\mathbf{A} = \mathbf{M} - \mathbf{N}$ leads to the iteration scheme,

$$\mathbf{x}^{(k+1)} = \mathbf{M}^{-1}\mathbf{N}\mathbf{x}^{(k)} + \mathbf{M}^{-1}\mathbf{b} \tag{7.15}$$

where $\mathbf{x}^{(0)}$ is given. Convergence to the desired solution is guaranteed if $\rho(\mathbf{M}^{-1}\mathbf{N}) < 1$ where ρ denotes the spectral radius of the matrix involved. Let matrices $\mathbf{D}$, $\mathbf{L}$, and $\mathbf{U}$ represent the diagonal, strictly lower triangular, and strictly upper triangular portions of $\mathbf{A}$. Then Gauss–Seidel is the result of setting $\mathbf{M} = \mathbf{D} - \mathbf{L}$ and $\mathbf{N} = \mathbf{U}$. Likewise, Jacobi follows from setting $\mathbf{M} = \mathbf{D}$ and $\mathbf{N} = \mathbf{L} + \mathbf{U}$. Both methods converge if $\mathbf{A}$ is strictly diagonally dominant.

Krylov subspace methods monotonically minimize the residual error using orthogonality conditions [36]. They are nonstationary in the sense that the computations involve information that changes each iteration. The prototypical example is CG, which is based on the assumption that matrix $\mathbf{A}$ is symmetric, positive-definite. The iterative update scheme is given by

$$\begin{aligned}
\alpha &= |\mathbf{r}^{(\mathrm{k})}|^2 / \mathbf{d}^{(\mathrm{k})}|_{\mathrm{A}}^2 \\
\mathbf{x}^{(\mathrm{k+1})} &= \mathbf{x}^{(\mathrm{k})} + \alpha\, \mathbf{d}^{(\mathrm{k})} \\
\mathbf{r}^{(\mathrm{k+1})} &= \mathbf{b} - \mathbf{A}\mathbf{x}^{(\mathrm{k+1})} \\
\beta &= |\mathbf{r}^{(\mathrm{k+1})}|^2 / |\mathbf{r}^{(\mathrm{k})}|^2 \\
\mathbf{d}^{(\mathrm{k+1})} &= \mathbf{r}^{(\mathrm{k})} + \beta\, \mathrm{d}^{(\mathrm{k})}
\end{aligned} \tag{7.16}$$

where $\mathbf{x}^{(0)}$ is given, $\mathbf{d}^{(0)} = \mathbf{r}^{(0)} = \mathbf{b} - \mathbf{A}\mathbf{x}^{(0)}$, $|\mathbf{r}|^2 = \mathbf{r}^{\mathrm{T}}\mathbf{r}$, and $|\mathbf{d}|_{\mathrm{A}}{}^2 = \mathbf{d}^{\mathrm{T}}\mathbf{A}\mathbf{d}$.

Whether a stationary method or a Krylov subspace method is used, the search for $\mathbf{x}$ is terminated either when the residual norm associated with the current iterate is deemed to be small enough, or when a fixed number of iterations have been executed.

In most imaging applications, the linear system is under-or overdetermined and a least squares problem must be solved instead. Let this be denoted by

$$\mathbf{x}^* = \operatorname{argmin}|\mathbf{Ax} - \mathbf{b}|^2 \ , \tag{7.17}$$

where $|\mathbf{Ax} - \mathbf{b}|^2 = (\mathbf{Ax} - \mathbf{b})^{\mathrm{T}}(\mathbf{Ax} - \mathbf{b})$. By taking the derivative of the squared residual norm with respect to $\mathbf{x}$ and equating the result to zero, we obtain the normal equations which form a linear system that can be solved by any of the above-mentioned methods, namely,

$$\mathbf{A}^{\mathrm{T}}\mathbf{A}\mathbf{x} = \mathbf{A}^{\mathrm{T}}\mathbf{b} \tag{7.18}$$

Other methods include SIRT, which is a stationary method for solving a weighted least squares problem that has been widely used in medicine and biology [15;18]) and LSQR, which is a CG-like method with better numerical properties [33].

The reconstructed image should not contain any negative values, as they have no physical interpretation. Their absence is only guaranteed if the least squares problem is constrained accordingly. This leads to,

$$\mathbf{x}^* = \operatorname{argmin}|\mathbf{A}\mathbf{x} - \mathbf{b}|^2 \text{subject to} \, \mathbf{x} \geq 0 \ . \tag{7.19}$$

Note that depending on the neutron imaging system, a highly scattering object can cause neutron detection in detectors just outside the shadow of the object under test. These additional neutrons can result in negative attenuation values. Therefore, modifications to the typical constraints may be required along these edges.

A number of methods exist for solving constrained quadratic optimization problems, one example being to embed the least squares algorithm in a projected gradient framework [5, 19]. Computationally, an outer algorithmic layer iteratively (i) uses an inexact line search based on safeguarded quadratic interpolation to determine the set of free variables, i.e., those that are strictly positive and (ii) invokes the least squares algorithm to compute a feasible solution in that subspace. Control is passed back from the least squares solver to the outer algorithmic layer either at convergence or when a variable is about to violate the nonnegativity constraint.

Computational steering can be introduced through the addition of a penalty term [26]. For example, suppose that the image is known to consist of regions of uniform macroscopic cross-sections. In that case, it might be advantageous to discourage neighboring voxels from taking on different values except near region boundaries. This could be achieved by applying an edge-preserving Markov random field that has smoothing properties [8]. Other structural constraints are possible. Generally speaking, the resulting optimization problem can be written as

$$\mathbf{x}^* = \operatorname{argmin}|\mathbf{A}\mathbf{x} - \mathbf{b}|^2 + \beta \mathrm{R}(\mathbf{x}) \text{ subject to} \, \mathbf{x} \geq 0 \tag{7.20}$$

where $\mathrm{R}(\mathbf{x})$ denotes the penalty term and β is a free parameter that controls the influence exerted thereby on the solution. From a computational point of view, linear penalty terms are preferable as they can be handled by straightforward row-expansion of **A** and **b**.

Bayesian reconstruction forms an alternative to algebraic methods. Commonly used in the field of medical imaging, this iterative approach is statistical by nature. The image formation is described probabilistically, thereby making reconstruction a matter of model parameter estimation. Specifically, let $p(\mathbf{x},\mathbf{b}|\Omega)$ denote the data likelihood for **x,** the unobservable image data, and **b,** the observable projection data, given Ω, the associated model parameters. As above, $\mathbf{Ax} = \mathbf{b}$, only now **A** is a matrix of detection probabilities. The goal is then to solve the penalized maximum likelihood problem

$$\Omega^* = \text{argmax } \log \text{ p}(\mathbf{x}, \mathbf{b}|\Omega) + \beta \text{R}(\Omega) \quad . \tag{7.21}$$

where **R,** strictly speaking, is a log-prior; although in practice a heuristic smoothing term is often used.

Methods for maximum likelihood estimation include gradient descent, conjugate gradient, and similar algorithms. These typically require first and/or second derivatives of the likelihood function to be evaluated. This is not the case for the two-step EM algorithm [11]. First, an E-step is carried out to compute the expectation of the data log-likelihood given the observed data and the current parameter estimate. Then, an M-step is used to re-estimate the model parameters by means of maximization of the expectation. The result is the following iteration scheme

$$\Omega^{k+1} = \text{argmax E}[\log \text{ p}(\mathbf{x}, \mathbf{b}|\Omega)|\mathbf{b}, \Omega^{(k)}] + \beta \text{R}(\Omega) \tag{7.22}$$

In some applications, e.g., maximum likelihood estimation for independent Poisson events, the M-step is trivial as it has a closed-form solution. This allows the E-step and the M-step to be combined. For other applications, the M-step has to be implemented as an iterative computation of its own. This may also be the case for penalized maximum likelihood estimation, as the penalty term may preclude true maximization from taking place. The M-step is then turned into an iterative parameter update computation that ensures a nondecrease of the likelihood [21].

There are several advantages associated with taking an iterative approach to image reconstruction. For one, it is possible to quite accurately model the system geometry including nonuniformities in the sampling pattern as well as sensitivity variations across the detector. As pointed out above, it is also possible to steer the computation away from structurally undesirable images. Poor count statistics tend not to be a problem either, as the objective of both algebraic and statistical reconstruction methods is to find a solution that best fits the data. That is, there is no built-in assumption of data consistency across projections. Iterative approaches also allow reconstruction with few projections and irregular projection patterns and enable the inclusion of more accurate physics models up to full Monte Carlo models. This flexibility comes at a high cost with respect to both implementation complexity and computational cost at run-time. However, if these can be overcome, the payoff is a high-quality image.

7.3.3 Computer Platforms

Many platforms exist that can provide the compute power needed in order for a volumetric image to be reconstructed in a reasonable amount of time. A relatively small cluster of multi-core PCs may suffice to solve even a rather large problem in a matter of minutes [18]. From a programming point of view this is conceptually straightforward as it merely involves writing multi-threaded code for a distributed memory environment. In recent years, commodity graphics processing units (GPUs) have seen a rapid increase in both capability and programmability to the point that they are now being used for general purpose computing and thus also tomographic image reconstruction [31]. Papers often report order-of-magnitude performance gains over optimized CPU applications. However, GPU programming is not for the novice. Also, floating-point arithmetic may not follow IEEE standards, which could be a problem. Field-programmable gate-arrays (FPGAs) provide a different but equally powerful platform from a data throughput point of view. However, FPGAs are notoriously difficult to program, especially since floating-point arithmetic may not be supported to the extent needed. Finally, we mention the cell processor, which conceptually is somewhat similar to a multi-core CPU only orders-of-magnitude faster. High-level programming is possible, e.g., in C, but some thought must be devoted to memory handling as the processing units only have a limited amount of memory available. Nonetheless, the cell may represent the fastest platform currently on the market for volumetric imaging [22].

7.4 Conclusions and Application

The ability to model, design, and measure the resolving power of a radiography system is critically important to the experimental design required to achieve relevant measurement characteristics of complex, material structures. The focus of this chapter has been on methods to design and characterize radiography systems according to experimental goals for volumetric measurements (i.e., CT recording and reconstruction methods). This focus has been primarily related to spatial resolution and the estimation of macroscopic material interaction cross-sections.

New or modified radiography facilities that generate cold neutrons [38] provide wavelength tunable monochromaters [39] or provide time-of-flight (TOF) measurements [27], facilitating the radiographic capture of images that take $I(\theta,\Sigma_T,x,y,z)$ to $I(E,\theta,\Sigma_T,t,x,y,z)$. The ability to better characterize the energy–time–space dependencies of materials, $\Sigma_T(E,t,x,y,z)$, will produce better material resolution.

As an example of how to adapt this discussion to higher resolution measurements, consider the example MTF shown in Fig. 7.7. Using Eq. (7.5), we have

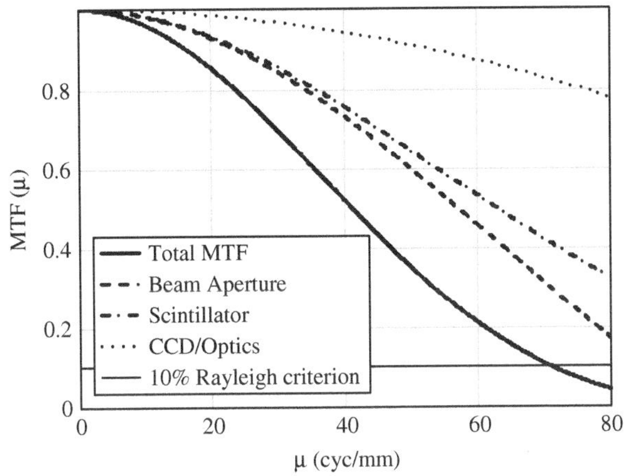

Fig. 7.7 MTF for a system design with $L/D = 750$, corresponding to an aperture size of 5 mm. the resolution (71 cyc/mm or 14 μm at the detector) is now comparable to the scintillator limit. How do we achieve sufficient neutron counting statistics?

hypothesized a radiography system with $L/D = 750$, corresponding to a beam aperture of 5 mm at a distance of 3.75 m from the object—a seemingly achievable design. For a scintillator material with an optical diffusion parameter of $\sigma = 10$ μm, the resolution of such a system would be 14 μm at the sensor, moving from the macro-analysis scale to the microscopy scale. Reducing the beam aperture from 25 mm (our earlier example) to 5 mm will increase system resolution by a factor of 4, yet it is likely that neutron-counting statistics and background noise will result in measurement challenges that are difficult to overcome.

Overcoming low neutron counts may be achievable by encoding the beam aperture, i.e., implementing an array of 5 ×5 mm (or smaller) apertures that constitute a significant total area (e.g., >25 ×25 mm). In such a case, the MTF of Eq. (7.5) could be replaced by a new geometric transfer function, $H_{CA}(\cdot)$ that describes a coded aperture [17], e.g.,

$$MTF(u) = \left| H_{CA}\left(\frac{\mathrm{d}\,\mathrm{M_{CCD}}}{L/D}u\right) \cdot Gauss(\sigma\, M_{CCD}u) \cdot \sin c(\Delta s\, u) \right| \qquad (7.23)$$

The point is that these analysis tools can be adapted to new, more stringent resolution requirements by applying creative ideas and methods.

For another example, the steady-state flux of a conventional reactor cannot readily interrogate time-dependent phenomena as deeply as the high-intensity and energy-dependent beams available at a spallation neutron source. Using a spallation source, the potential exists to develop new time-synchronized, energy-resolved methods to better resolve subtle material differences. This is potentially achievable by taking advantage of the Bragg edges in materials [27] or of the fact that absorption cross-sections, σ_a at low neutron energies are typically proportional to $1/\sqrt{E}$. Using prior knowledge of the material characteristics under test should facilitate the collection of a series of energy-windowed images, $I_{\Delta E}(\theta,\Sigma_T,x,y,z)$, whose linear combinations, e.g., using principle component or linear discriminant analysis [12], may provide improved contrast (i.e., CDF) between materials.

Finally, any of these methods that compartmentalize the total neutron beam into small spatial- or energy-dependent units will make volumetric recovery of material characteristics more challenging due to reduced signal-to-noise and the corresponding artifacts that conventional CT methods produce in low-signal environments. In these situations, the iterative reconstruction methods introduced in Section 7.3.2 become more relevant to achieving adequate signal recovery and reconstruction.

References

1. R. Accorsi, F. Gasparini, et al., A Coded aperture for high-resolution nuclear medicine imaging with conventional anger camera. *IEEE Trans. Nucl. Sci.* **48**:6, 2411 (2001).
2. H. Asano, N. Takenaka, et al., Visualization and void fraction measurement of gas-liquid two-phase flow in a commercial plate heat exchanger by thermal neutron radiography. *IEEE Trans. Nucl. Sci.* **52**:1, 280–284 (2005).
3. M. Balasko, E. Svab, et al., Comparison of neutron radiography with other nondestructive methods. *IEEE Trans. Nucl. Sci.* **52**:1, 330–333 (2005).
4. L.G.I. Bennett, T.R. Chalovich, et al., Comparison of neutron radiography with other non-destructive techniques for the inspection of cf188 flight control surfaces. *IEEE Trans. Nucl. Sci.* **52**:1, 334–337 (2005).
5. M. Bierlaire, P.L. Toint, et al., On iterative algorithms for linear least squares problems with bound constraints. *Linear Algebra Appl.* **143,** 111–143 (1991).
6. P. Bingham, L. Arrowood, et al., Calibration and performance testing for reconfigurable computed tomography systems. *Mat. Eval.* **65,** 1102–1107 (2007).
7. G.D. Boreman, *Modulation Transfer Function in Optical and Electro-Optical Systems.* Bellingham, WA, SPIE Press (2001).
8. C.Bouman and K. Sauer, A generalized Gaussian image model for edge-preserving MAP estimation. *IEEE Trans. Image Process.* **2**, 296–310 (1993).
9. S. Casalta, G.G. Dawuino et al., Digital image analysis of X-ray and neutron radiography for the inspection and the monitoring of nuclear materials. *NDT&E Int.* **36**, 349–355 (2003).
10. J.M. Cimbala, J.S. Brenizer, et al., Study of a loop heat pipe using neutron radiography. *Appl. Radiat. Isot.* **61**, 701–705 (2004).
11. A. Dempster, N. Laird, et al., Maximum likelihood from incomplete data via the EM algorithm. *J. R. Stat. Soc. B* **39**, 1–38 (1977).
12. R.O. Duda, P.E. Hart, et al., *Patt. Class.*. New York, John Wiley & Sons, Inc. (2001).
13. L.A. Feldkamp, L.C. Davis, et al., Practical cone beam algorithm. *J. Opt. Soc. Am. A* **1**, 612–619 (1984).
14. J.D. Gaskill, *Linear Systems, Fourier Transforms, and Optics*, John Wiley & Sons, Inc. (1978).
15. P. Gilbert, Iterative methods for the three-dimensional reconstruction of an object from projections. *J. Theor. Biol.* **36**, 105–117 (1972).
16. J.W. Goodman, *Introduction to Fourier Optics*, Roberts & Company Publishers (2004).
17. S.R. Gottesman, and E.E. Fenimore, New family of binary arrays for coded aperture imaging. *Appl. Opt.* **28**:20, 4344 (1989).
18. J. Gregor and T. Benson, Computational analysis and improvement of SIRT. *IEEE Trans. Med. Imaging* **27**, 918–924 (2008).

19. J. Gregor and F.R. Rannou, *Least squares framework for projection MRI reconstruction.* Medical Imaging, San Diego, CA, Proceedings of SPIE (2001).
20. A.A. Harms and D.R. Wyman, *Mathematics and Physics of Neutron Radiography.* Boston, D. Reidel Publishing Company (1986).
21. T. Hebert and R. Leahy, A generalized EM algorithm for 3-D Bayesian reconstruction from Poisson data using Gibbs priors. *IEEE Trans. Med. Imaging* **8**, 194–202 (1989).
22. M. Kachelriess, M. Knaupp, et al., Hyperfast parallel-beam and cone-beam backprojection using the Cell general purpose hardware. *Med. Phys.* **34**, 1471–1486 (2007).
23. A. Kak and M. Slaney. *Principles of Computerized Tomographic Imaging*, IEEE Press (1988).
24. G.F. Knoll, *Radiation Detection and Measurement*, John Wiley & Sons, Inc. (2000).
25. J.R. Lamarsh, *Introduction to Nuclear Engineering*, Addison-Wesley Publishing Company (1983).
26. C.L. Lawson and R.J. Hanson, *Solving Least Squares Problems*, Prentice-Hall (1974).
27. E. Lehmann, P. Vontobel, et al., Neutron imaging – present status and options with tof methods – TOF neutron imaging. *J. Neutron Res.* **13**, 27–31 (2005).
28. U. Matsushima, Y. Kawabata, et al., Measurement of changes in water thickness in plant materials using very low-energy neutron radiography. *Nucl. Instrum. Meth. Phys. Res. A* **542**, 76–80 (2005).
29. S.L. Meeks, J.F. Hamon, et al., Performance characterization of megavoltage computed tomography imaging on a helical tomotherapy unit. *Med. Phys.* **32**, 2673–81 (2005).
30. J.J. Milczarek, A. Czachor, et al., Dynamic neutron radiography observations of water migration in porous media. *Nucl. Instrum. Meth. Phys. Res. A* **542**, 232–236 (2005).
31. K. Müller and F. Xu, Accelerating popular tomographic reconstruction algorithms on commodity PC graphics hardware. *IEEE Trans. Nucl. Sci.* **3**, 654–663 (2005).
32. M. Ohkubo, S. Wada, et al., An effective method to verify line and point spread functions measured in computed tomography. *Med. Phys.* **33**, 2757–64 (2006).
33. C.C. Paige and M.A. Saunders, LSQR: An algorithm for sparse linear equations and sparse least squares. *ACM Trans. Math. Software* **8**, 43–71 (1982).
34. D. Parker, Optimal short-scan convolution reconstruction for fanbeam CT. *Med. Phys.* **9**, 254–257 (1982).
35. A. Rosenfeld and A. Kak, *Digital Picture Processing*, Academic Press (1976).
36. Y. Saad, *Iterative Methods for Sparse Linear Systems*, PWS Publishing Co (1996).
37. E. Samei and M.J. Flynn, A method for measuring the presampled MTF of digital radiographic systems using an edge test device. *Med. Phys.* **25**, 102–113 (1998).
38. M. Tamaki, Conceptual monochromatic digital neutron radiography using continuous cold neutron beam. *Nucl. Instrum. Meth. Phys. Res. A* **542,** 32–37 (2005).
39. W. Treimer, M. Strobl, et al., Wavelength tunable device for neutron radiography and tomography. *Appl. Phys. Lett.* **89,** 203504 (2006).
40. H. Tuy, An inversion formula for cone-beam reconstruction. *SIAM J. Appl. Math.* **43,** 546–552 (1983).

Chapter 8
Neutron Phase Imaging

F. Pfeiffer

Abstract This chapter is dedicated to neutron phase imaging. We first review the underlying physical principles describing the phase shift that neutrons experience when they propagate through matter. Based on these principles we describe the three main approaches for neutron phase imaging and discuss their extension to three-dimensional neutron phase tomography. The chapter ends with a final discussion of the concept of coherence and its implications concerning practical neutron phase-imaging applications.

Keywords Neutron phase-contrast tomography · Neutron phase-contrast imaging · Neutron phase imaging

8.1 Introduction

From a classical physics point of view, neutrons are small massive particles with a confinement radius of about 10^{-15} m and a distinct internal quark-gluon structure. In quantum mechanics, neutrons are described by de Broglie [1] wave packets whose spatial extent may be large enough to show interference effects similar to what can be observed with visible laser light or highly brilliant X-rays from synchrotron sources. Measurements of the neutron wave packet's phase shift induced by different interaction potentials have a long and distinguished history in the exploration of the fundamental principles of quantum mechanics [2, 3]. If such phase-sensitive measurements are further combined with neutron-imaging approaches, two or even three dimensionally resolved spatial information on the quantum mechanical interactions of massive particles with matter can be obtained.

In the following, the term "phase imaging" is used to describe not only phase contrast methods, which simply record some form of contrast generated due to a phase shift induced by the object, but also methods that determine the phase

F. Pfeiffer (✉)
Department of Physics (E17), Technical University Munich,
85747 Garching/Munich, Germany
e-mail: franz.pfeiffer@pri.ch

I.S. Anderson et al. (eds.), *Neutron Imaging and Applications*,
Neutron Scattering Applications and Techniques, DOI 10.1007/978-0-387-78693-3_8,

shift quantitatively. The research activities in the direction of quantitative phase imaging are particularly interesting because they open the way for more sophisticated methods such as neutron phase tomography or quantitative phase imaging of fundamental quantum mechanical interactions.

Conventional attenuation-based neutron computed tomography (CT) is well developed as a three-dimensional (3D) imaging method. Neutron phase tomography was attained by introducing the technique of neutron phase-sensitive imaging into neutron tomography. In this case, because the neutron phase shift is expressed as a projection of the real part of the refractive index, a reconstructed 3D image corresponds to a map of the real part of the refractive index. Neutron phase tomography, of course, inherits its high and complementary sensitivity from neutron phase imaging.

Since phase information is generally lost in measuring neutron transmission images, several techniques have been developed to convert the neutron phase shift to neutron intensity modulation. Because of the short wavelength of neutrons, comparatively low coherence of neutron beams, and limited numerical aperture of neutron optical elements, the construction of phase-sensitive optics is not as flexible as in the visible-light region. Nevertheless, recent developments, specifically in neutron optics, have opened the door for several approaches to carry out neutron phase imaging. In this chapter we describe the basic characteristics and principles of phase-sensitive imaging and tomographic methods for extending quantitative phase imaging into three dimensions and present an overview of the existing experimental approaches for phase-sensitive imaging and tomography.

8.2 Principles

8.2.1 Phase Shift Versus Attenuation

Let us consider an object characterized by a complex neutron refractive index distribution

$$n(x,y,z) = 1 - \delta(x,y,z) + i\beta(x,y,z) \quad , \tag{8.1}$$

where δ (x,y,z) describes the decrement from unity of the real part of the refractive index and β (x,y,z) is the imaginary part of the refractive index. A conventional neutron attenuation-based projection image maps the imaginary part of n and is given by

$$\frac{I(y,z)}{I_0(y,z)} = \exp\left[-\int \Sigma(x,y,z)dx\right] = \exp\left[-\int 4\pi\beta(x,y,z)/\lambda dx\right] \quad , \tag{8.2}$$

where λ is the wavelength of the neutron radiation. Moreover, we have assumed that the neutron beam projects the object along the x-direction onto the (y,z) detector plane and that the linear coefficient Σ is related to the imaginary part of

n with $\beta = \lambda/4\pi\,\Sigma$. Generally Σ contains not only the specific neutron capture cross-section of the material, but also the effect that neutrons are removed from the beam traversing the specimen to coherent or incoherent scattering processes.

In analogy to the description of attenuation-based projections, the projected phase shift of an object can be written as [3]

$$\Phi(y,z) = \frac{2\pi}{\lambda}\int \delta(x,y,z)dx \quad . \tag{8.3}$$

A phase shift projection image thus can give direct information on the real part of the refractive index distribution in the object and carries additional and complementary information if compared to an attenuation image.

Unlike X-rays, where the values of δ and β are linked through the X-ray optical properties of the materials, the corresponding values for neutrons can be very different. Table 8.1 shows values for several materials for a neutron wavelength of 4Å. It is interesting to note that, for example, titanium and copper have very similar β values, but opposite signs in δ. Moreover, we observe that aluminum and silicon are characterized by very small β values and thus yield very little contrast in a conventional attenuation image. Their relatively large δ values, on the other hand, make them easily detectable in phase-sensitive imaging (see below). A graphical overview of the distribution of δ values and corresponding attenuation coefficients Σ is shown in Fig. 8.1.

Table 8.1 Values for the complex neutron refractive index $n = 1\text{-}\delta + i\beta$ of several representative materials for neutrons with a wavelength of 4Å [4, 5]

Material	δ	β	δ/β
Al	5.3×10^{-6}	9.8×10^{-11}	5.4×10^{4}
Si	5.3×10^{-6}	6.0×10^{-11}	8.7×10^{4}
Ti	-5.0×10^{-6}	2.4×10^{-9}	-2.0×10^{3}
Mn	-7.8×10^{-6}	7.7×10^{-9}	-1.0×10^{3}
Cu	1.7×10^{-5}	2.3×10^{-9}	7.3×10^{3}
Gd	5.0×10^{-6}	1.1×10^{-5}	4.8×10^{-1}
Pb	7.9×10^{-6}	4.0×10^{-11}	2.0×10^{5}

In a more fundamental way, the decrement of the real part of the neutron refractive index is related to the interaction potential that causes the phase shift. For most materials, the nuclear interaction potential produces the strongest phase shift and thus determines the δ value. For an idealized slab of homogenous material with a thickness d, the phase shift due to the nuclear interaction potential can be expressed by [3]

$$\Delta\Phi = Nb_c\lambda d \quad , \tag{8.4}$$

where N is the atom number density and b_c the neutron nuclear scattering length. This is consistent with Eq. (8.3), when δ is expressed by

$$\delta = \frac{\lambda^2 N b_c}{2\pi} \quad . \tag{8.5}$$

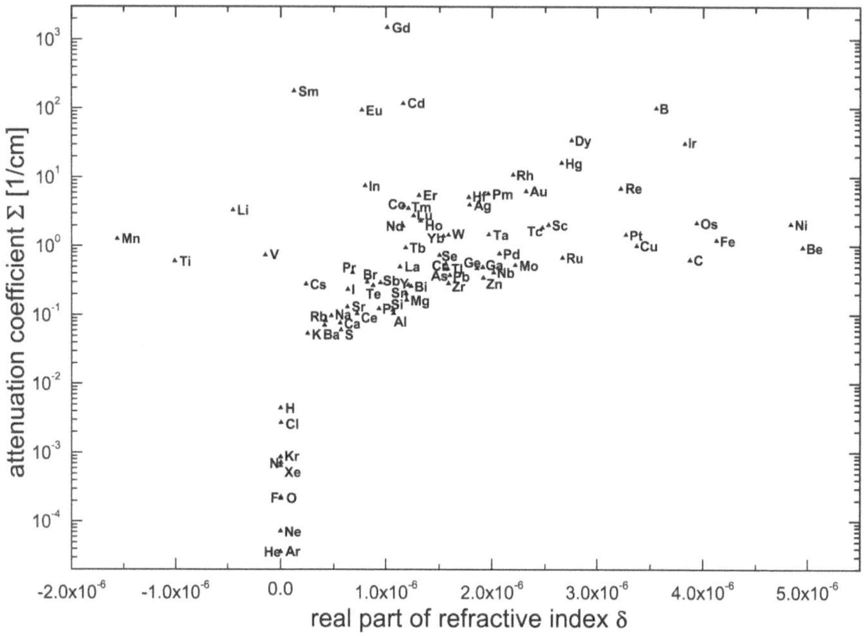

Fig. 8.1 Semilogarithmic plot of linear attenuation coefficient Σ versus the real part of the refractive index δ for different elements for thermal neutrons with a wavelength $\lambda = 1.8$Å [4, 5] (courtesy of K. Lorenz)

The second most important interaction potential causing a phase shift is the magnetic interaction. In this case $\Delta\Phi$ can be written as

$$\Delta\Phi = \pm \frac{\mu B m \lambda d}{2\pi\hbar^2} \quad , \tag{8.6}$$

where μ is the neutron dipole moment, B the applied magnetic field, m the neutron mass, and d the sample thickness.

In addition to the phase shift effect caused by the nuclear and magnetic interactions, several other interaction potentials exist that cause a measurable phase shift. Among those are the phase shift effects caused by the Coriolis [6], the Aharanov–Casher [7], the scalar Aharanov–Bohm [8], the magnetic Josephson [9], the Fizeau [10], and the geometric (Berry) [11] interactions.

8.2.2 Phase Tomography

Computed tomography enables the reconstruction of a 3D object from projection data measured over a range of projection angles. In conventional attenuation-based neutron CT, a projection image corresponds essentially to a projection line integral over the 3D distribution of the imaginary part of the neutron refractive index of the object (Chapter 6).

In phase tomography, the aim is to quantitatively reconstruct the 3D distribution of the real part of the refractive index of the object. To do so, a series of projection images that exhibit some sort of phase contrast are recorded and used as input for adequately matched CT reconstruction algorithms. The high sensitivity or complementary information present in the phase contrast images is of course inherited.

There are essentially three modalities for measuring phase contrast projection images. On the fundamental physics level, they differ in the way they represent a phase change induced by the object in the measured projection image. They can be classified into (1) methods measuring the phase shift Φ directly, (2) methods that yield one or both components of the phase gradient $\nabla\Phi$, or (3) methods that measure a projection image containing information on the Laplacian of the wave front phase $\nabla^2\Phi$. In the following sections we will briefly discuss how tomographic reconstruction can be achieved in all three cases.

8.2.2.1 Reconstruction from Phase Shift Projections

Consider a 3D object described by $g\,(x,y,z)$ as shown in Fig. 8.2. The projections in a plane defined by $z = z_0$ through the object at an angle θ to the x-axis are given by the Radon transform [12]:

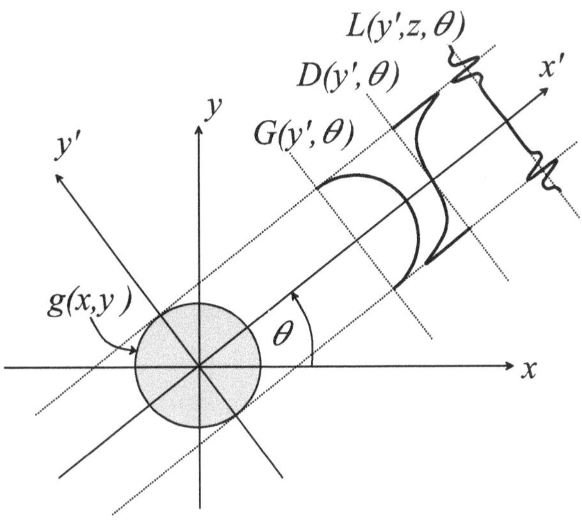

Fig. 8.2 Projection geometry for tomographic reconstruction. The three different projection functions correspond to the three main modalities for obtained phase contrast images. $G\,(y',\theta)$: direct measurement on the projected phase shift Φ induced by the object, $D\,(y',\theta)$: measurement of the gradient of the phase shift $\nabla\Phi$, and $L\,(y',\theta)$: measurement of a projection image containing information on the Laplacian of the wave front phase $\nabla^2\Phi$

$$\mathcal{G}(y',\theta)=\int_{-\infty}^{\infty} g(x',y')dx' \ , \tag{8.7}$$

where x' and y' denote a coordinate system which is rotated by an angle θ around the z-axis with respect to x and y. Note that we have omitted the variable z because it does not affect further derivation. A fast and accurate algorithm for reconstructing the original object function $g\,(x,y)$ from the Radon transform $\mathcal{G}(y',\theta)$ is the filtered backprojection (FBP) algorithm, which can be represented by the convolution backprojection integral [13, 14]:

$$g(x,y)=\int_0^{\pi} \mathcal{FT}^{-1}[\tilde{\mathcal{G}}(v',\theta)\cdot \tilde{k}(v')]d\theta \ , \tag{8.8}$$

where $\tilde{\mathcal{G}}(v',\theta)$ is the Fourier transform of $\mathcal{G}(y',\theta)$, v' the reciprocal space coordinate corresponding to $y'=-x\sin\theta+y\cos\theta$, $\tilde{k}(v')$ the Fourier representation of the filter function in the FBP, and $\mathcal{FT}^{-1}$ the inverse Fourier transform. For continuously sampled projection functions $\mathcal{G}(y',\theta)$, the filter is defined as $\tilde{k}(v')\equiv|v'|$. In practice, where the projections are sampled with a finite resolution, $\Delta y'$, the filter should be truncated at the Nyquist frequency, $1/2\Delta y'$, and becomes [13]

$$\tilde{k}(v')\equiv\begin{cases}|v'|, & |v'|\leq 1/2\Delta y'\\ 0, & |v'|>1/2\Delta y'\end{cases} \ . \tag{8.9}$$

(See also Fig. 8.3a.)

8.2.2.2 Reconstruction from Phase Gradient Projections

Let us now consider the case where the experimental arrangement does not yield the ordinary line projection of the object function as defined by Eq. (8.7), but the line projection of the partial derivative of the object function

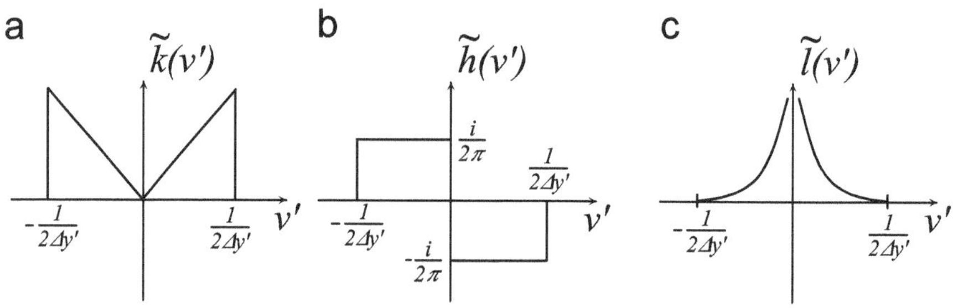

Fig. 8.3 Backprojection filter functions for (**a**) simple projection line integrals, (**b**) gradient projection integrals, and (**c**) projection image containing information on the Laplacian of the object

$$\mathcal{D}(y',\theta) = \int_{-\infty}^{\infty} \frac{\partial g(x',y')}{\partial y'} dx' \quad . \tag{8.10}$$

Then tomographic reconstruction based on the conventional FBP using the above-mentioned linear filter function [Eq. (8.9)] will not result in a correct reconstruction of the original object function $g\ (x,y)$. In analogy to what has been suggested by Faris and Byer [15], in the case of deflection tomography using visible laser light, the problem can be solved by adapting the filter function in the FBP algorithm accordingly. Using the Fourier derivative theorem, we find $\widetilde{\mathcal{D}}(v',\theta) = 2\pi i v' \times \widetilde{\mathcal{G}}(v',\theta)$, where $\widetilde{\mathcal{D}}$ is the Fourier transform of $\mathcal{D}$. Changing the integration and differentiation order in Eq. (8.10) and substituting into Eq. (8.8), we obtain

$$g(x,y) = \int_0^{\pi} \mathcal{FT}^{-1}[\widetilde{\mathcal{D}}(v',\theta) \cdot \widetilde{h}(v')]d\theta \quad , \tag{8.11}$$

with the differential phase contrast (DPC)-FBP filter function:

$$\widetilde{h}(v') \equiv \begin{cases} \frac{1}{2\pi i} \cdot sgn(v'), & |v'| \le 1/2\Delta y' \\ 0, & |v'| > 1/2\Delta y' \end{cases} \quad , \tag{8.12}$$

where $sgn(v')$ is the sign function (Fig. 8.3b).

8.2.2.3 Reconstruction from Laplacian Phase Projections

In the case where the projection images contain information on the second derivative, or more precisely the Laplacian of the phase shift induced by the object, $\nabla^2\Phi$, the projections are given by

$$\mathcal{L}(y',z,\theta) = \int_{-\infty}^{\infty} \left(\frac{\partial^2 g(x',y',z)}{\partial^2 y'} + \frac{\partial^2 g(x',y',z)}{\partial^2 z} \right) dx' \quad . \tag{8.13}$$

We note that the z-axis, which was omitted for reasons of simplicity for the treatment above, was reintroduced here. This is because projection images containing the Laplacian of the phase shift cannot simply be factored into single tomographic slices. The filter applied to the individual projections before the back projection intrinsically becomes two-dimensional (2D) in this case.

Analogous to the derivation of the correct filter function, using the Fourier derivative theorem above, we find that the tomographic reconstruction of the object is given by

$$g(x,y,z) = \int_0^{\pi} \mathcal{FT}^{-1}[\widetilde{\mathcal{L}}(v',w,\theta) \cdot \widetilde{l}(v',w)]d\theta \quad , \tag{8.14}$$

where $\widetilde{\mathcal{L}}(v', w, \theta)$ is the 2D Fourier transform of $\mathcal{L}(y', z, \theta)$ and w the reciprocal counterpart of the z-coordinate. The filter function $\widetilde{l}(v', w)$ is of the form

$$\widetilde{l}(v', w) \equiv \begin{cases} \frac{1}{4\pi^2} \cdot \frac{|v'|}{v'^2 + w^2} & |v'| \leq 1/2\Delta y' \\ 0, & |v'| > 1/2\Delta y' \end{cases}, \quad (8.15)$$

where $\mathrm{sgn}(v')$ is the sign function (Fig. 8.3c).

8.3 Experimental Methods

8.3.1 Crystal Interferometer

The most direct way of measuring a phase shift is interferometry, where a coherent reference beam interferes with a beam that has been passed through the sample. Because of the short wavelength of neutrons, however, the construction of neutron interferometers is not straightforward. To form neutron beam paths generating a stable interference, optical elements of a neutron interferometer should be sufficiently stable so that the deviation of the optical path length is smaller than the wavelength. Although it is a challenging task, various neutron crystal interferometers have been developed and used for neutron phase imaging. The stability of a crystal interferometer is ensured by cutting its entire body monolithically out of a single-crystal ingot.

Experimentally, the first single-crystal X-ray interferometer was reported by Bonse and Hart in 1965 [16], and the technology was later transferred to neutrons [17, 18]. As can be seen in Fig. 8.4, these devices typically consist of three parallel lamellae separated by a constant spacing. The lamellae function as beam splitters when a neutron beam is incident at the Bragg diffraction condition on a lattice plane perpendicular to the surface of the lamellae. The amplitude of the incident neutron wave front is coherently divided into diffracted and

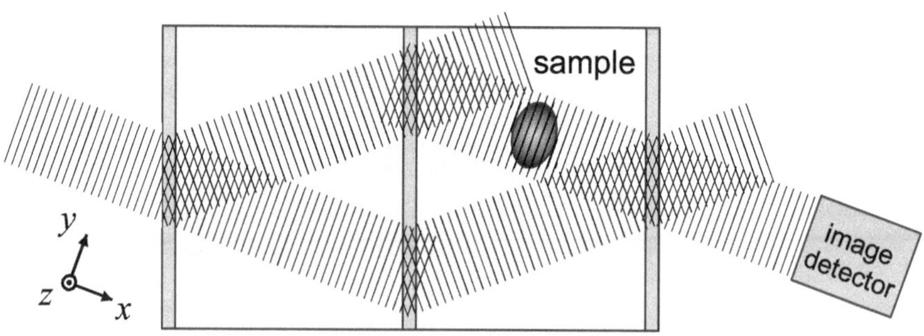

Fig. 8.4 Sketch of a monolithic crystal interferometer

forward-diffracted beams outgoing from the opposite side of the lamella. Neutrons thus divided by the first lamella are divided again by the second lamella in the same manner. Two beams overlapping at the third lamella are also divided and interference is observed in the beams outgoing from the third lamella. Because of its monolithic configuration, no mechanical tuning is needed except in the arrangement of the interferometer so that the Bragg diffraction condition is satisfied.

When an object is placed into one arm of the interferometer, it generates an interference pattern, which in general is given by

$$I(y,z) = a_0(y,z) + a_1 \cos[\Phi(y,z) + \Phi_{\mathrm{inst.}}(y,z)] \quad , \tag{8.16}$$

where a_0 and a_1 are the average intensity and fringe contrast, respectively. In practice the interferometer has some residual strain that normally generates a built-in fringe pattern, and the influence is involved in $\Phi_{\mathrm{inst.}}$. Often this effect is not negligible, making the interference pattern too complicated for grasping structural information by eye. One way to separate the effect of the sample from the characteristics of the interferometer is using a "phase-stepping"approach. This phase-stepping method processes multiple interference patterns observed by varying the phase difference between the two arms, for instance by using a tunable phase shifter. When interference patterns

$$\begin{gathered} I_m(y,z) = a_0(y,z) + a_1(y,z)\cos[\Phi(y,z) + \Phi_{inst.} + 2\pi m/M], \\ (m = 1, 2, \ldots, M) \ , \end{gathered} \tag{8.17}$$

are obtained with $M \geq 3$, both with and without the sample in place, $\Phi(y,z)$ and $\Phi_{\mathrm{inst.}}$ can uniquely be processed out from the data set. For phase tomography, this measurement is repeated while rotating the sample.

Figure 8.5 shows an example [19] where an aluminum screw has been imaged in an aluminum block of slightly different composition by using neutron crystal interferometry. From the image without sample (Fig. 8.5a), we can deduce that the phase of the empty interferometer is reasonably constant over the total projection area of about 10 mm width and 25 mm height. With sample (Fig. 8.5b), the modulation of the phase induced by the sample is clearly recognizable.

By essentially taking the difference and correcting for phase-wrapping errors, one obtains a projected phase map of the specimen, as shown in Fig. 8.6a. Because the measured phase projection images are of a quantitative nature, phase tomography reconstructions can be performed using Eq. (8.8). One such reconstruction is shown in Fig. 8.6b, in the form of a 3D rendering of the object (see [19] for more details). Other examples of the application of neutron interferometry for phase imaging are reported in [20, 21, 22, 23, 24].

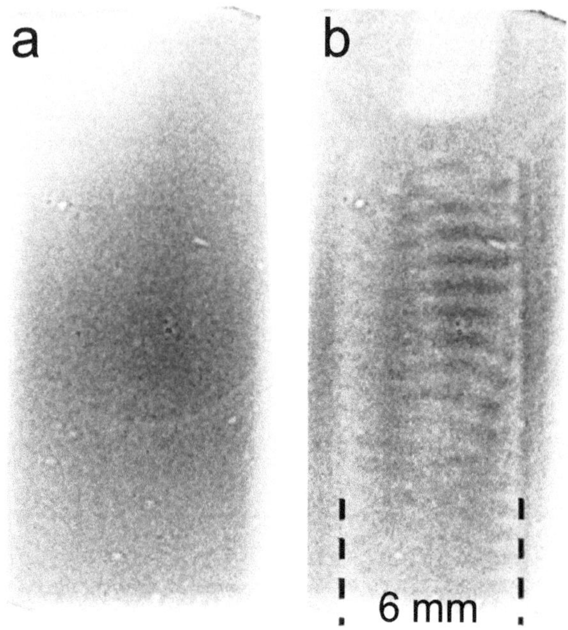

Fig. 8.5 Raw images behind a neutron crystal interferometer of an aluminum screw in an aluminum block. (**a**) Reference image without sample and (**b**) interferometer image with sample in the beam path (images from [19])

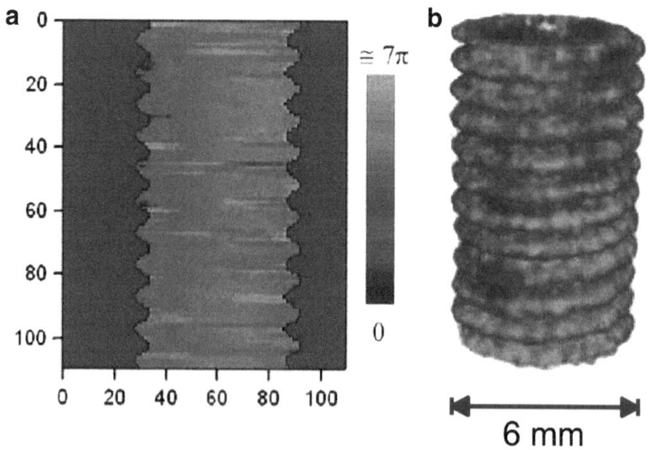

Fig. 8.6 Processed neutron interferometer images of an aluminum screw in an aluminum block. (**a**) Projection phase map of the sample after processing the data. (**b**) 3D rendering of the reconstructed phase object (images from [19])

8.3.2 Refraction-Based Techniques

8.3.2.1 Crystal Analyzer-Based Techniques

Similar to the refraction of visible light that is governed by Snell's law, the refraction of a neutron beam can be caused by a spatially varying phase shift induced by a specimen. Thus, by measuring these deflection angles α, a contrast related to the phase shift gradient (DPC) is generated, which can be expressed by [25]

$$\alpha \propto \frac{\lambda}{2\pi} \nabla \Phi \quad , \tag{8.18}$$

where ∇ denotes the gradient operator.

The deflection angle caused by refraction is extremely small, and neutrons are assumed conventionally to go straight through materials. To generate the contrast, an angular sensor sufficiently sensitive to the neutron beam refraction is needed, and the incident neutron beam should furthermore be correspondingly collimated. Perfect crystal analyzers are well suited for that purpose because the angular width of the Bragg diffraction is on the order of 10–100 μrad. Figure 8.7 shows a typical arrangement for generating a contrast based on the refraction of neutron beams using perfect crystals. When a neutron beam collimated by a crystal passes through a sample, the wave front is deformed by the phase shift at the sample and the neutron propagation direction varies depending on the phase gradient. The analyzer crystal placed downstream of the sample reflects only the neutrons that meet the Bragg diffraction condition. As a result, a contrast can be seen in the reflected beam. By changing the angular setting of the analyzer slightly, the contrast can be varied in a very sensitive way.

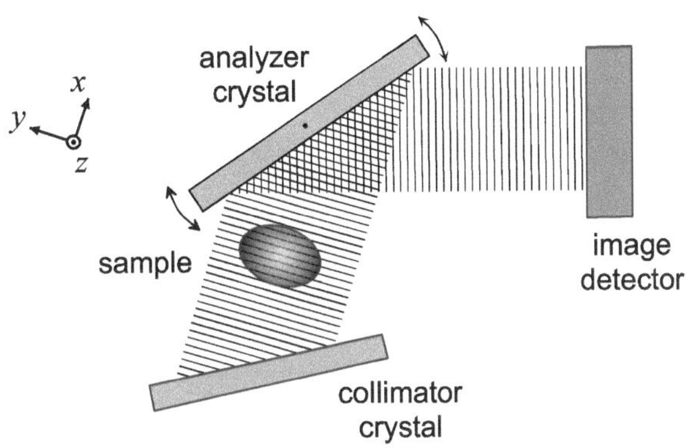

Fig. 8.7 A typical setup for analyzer-based phase contrast imaging and tomography

Given the reflection curve $R(\gamma)$ of the analyzer and the complex transmission function $\sqrt{I(y,z)/I_0(y,z)}\exp\,[-i\Phi(y,z)]$ of the specimen, the image $I(y,z,\gamma_0)$ of the beam reflected by the analyzer is given by

$$I(y,z,\gamma_0) = \sqrt{I(y,z)/I_0(y,z)}R[\gamma_0 - \alpha(y,z)] \quad , \tag{8.19}$$

where γ_0 is the incident angle without the sample. Similar to the phase-stepping procedure in single-crystal interferometry, the values α (y,z) can be determined more precisely by measuring several points along the crystal rocking curve and subsequent data processing. Tomographic reconstruction of an object from a data set containing measured projections of γ (y,z) is also possible. Since these projections basically represent gradient projections of the object phase shift, CT reconstruction can be achieved by applying Eq. (8.11).

An example of neutron phase tomography using such a crystal analyzer-based method is shown in Fig. 8.8 (from [26]). We see tomographic images of an aluminum test sample imaged with conventional attenuation-based neutron tomography (Fig. 8.8a) and in phase tomography (Fig. 8.8b). Because aluminum exhibits only a very small neutron capture cross-section, it produces negligible contrast in absorption. In the phase reconstruction, however, the internal details are clearly visible. Moreover, it has recently been shown that these reconstructions are in quantitative agreement with the expected literature values for the neutron refractive index of materials (see [27] for more details). Other examples and possible future applications of analyzer-based neutron phase imaging are reported in [28, 29, 30, 31].

8.3.2.2 Grating Interferometers

Similar to the crystal analyzer-based method described, grating interferometers can also be classified as a refraction-based technique. The main advantage, however, is that grating interferometers are considerably more efficient because they can accept a much larger neutron energy spread and incoming divergence.

Figure 8.9 shows the basic setup of a grating interferometer. It consists of a source grating G0, a phase grating G1, and an analyzer absorption grating G2.

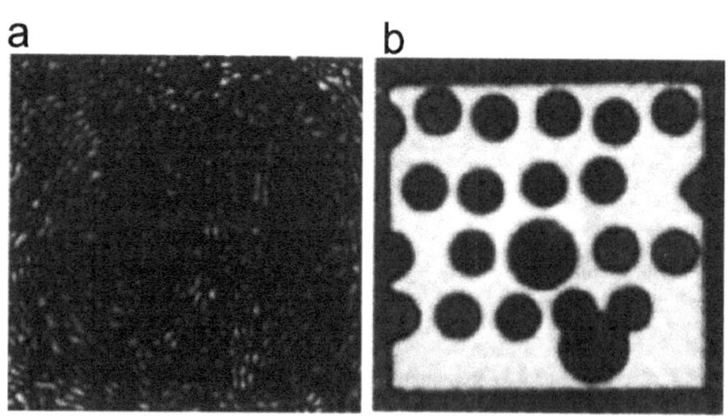

Fig. 8.8 Reconstructed tomographic slices obtained for an aluminum test sample with holes. (**a**) Conventional attenuation-based reconstruction; (**b**) phase contrast-based reconstruction (images from [26]). The size of the sample is approximately 25 × 25 mm^2

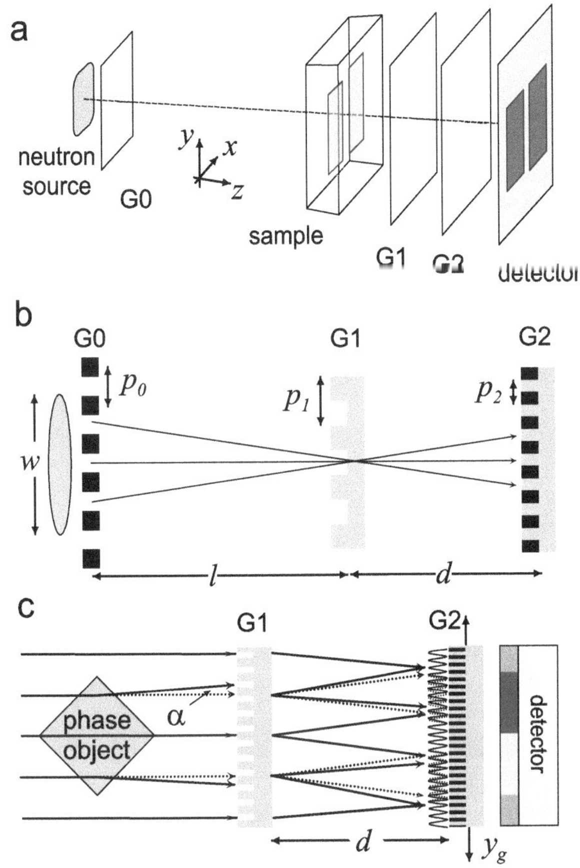

Fig. 8.9 (**a,b**) Principle setup of a grating-based neutron shearing interferometer. (**c**) The phase grating (G1) forms a periodic interference pattern in the plane of the analyzer grating. A phase object in the incident beam will cause a slight refraction, which results in changes of the locally transmitted intensity through the analyzer

The source grating (G0), an absorbing mask with transmitting slits, typically placed close to the neutron beam exit port, creates an array of individually coherent but mutually incoherent sources. Each individual line source provides enough spatial coherence for the image formation process [32, 33, 34]. As the source mask G0 can contain a large number of individual lines, each creating a virtual source, efficient use can be made of typical neutron source sizes of more than a few square centimeters. To ensure that each of the line sources contributes constructively to the image formation process, the geometry of the setup should satisfy the condition $p_0 = p_2\, l/d$, where p_0 (p_2) is the period of G0 (G2), l the distance between G0 and G1, and d the distance between G1 and G2 (Fig. 8.9b).

The DPC image formation process achieved by the two gratings G1 and G2 is similar to *Schlieren* imaging [25] or to a Shack–Hartmann wave front sensor [35]. The second grating (G1) acts as a beam splitter and divides the incoming beam essentially into the two first diffraction orders. Since the wavelength λ of the illuminating neutron radiation ($\approx 10^{-10}$ m) is much smaller than the grating pitch ($\approx 10^{-6}$ m), the angle between the two diffracted beams is so small that they overlap almost completely.

Downstream of the beam-splitter grating, the diffracted beams interfere and form linear periodic fringe patterns in planes perpendicular to the optical axis [33, 36]. For a phase grating with a phase shift of π, the periodicity p_2 of the fringe pattern equals half the period of G1. Neither the period nor the lateral position of these fringes depends on the wavelength of the radiation used. Perturbations of the incident wave front, such as those induced by refraction on a phase object in the beam, lead to local displacement of the fringes (Fig. 8.9c).

The fundamental idea of the method presented here is to detect the local positions of the fringes and determine from these the phase shift induced by the object. However, since the spacing of the interference fringes does not exceed a few microns, an area detector placed in the detection plane will generally not have sufficient resolution to resolve the fringes, let alone the exact position of their maxima. Therefore, a grating G2 with absorbing lines and the same periodicity and orientation as the fringes is placed in the detection plane, immediately in front of the detector. This analyzer grating acts as a transmission mask for the detector and transforms local fringe positions into signal intensity variations. The detected signal profile thus contains quantitative information about the phase gradient of the object.

To separate this phase information from other contributions to the signal such as attenuation in the sample, inhomogeneous illuminations, or imperfections of the gratings, the phase-stepping approach used in other interferometry methods [3] was adapted to this setup. When one of the gratings is scanned along the transverse direction y_g (Fig. 8.9c), the intensity signal I (y,z) in each pixel (y,z) in the detector plane oscillates as a function of y_g (Fig. 8.10e). The interferometer phases $\vartheta(y,z)$ of the intensity oscillations in each pixel are related to the wave front phase profile $\Phi(y,z)$, the neutron wavelength λ, and the period p_2 of the absorption grating by [25]

$$\vartheta = \frac{\lambda d}{p_2}\frac{\partial \Phi}{\partial y} \quad . \tag{8.20}$$

Since ϑ contains no other contributions, particularly no attenuation contrast, the phase profile of the object can be retrieved from $\vartheta(y,z)$ by a simple one-dimensional [1D] integration.

Figure 8.11 shows processed transmission (a), DPC (b), and integrated phase (c) images of a sample comprising titanium (Ti), lead (Pb), molybdenum (Mo), and magnesium (Mg) metal rods with square profiles (Fig. 8.9c). The results

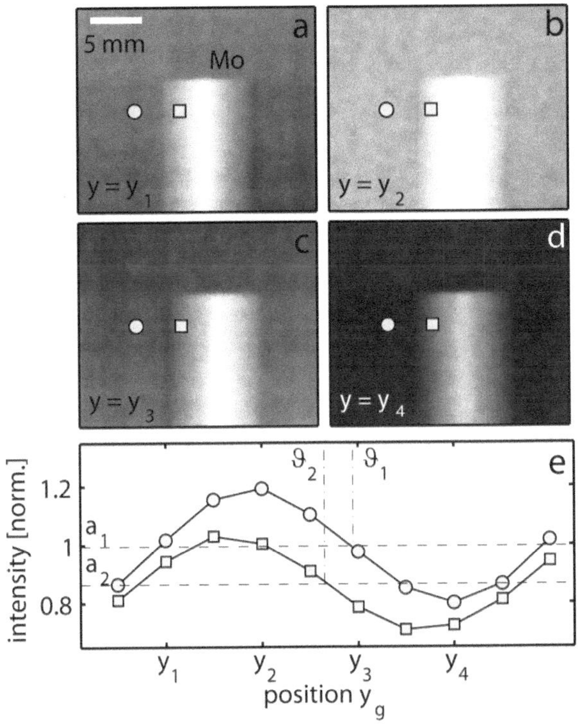

Fig. 8.10 Principle of phase stepping. (**a–d**) Raw image data of a square molybdenum (Mo) metal rod, taken at different positions $y_g = y_1, \ldots, y_4$ of the analyzer grating *G2*. (**e**) Intensity oscillation in two detector pixels as a function of y_g (images from [33])

were obtained at the ICON beam line of the SINQ (Paul Scherrer Institut, Switzerland). Due to their specific neutron capture cross-sections and incoherent scattering cross-sections, differences in the transmitted intensity through the rods are observed in Figure 8.11a. Because the vector component of the phase gradient in the direction perpendicular to the grating lines is constant, but of opposite signs for both sides of the rods, the rods appear black and white in the DPC image. It is interesting to note that titanium has a negative neutron scattering length density [5], and consequently, a negative phase shift is measured in the material (see Fig. 8.11c). Furthermore, we note that particularly for weakly absorbing materials, like magnesium and lead (Fig. 8.11a), the DPC (Fig. 8.11b) and the integrated phase shift (Fig. 8.11c) signal yield a higher contrast in the image.

As the phase contrast obtained with a grating interferometer can be described quantitatively, it can further be used for neutron phase tomography. Figure 8.12 shows one exemplary processed DPC image (a), a 3D rendering of the reconstructed coherent scattering length density distribution (b), and 2 (out of 512) reconstructed tomographic slices (c,d) of a sample comprising titanium

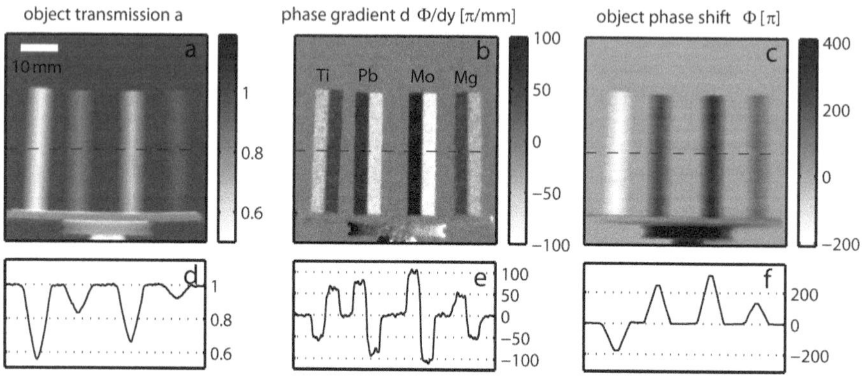

Fig. 8.11 Linear contour plots and section profiles of processed data for a test object comprising titanium (Ti), lead (Pb), molybdenum (Mo), and magnesium (Mg) metal rods with a square profile. (**a**) Conventional neutron transmission image. (**b**) Differential phase contrast image d Φ/d y. (**c**) Integrated phase shift $\Phi(y,z)$ retrieved from dΦ/d y by integration. (**d,e,f**) Section profiles through the corresponding image data (images from [33])

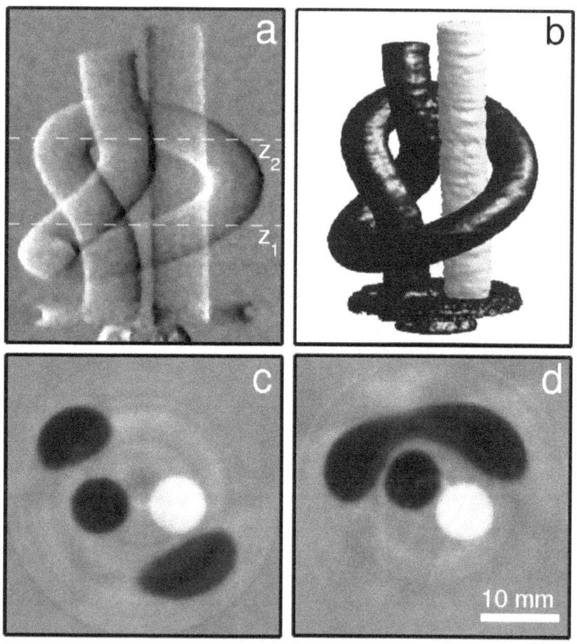

Fig. 8.12 Projection images and tomographic reconstruction for a sample comprising lead (*black*) and titanium (*white*) rods with diameters of 6.35 mm. (**a**) Differential phase contrast projection image. (**b**) 3D rendering of the reconstructed coherent scattering length distribution in the sample. (**c,d**) 2 (out of 512) reconstructed tomographic slices (images from [33])

and lead rods of 6.35 mm diameter. The corresponding CT reconstruction algorithm used 180 differential phase projections and 1 reference projection without the sample as input (see Eq. (8.12) and [37]).

8.3.3 Propagation-Based Techniques

In the classical sense, the distance between an object and a neutron image detection plane should be as short as possible to avoid blurring. However, under (partially) coherent illumination, Fresnel or Fraunhofer diffraction is observed at a distance away from an object even with neutrons. Then, an edge-enhanced contrast is observed without optical elements even for a pure phase object, as illustrated in Fig. 8.13. The choice of the propagation distance has a direct influence on the characteristics of the observed fringe contrast produced by the edges of the object.

The effect can be described by considering the propagation of a wave field along a certain direction in space (Fig. 8.13). It can be written under the paraxial approximation by [38, 39]

$$I(x, y, z) = 1 + \frac{\lambda x}{2\pi} \nabla^2 \Phi(0, y, z) \quad , \tag{8.21}$$

where we have assumed a propagation along the x direction and ∇ denotes the 2D gradient operator acting in the (y,z) plane. This expression can be derived from the so-called transport of intensity equation [38, 39], assuming a nonabsorbing, weak-phase object. The character of the Laplacian in Eq. (8.21) suggests the generation of the contrast particularly outlining surfaces and structural boundaries, where the refractive index changes abruptly.

In propagation-based phase contrast radiography, the distance between the sample and the image detector is large so that the weak lateral deflections due to

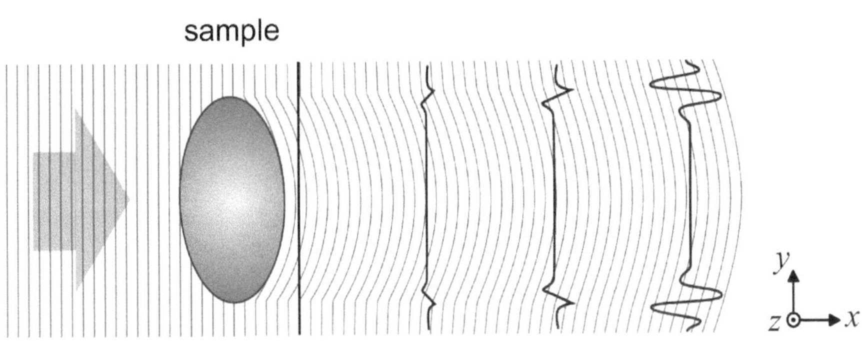

Fig. 8.13 Illustration showing the generation of an edge-enhanced contrast of a phase object due to propagation

nonabsorbing, refractive sample components can form, bringing these fine structures into sharp relief. For this method, the collimation ratio, or angular spread of incident beam, is necessarily on the order of 10^4–10^5, so that the detail is not washed out in source size blurring.

The development of phase contrast with propagation is shown in Fig. 8.14. Here a hollow lead sinker has been mounted on an aluminum screw. In the contact image (at 20 mm), scattering/absorption within the microcrystalline structure of the lead leads to little image contrast, but affords no detail. The aluminum screw is invisible. Images for propagation steps of approximately 220, 450, 900, and 1800 mm are then shown. In the final phase contrast image, the hollow structure and flaws within the sinker, the aluminum screw, and shavings in the axial hole from the threading process are all clearly visible [40]. Other examples of the application of propagation-based phase contrast imaging with neutrons are reported in [41, 42, 43, 44, 45, 46, 47].

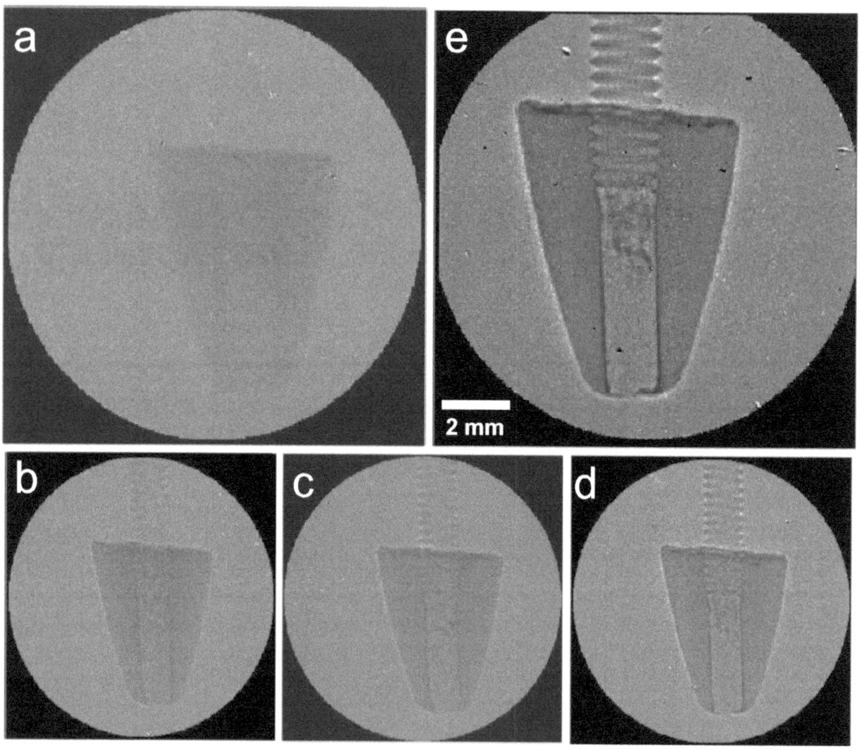

Fig. 8.14 A series of neutron radiographs showing the development of the phase contrast of a lead sinker mounted on an aluminum screw. The images, counterclockwise from top left to top right (**a–e**), correspond to propagation distances of approximately 0.02, 0.22, 0.45, 0.90, and 1.8 m, respectively. The visibility of the edge detail within the sinker, screw, and shavings trapped in the axial hole are greatly enhanced (images from [40])

It is conceivable that a number of these projections could be used to generate a quantitative tomographic reconstruction based on Eq. (8.15) in a manner similar to how it is done in the case of X rays [48].

8.4 Coherence Requirements

To generate contrast with phase information, a wave front with a defined phase is required. This means that a certain degree of coherence is required in the neutron beam. The coherence can be evaluated perpendicular to the optical axis and along the optical axis. The first case refers to the so-called *spatial* or *transverse* coherence and the second to the *temporal* or *longitudinal* coherence [49, 50].

Let us first consider the coherence properties of a source in a plane *transverse* to the propagation direction of the radiation. A good measure for the degree of coherence is the interference contrast recorded in the far-field diffraction pattern of an object in the beam. Let us take as an object Young's double slit arrangement [25]. Illuminating the double slit with a monochromatic point source results in the far field in a set of interference fringes with fully developed maxima and zero intensity in the minima, corresponding to a fringe visibility of 100%.

In practice, however, the source is not point-like, but has some lateral extension. It is easy to see that this reduces the visibility of the interference fringes. Although the fringe patterns resulting from different points of the extended source are identical, they emerge at slightly shifted diffraction angles. Their superposition, therefore, causes the observed pattern to be somewhat washed out (i.e., the fringe visibility to be reduced). The effect of the extension of the source can be readily quantified. Let us consider two narrow slits at distance d, illuminated by a uniform line source of height w (Fig. 8.15).

The interference fringes produced by an infinitesimally small emitting source element located on the optical axis appear at diffraction angles $\alpha = m\lambda/d$, with $m = 0, \pm 1, \pm 2, \ldots$ for the maxima and $m = \pm\frac{1}{2}, \pm\frac{3}{2}, \pm\frac{5}{2}\cdots$ for the minima. Here, the small-angle approximation has been made. The fringes produced by an emitting element at the border of the source are found at the shifted angles $m\lambda/d + w/2R$, with R the distance between the source and the slits. It is easy to see that the maxima from a border element of the source coincide with the minima from the central element at a slit separation $d = \lambda R/w$. This distance is defined as the transverse coherence length

$$\xi_t = \frac{\lambda R}{w} \quad . \tag{8.22}$$

Considering now a uniform rectangular source with horizontal and vertical widths w_h and w_v, which illuminates two pairs of pinholes at right angles at distances d_h and d_v, we arrive at the corresponding transverse coherence lengths $\xi_h = \lambda R/w_h$ and $\xi_v = \lambda R/w_v$.

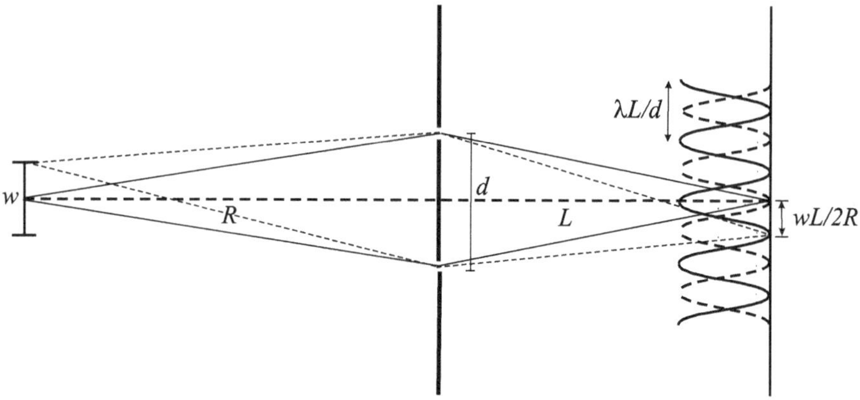

Fig. 8.15 Diffraction patterns from two narrow slits at distance d, originating from the central part of the source (*solid curve*) and from the edge of the source at height $w/2$ (*dashed curve*). The slit distance d is such that the two patterns are in antiphase. This occurs for $d=\lambda\ R/w$. Angles and distances are not to scale (adapted from [50])

The degree of coherence of the radiation along its propagation direction (i.e., its *longitudinal* [or *temporal*] coherence) enters the imaging process in a different way. Let us consider two wave fronts, one at wavelength λ and the other at a slightly different wavelength $\lambda + \Delta\lambda$, which simultaneously depart from a single-point. Let us assume that after some distance ξ_l both wave fronts are in antiphase (Fig. 8.16).

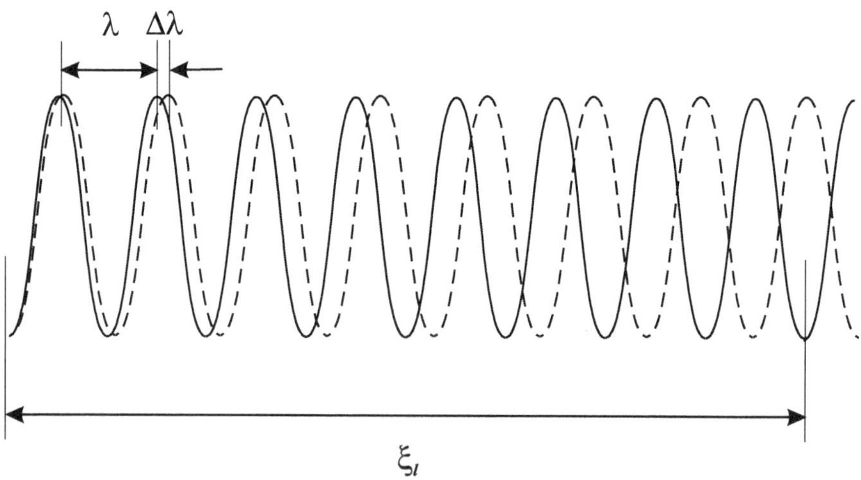

Fig. 8.16 Propagation of two waves with wavelengths λ and $\lambda + \Delta\lambda$. The longitudinal coherence length ξ_l is defined as the distance over which the phase difference of the two waves has become π (adapted from [50])

If the first wave has made N oscillations over that distance, the second wave must have made $N - \frac{1}{2}$ oscillations. One therefore has $N\lambda = (N - \frac{1}{2})\ (\lambda + \Delta\lambda)$. Solving for N and substituting in $\xi_l = N\lambda$ we find for this distance

$$\xi_l \simeq \frac{1}{2} \cdot \frac{\lambda^2}{\Delta\lambda} \quad . \tag{8.23}$$

The precise prefactor is dependent on the spectral power density of the source; for a Lorentzian spectrum one has $\xi_l \simeq (2/\pi) \cdot \lambda^2/\Delta\lambda$ [51].

In practice the intrinsic coherence properties of a neutron spallation or reactor source are relatively low. This is because such sources typically are a few centimeters in size and emit a broad polychromatic spectrum. But the coherence can be improved by adapting the experimental setup accordingly.

As can easily be seen from Eq. (8.22), the transverse coherence can be increased by inserting a small aperture into the neutron beam at a large distance upstream from the sample. For example, if we insert an aperture with a 1 mm diameter at a distance of 10 m upstream from the sample, we will obtain a transverse coherence length on the order of a few microns for thermal neutrons with a typical wavelength of a few angstroms. The propagation-based phase contrast imaging method, for example, typically requires this amount of transverse coherence (see also Section 8.3.3).

As the longitudinal coherence depends on the wavelength spread of the neutron energy spectrum [Eq. (8.23)], it can be improved by reducing the bandwidth of the wavelength distribution. This can be achieved, for example, by using a double-crystal silicon monochromator. With a typical bandwidth of $\Delta\lambda/\lambda \approx 10^{-4}$, a longitudinal coherence length of a few microns can be achieved for thermal neutrons with a typical wavelength of a few angstroms. Such a beam preparation is necessary, for example, in the case of the crystal analyzer-based method (see Section 8.3.2.1).

Table 8.2 contains a comparison of the coherence requirements for the phase imaging methods described above.

Table 8.2 Comparison of the coherence requirements of several phase-sensitive imaging methods. Note that, strictly speaking, the *requirements* of a single-crystal interferometer on the coherence of the incident neutron beam are relatively low. It is more an intrinsic *property* of the device that it only accepts the spatially and temporally coherent fraction of the incident beam

Method	Requirements on spatial coherence	Requirements on longitudinal coherence
Crystal interferometer	High	High
Crystal analyzer	High	High
Grating interferometer	Low	Moderate
Propagation-based imaging	High	Low

References

1. A. F. Kracklauer, *On the theory of quanta* (a translation of the thesis of de Broglie, "Recherches sur la théorie des quanta"), *Ann. de Phys.* **10**, t. III (1925).
2. A. G. Klein, S. A. Werner, *Rep. Prog. Phys.* **46**, 259 (1983).
3. H. Rauch, S. A. Werner, *Neutron Interferometry*, Oxford University Press, Oxford, (2000).
4. http://www.ncnr.nist.gov/resources/sldcalc.html.
5. V. F. Sears, *Neutron News*, **3**, 26 (1992).
6. M. Zawisky, U. Bonse, F. Dubus, et al., *Europhys. Lett.* **68**, 337–343 (2004).
7. A. Cimmino, G. I. Opat, A. G. Klein, H. Kaiser, *Phys. Rev. Lett.* **68**, 380 (1989).
8. B. E. Allman, A. Cimmino, A. G. Klein, *Phys. Rev. Lett.* **68**, 2409 (1992).
9. G. Badurek, H. Rauch, D. Tuppinger, *Phys. Rev. A* **34**, 2600 (1986).
10. A. G. Klein, G. I. Opat, A. Cimmino, et al., *Phys. Rev. Lett.* **46**, 1551 (1981).
11. A. G. Wagh, V. C. Rakhecha, J. Summhammer et al., *Phys. Rev. Lett.* **78**, 755 (1997).
12. J. Radon, Ber. Verh. Sächs. Akad. Wiss. Leipzig, *Math–Nat.* **69**, 262 (1917).
13. G.T. Herman, *Image reconstruction from projections*, Academic Press, New York (1980).
14. A. C. Kak M. Slaney, *Principles of Computerized Tomography* IEEE Press, New York, (1987).
15. G. W. Faris, R. L. Byer, *Appl. Opt.*, **27**, 5202–5212 (1988).
16. U. Bonse, M. Hart, *Appl. Phys. Lett.* **6**, 155 (1965).
17. H. Rauch, W. Treimer, U. Bonse, *Phys. Lett. A* **47**, 369–371 (1974).
18. W. Treimer, *Krist. Tech. Crystal Res. Technol.* **13**, 1105–1116 (1978).
19. F. Dubus, U. Bonse, M. Zawisky, et al., *IEEE Trans. Nucl. Sci.* **52**, 364–370 (2005).
20. G. Badurek, M. Hochhold, H. Leeb, *Physica B* **234**, 1171–1173 (1997).
21. G. Badurek, R. J. Buchelt, H. Leeb, et al., *Physica B* **335**, 114–118 (2003).
22. E. Jericha, R. Szeywerth, H. Leeb, et al., *Physica B* **397**, 159–161 (2007).
23. J. Rehacek, Z. Hradil, M. Zawisky, et al., *Phys. Rev. A*, **71**, 023608 (2005).
24. S. A. Werner, J. L. Staudenmann, R. Colella, *Phys. Rev. Lett.* **35**, 1053 (1979).
25. M. Born, E. Wolf, *Principles of Optics*, Pergamon Press, Oxford (1980).
26. W. Treimer, M. Strobl, A. Hilger, et al., *Appl. Phys. Lett.* **83**, 398–400 (2003).
27. M. Strobl, K. Staack, W. Treimer, et al., *Meas. Sci. Technol.* **19**, 034020 (2008).
28. M. Strobl, W. Treimer, A. Hilger, *Nucl. Instrum. METHODS Phys. Res. B* **222**, 3–4 (2004).
29. M. Strobl, W. Treimer, A. Hilger, *Phys. B*, **385–386**, 1209–1212 (2006).
30. M. Strobl, W. Treimer, P. Walter, et al., *Appl. Phys. Lett.* **91**, 254104 (2007).
31. W. Treimer, M. Strobl, A. Hilger, et al., *IEEE Trans. Nucl. Sci.*, **52**, 386–388 (2005).
32. F. Pfeiffer, O. Bunk, C. Schulze–Briese, et al., *Phys. Rev. Lett.* **94**, 164801 (2005).
33. F. Pfeiffer, C. Grunzweig, O. Bunk, et al., *Phys. Rev. Lett.*, **96**, 215505 (2006).
34. F. Pfeiffer, T. Weitkamp, O. Bunk, et al., *Nat. Phys.* **2**, 258–261 (2006).
35. J. Hartmann, *Z. Instrumentenkd.* **8**, 2 (1990).
36. T. Weitkamp, A. Diaz , C. David, et al., *OPT. Express*, **13**, 6296–6304 (2005).
37. F. Pfeiffer, C. Kottler, O. Bunk, et al., *Phys. Rev. Lett.* **98**, 108105 (2007).
38. J. M. Cowley, *Diffraction Physics*, Elsevier, Amsterdam (1995).
39. M. R. Teague, *J. Opt. Soc. Am.* **73**, 1434 (1983).
40. B. E. Allman, K. A. Nugent, *Physica B*, **385–86**, 1395–1401 (2006).
41. B. E. Allman, P. J. McMahon, K. A. Nugent, et al., *Nature* **408**, 158–159 (2000).
42. J. W. Chen, H. Y. Gao, H. F. Zhu, et al., *Acta Phys. Sin.* **54**, 1132–1135 (2005).
43. F. Fiori, A. Hilger, N. Kardjilov, et al., *Meas. Sci. Technol.* **17**, 2479–2484 (2006).
44. D.L. Jacobson, B.E. Allman, J. McMahon, et al., *Appl. Radiat. Isot.* **61**, 547–550 (2004).
45. E. Lehmann, K. Lorenz, E. Steichele, et al., *Nucl. Instrum. Methods Phys. Res. A* **542**, 95–99 (2005).
46. P. J. McMahon, B. E. Allman, D. L. Jacobson, et al., *Phys. Rev. Lett.* **91**, 145502 (2003).
47. B. Schillinger, E. Calzada, K. Lorenz, *Mat. Trans., Proc.* **112**, 61–71 (2006).

48. S. C. Mayo, T. J. Davis, T. E. Gureyev, et al. *Opt. Express* **11**, 2289–2302 (2003).
49. J. Als-Nielsen, D. McMorrow, *Elements of Modern X–Ray Physics*, John Wiley & Sons Ltd, New York (2001).
50. F. van der Veen, F. Pfeiffer, *J. Phys. Condens. Matter* **16**, 5003–5030 (2004).
51. J. W. Goodman, *Statistical Optics*, Wiley, New York, (1985).

Chapter 9
Thermal Neutron Holography

B. Sur, R.B. Rogge, V.N.P. Anghel, and J. Katsaras

Abstract X-ray and neutron diffraction techniques have for almost a century produced results that provide important insights into materials of interest to a wide range of scientific and technological disciplines. However, traditional diffraction techniques have their limits, and these limits are best exemplified by the fact that certain important materials (e.g., integral proteins) are difficult if not impossible to crystallize — diffraction techniques usually require high-quality single crystals. Recently developed atomic resolution X-ray and neutron holography techniques offer the promise to resolve the structures of difficult-to-crystallize materials to atomic resolution. This chapter will discuss the latest developments in neutron holography and the challenges that must be overcome to make the technique a viable tool.

Keywords X-ray holography · Neutron holography · Inside source holography · Inside detector holography · Protein crystallography

9.1 Introduction

Holography can be traced to Bragg's [1] X-ray work and Gabor's [2] electron interference microscope. Although the technique lay dormant for a number of decades, there has been increasing interest in its development as implied by the increasing number of publications on atomic-resolution holography using either electrons [3–7] or hard X rays [8–13]. Moreover, the feasibility of atomic-resolution thermal neutron holography, either the "inside source" or "inside detector" type, was experimentally demonstrated [14–16], and a theory was developed showing that the diffraction pattern of plane waves incident on a sample with a uniformly random distribution of incoherent scatterers

J. Katsaras (✉)
National Research Council, Chalk River Laboratories, Chalk River, ON K0J 1J0, Canada
e-mail: john.katsaras@nrc.gc.ca

I.S. Anderson et al. (eds.), *Neutron Imaging and Applications*,
Neutron Scattering Applications and Techniques, DOI 10.1007/978-0-387-78693-3_9,

(e.g., ^{1}H in a polymeric material) per unit cell is the same as that from a sample with a single incoherent scatterer per unit cell [17]. There is interest in the possibility of using neutron holography to reconstruct, to atomic resolution, the three– dimensional (3D) structure of materials rich in hydrogen that are presently difficult to crystallize (e.g., membrane-associated proteins).

Dennis Gabor (1900–1979) developed holography as a technique to improve the resolution of the electron microscope while working for the British Thomson-Houston Company (Rugby, England). A patent was filed in 1947 [18], but the general field did not advance until the invention of the laser [19, 20] (the process of stimulated emission was first theorized by Einstein in 1917) [21]. The first 3D holograms were made by Denisyuk [22] in 1962 (reflection hologram) and later that year by Leith and Upatnieks at the University of Michigan (transmission hologram) [23]. Since then, optical holography has been extensively used in optical devices (e.g., holographic lens for heads-up displays), security (e.g., credit cards), scanners, computer memory storage, medicine, and a variety of other applications. For example, holographic interferometry is a precise technique for measuring changes in the dimensions of an object, and digital cameras with holographic capabilities are able to focus in poor contrast conditions.

For many decades, X-ray diffraction has been relied upon to determine the 3D structure of a number of crystalline materials, the prerequisite being a very high-quality single crystal. However, even when the requisite sample was made available, the technique suffered from the so-called "phase problem" — that is, the phase information was not recorded. As a result, a number of isomorphous crystals had to be grown to allow for the direct solution of molecular structure — at increased cost and time. It should also be pointed out that, in some cases, isomorphous derivatives were impossible to produce. Recently, however, synchrotron sources with their tunable X-ray beams have revolutionized crystallography by allowing the development of multiple-wavelength anomalous dispersion (MAD), a technique used to obtain reliable phase information. Now a single crystal serves the purpose of several different metal-containing crystals and has the added advantage of faster data collection times. Despite these advances, the need for high-quality crystals remains.

Both holography and crystallographic techniques, such as MAD, solve the phase problem in the case of 3D materials. Holographic techniques, however, have the added advantage that only orientational order is required; that is, translational symmetry is not necessary. In addition to imaging crystalline materials, holography is also suitable for imaging noncrystalline materials (e.g., nematic liquid crystals) or poorly crystallized membrane-associated proteins.

Although the idea of atomic-resolution holography has its roots in Bragg's [1] X-ray work and Gabor's [2] electron interference microscope, atomic-resolution holography did not evolve until Szöke [24] pointed out that photo-excited atoms within a sample emit highly coherent outgoing electrons or fluorescent X rays. The interference between the unperturbed wave from the photoexcited atom (reference wave) and waves scattered from neighboring atoms (object waves) thus permitted the realization of atomic-resolution holography (Fig. 9.1). Of course, evidence for atomic-resolution holography existed as far back as the early twentieth century in the experiments of Kikuchi [25] and Kossel [26].

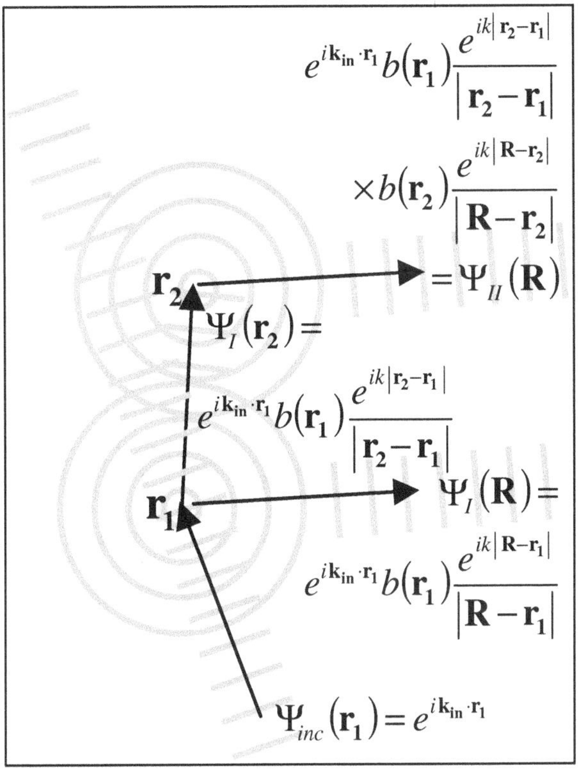

Fig. 9.1 (A) Schematic depiction of second-order plane-wave scattering. The incident neutron plane wave Ψ_{inc} interacts with an atom at $\mathbf{r}_1$, producing a primary spherical wave (*S* wave) Ψ_I. This primary *S* wave interacts with a second atom at $\mathbf{r}_2$, producing a secondary *S* wave Ψ_{II}. The interference between the primary and secondary *S* waves leads to an intensity modulation (hologram) detected at $\mathbf{R}$. The detected intensity (evaluated for incident and detected wave vectors, $\mathbf{k}_{\text{in}}$ and $\mathbf{k}_{\text{out}}$) is given as $I = \Psi_I^*\Psi_I + 2\text{Re}(\Psi_I^*\Psi_{II}) + \Psi_{II}^*\Psi_{II}$. Figure adapted from Sur et al. [17]. See Ref. [17] or Sect. 9.5 of this paper for details

9.2 Atomic-Resolution Holography

There are two kinds of holographic methods capable of atomic resolution: inside source and inside detector. The inside source method [24] uses some of the atoms making up the sample as independent sources of coherent illumination. In the case of X-ray holography and certain types of electron holography (e.g., Auger holography), the radiation generated by atomic de-excitation manifests itself in the form of nearly spherical waves (*S* waves). This radiation can either pass through the sample unperturbed (reference wave) or be scattered by the surrounding atoms (object wave). The interference between the reference and object beams produces the hologram.

In the inside detector method developed by Gog et al. [9, 27], the inside source holography beam paths can be reversed and the source and detector exchanged. In other words, an incident plane wave can reach the photoexcitable atom in a crystal unperturbed (reference wave) or after it has scattered off other atoms in the sample (object wave). The interference between these two beams dictates the fluorescence intensity given off by the photoexcitable atom, which is now the detector. When the incident beam is scanned, the phase relation between the reference and object waves leads to oscillations of the fluorescent intensity [28] and forms a hologram.

9.3 Neutrons

In 1932, James Chadwick [29, 30] discovered the neutron — first postulated by Rutherford [31]—a neutral, subatomic elementary particle found in all atomic nuclei except hydrogen nuclei. A neutron's mass is similar to that of a proton, but a neutron possesses a nuclear spin of ½ and a magnetic moment. Neutrons are stable only when bound in an atomic nucleus; an unstable free neutron has a mean lifetime of approximately 900 s and decays into a proton, an electron, and an antineutrino [32].

Even though the neutron's interaction with atomic nuclei is weak, the scattering power (cross section) of an atom is not related to its atomic number. Therefore, neighboring elements in the periodic table can have substantially different scattering cross sections. It is important to note that the interaction of a neutron with the nucleus of an atom allows for an element's isotopes to be differentiated. An example is the isotopic substitution of ^{1}H (hydrogen) for ^{2}H (deuterium), a technique commonly used in the study of polymeric and biologically relevant materials. Hydrogen has a negative coherent scattering length ($b_{coh} = -3.74 \times 10^{-15}$ m), lending it "contrast" when surrounded by other, positive-scattering-length atoms. For biological samples, intrinsically rich in hydrogen, judicious substitution of ^{2}H ($b_{coh} = +6.67 \times 10^{-15}$ m) for ^{1}H provides a powerful method for selectively tuning the "contrast" of a given macromolecule. One can therefore accentuate or nullify the scattering from particular parts of a macromolecular complex by selective deuteration [33]. Of importance to holography is that ^{1}H has a large incoherent cross section for thermal neutrons ($\sigma_i = 80$ barns), producing *S* waves similar to the de-excitation of a photoexcitable atom in X-ray holography. Unlike forms of electromagnetic radiation that interact primarily with electrons, neutrons interact directly with nuclei and are, to first order, equally capable of imaging light atoms and heavy atoms. Consequently, atomic-resolution holographic images of materials comprising entirely light atoms can be obtained, a great advantage in the study of polymers and biologically relevant materials.

Neutrons suitable for scattering experiments are presently being produced either by nuclear reactors (e.g., the 120-MW National Research Universal [NRU], Chalk River, Ontario) via the fission of uranium nuclei or by spallation sources (Spallation Neutron Source, Oak Ridge, Tennessee), in which accelerated subatomic particles (e.g., protons) strike a target (e.g., tungsten, lead, or liquid mercury), spalling neutrons from the target material's nuclei (for further details, refer to the instrumentation chapter, which is online at http://www.springer.com/series/8141).

9.4 Neutron Holography and *K* Lines

The feasibility of atomic-resolution neutron holography was discussed by Cser et al. [34] on the basis of the large incoherent thermal neutron cross section exhibited by ^{1}H atoms. Experimentally, neutron holography was demonstrated by Sur et al. [14] using a single crystal of natural simpsonite, a rare oxide mineral of aluminum and tantalum first discovered in western Australia [35] in 1939. The mineral, with the chemical formula $Al_4Ta_3O_{13}(OH)$, was examined using X-ray diffraction [36] in the same year as its discovery; more recently, X-ray diffraction has been used to determine the crystal structure in the trigonal space group *P3* with unit cell parameters $a = 7.386(1)$ Å and $c = 4.516(1)$ Å [37]. More important, simpsonite contains only one ^{1}H atom per crystallographic unit cell and was shown to lend itself ideally to holographic reconstruction.

The following year, Sur et al. [15] reported the observation of *K* lines (representing, collectively, Kossel and Kikuchi lines) produced by monochromatic thermal neutrons interacting with a potassium dihydrogen phosphate (KDP) single crystal. *K* lines contain phase information, further establishing the experimental basis for direct crystallographic phasing of atomic structures containing incoherent scatterers via thermal neutron inside source holography.

The interaction of single-frequency plane waves with a periodic potential, formulated in the Born approximation, produces diffraction intensities in discrete directions commonly known as Bragg reflections. Similarly, single-frequency *S* waves interacting with a 3D periodic potential (e.g., an atomic lattice) give rise to sharp conical intensity variations. Viewed on a surface, the conic sections appear as lines.

Nearly spherical single-frequency electromagnetic waves are obtained by electronic de-excitations of atoms (e.g., fluorescence, photoemission) within the crystal sample itself, an "inside source," or a source in close proximity to the sample. In 1922 Clark and Duane [38] predicted the resulting line structures for X rays, which were experimentally observed using a single crystal of copper in 1934 by Kossel, Loeck, and Voges [26]. For X rays, such lines are therefore commonly referred to as Kossel lines. A theoretical explanation was subsequently provided in 1935 by von Laue [39]. Spherical waves can also be generated by dynamical effects (e.g., multiple and inelastic scattering); the resulting lines were first observed by Kikuchi [25] in electron scattering studies using mica plates of varying thicknesses. Thus such lines are generally referred to as Kikuchi lines. In the present case, thermal neutron *S* waves are generated by incoherent elastic scattering from hydrogen. The fundamental scattering process that gives rise to the characteristic lines — *S* wave interaction with the crystal lattice — is the same as for Kossel and Kikuchi lines; therefore, they are collectively referred to as *K* lines.

The *K* line intensity pattern depends on the relative position of the periodic scatterers with respect to the *S* wave source (i.e., the crystallographic phase of the scattering structure function). Thus *K* lines provide the signal for direct 3D, atomic-resolution imaging techniques (holography) for single crystals via the inside source or inside detector concept.

Atomic nuclei are, to a very good approximation, point or pure *S* wave scatterers of thermal neutron waves. Hydrogen atoms have a large bound, spin-incoherent cross section, $\sigma_i = 80$ barns, for unpolarized thermal neutrons. An unpolarized, parallel, thermal neutron beam interacting with a single crystal containing hydrogen atoms should therefore give rise to both Bragg reflections, from the incident plane-wave neutron beam, and *K* lines, from the *S* waves produced by the incoherent scattering from the ^{1}H atoms.

9.5 Theory

To gain an understanding of the process, consider a single-point *S* wave source of amplitude *a* at the origin surrounded by a distribution of weak point or purely *S* wave scatterers described by a coherent scattering length density $b(\mathbf{r})$. Then, in the Born approximation, the unperturbed or reference wave is

$$\Psi_{(\mathrm{ref})}(\mathbf{r}) = \frac{a\mathrm{e}^{\mathrm{i}|\mathbf{k}||\mathbf{r}|}}{|\mathbf{r}|}, \tag{9.1}$$

and the scattered or object wave is

$$\Psi_{\mathrm{obj}}(\mathbf{r}) = a\int_{r_0} \frac{b(\mathbf{r}_0)\mathrm{e}^{\mathrm{i}|\mathbf{k}|(|\mathbf{r}_0|+|\mathbf{r}-\mathbf{r}_0|)}}{|\mathbf{r}_0||\mathbf{r}-\mathbf{r}_0|}\mathrm{d}\,\mathbf{r}_0\,. \tag{9.2}$$

The total wave amplitude, Ψ_{tot} for a detector at $\mathbf{R}$ is given by $\Psi_{\mathrm{ref}}(\mathbf{R}) + \Psi_{\mathrm{obj}}(\mathbf{R})$. The detected intensity, I_{1S} $(\mathbf{R})$, is then $\Psi^*_{\mathrm{tot}}\ \Psi_{\mathrm{tot}}$. Using the far field diffraction condition (i.e., $R >> \hat{r}$) and defining the outgoing wave vector in the detector direction as $\mathbf{k}_{out} = k\,\mathbf{R}$, the detected intensity is then given as

$$I_{1S} \cong \frac{a^*a}{R^2}\left[1 + 2\mathrm{Re}[\chi(\mathbf{k}_{\mathrm{out}})] + |\chi(\mathbf{k}_{\mathrm{out}})|^2\right], \tag{9.3}$$

where the intensity modulation function is

$$\chi(\mathbf{k}_{\mathrm{out}}) = \int\limits_{r_0} \frac{b(\mathbf{r}_0)e^{i(kr_0-\mathbf{k}_{\mathrm{out}}\cdot\mathbf{r}_0)}}{r_0} d\,\mathbf{r}_0 \;. \tag{9.4}$$

In the case of inside source holography, the scattering length distribution $b(\mathbf{r})$ is reconstructed by a Fourier inversion of the observed modulation intensity [14].

To gain an understanding of K lines, consider this expression [Eq. (9.4)]. In reciprocal space, the modulation function (aside from constants) is given by

$$\chi(\mathbf{k}_{\mathrm{out}}) = \int \frac{B(\mathbf{q})}{|\mathbf{q}|^2 - 2\mathbf{k}_{\mathrm{out}}\cdot\mathbf{q}} d\mathbf{q} \;, \tag{9.5}$$

where $B(\mathbf{q})$ is the Fourier transform of $b(\mathbf{r})$.

In the case of a crystal lattice, the 3D scattering potential is completely specified by the magnitude and phase of the structure function F_{hkl} at discrete points in reciprocal space. The reciprocal lattice points are given by $\boldsymbol{\tau}_{hkl} = h\mathbf{b}_1 + k\mathbf{b}_2 + l\mathbf{b}_3$, where $\mathbf{b}_1$, $\mathbf{b}_2$, and $\mathbf{b}_3$ are three reciprocal lattice basis vectors. The integral in Eq. (9.5) then becomes a sum over all hkl, namely

$$\chi(\mathbf{k}_{\mathrm{out}}) = \sum_{hkl} \frac{F_{hkl}}{|\tau_{hkl}|^2 - 2\mathbf{k}_{\mathrm{out}}\cdot\tau_{hkl}} \;. \tag{9.6}$$

For every pair of discrete points $\pm\,\boldsymbol{\tau}_{hkl}$ located within a sphere of radius $2k$, the zeros in the denominator of Eq. (9.6) cause the modulation to go to infinity in $\mathbf{k}_{\mathrm{out}}$ directions whose locus is a cone with axis along $\pm\,\boldsymbol{\tau}_{hkl}$ and full opening angle 2θ, satisfying the relation $2k\cos\theta = \tau_{hkl}$. Viewed on the surface of a hemisphere of radius k_{out}, the intersections of the cones with the hemisphere are K line rings centered on $\boldsymbol{\tau}_{hkl}$. These K line rings can readily be calculated for any hkl relative to the crystal coordinate system of a known structure. For example, a selection of K lines expected for the structure of KDP in a particular sample orientation is shown in Fig. 9.2A on the surface of a hemisphere $\mathbf{k}_{\mathrm{out}}/|\mathbf{k}_{\mathrm{out}}|$ with $|\mathbf{k}_{\mathrm{out}}| = 4.8$ Å^{-1} [15]. The incident plane wave, $\mathbf{k}_{in}$, will also produce Bragg peaks, which will appear as discrete points on the surface of the hemisphere, when condition $\mathbf{k}_{\mathrm{out}} - \mathbf{k}_{\mathrm{in}} = \boldsymbol{\tau}_{hkl}$ is satisfied (Fig. 9.2A).

In a high-resolution experiment, K lines are readily identified by their characteristic fine structure [27]. However, K line fine structure is affected by factors such as

- sample size and shape;
- dynamics or higher-order partial wave scattering in both the coherent and incoherent scatterers, leading to a broadening of the signal and a nonuniform reference wave intensity; and
- wavelength spread and angular resolution.

All of these effects conspire to mute the sharp modulations characteristic of K lines, but they still produce modulations along the lines (Fig. 9.2B).

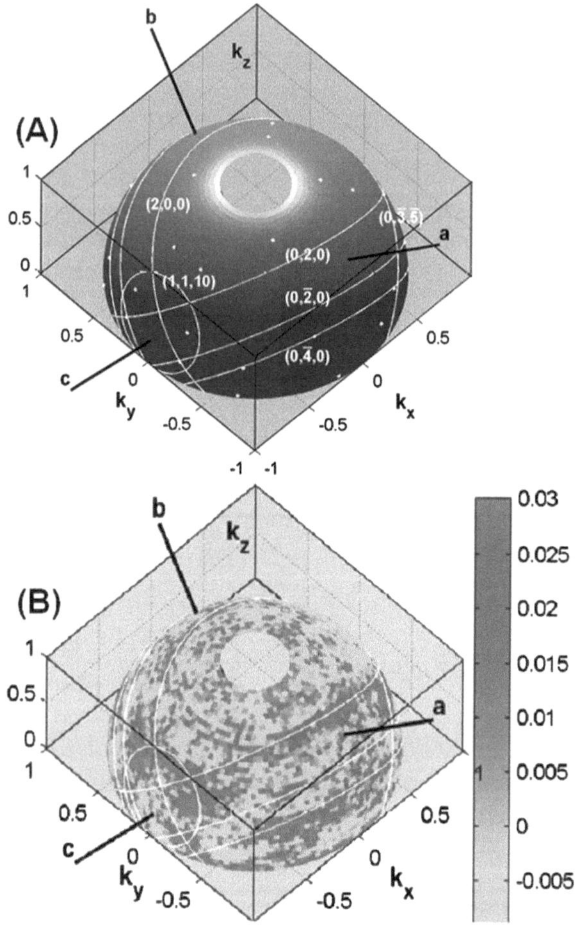

Fig. 9.2 (**A**) Calculated *K* lines for the various KDP lattice planes for $\lambda = 1.30$ Å neutrons. In addition to the *K* lines, the locations of Bragg reflections are shown as white spots. The orientation of the crystal with respect to the sphere-of-scattering ($\mathbf{k}_{\text{in}} \parallel k_z$) is depicted by the crystallographic axes *a*, *b*, and *c*. (**B**) KDP holographic data corrected for background, Debye-Waller factor, and sample shape. *K* lines appear as bands of scattering corresponding to the *white lines* in (A). Figure adapted from Sur et al. [15]

9.6 Reconstruction of the Diffraction Patterns from *S* Wave Scatterers

According to Eqs. (9.3) and (9.4), the experimentally observed hologram modulation function $\chi(\mathbf{k})$ produced by the interaction of a weak *S* wave coherent scattering length distribution $b(\mathbf{r})$ with either a single incoherent point scatterer or with a uniform distribution of incoherent scatterers is given, aside from constants, by

$$\chi(\mathbf{k}) \sim \int_{\text{all space}} \operatorname{Re}\left(\frac{b(\mathbf{r})}{r} e^{[i(kr-\mathbf{k}\cdot\mathbf{r})]}\right) d\mathbf{r} + O(b^2) . \tag{9.7}$$

The conditions for pure *S* wave weak scattering are particularly well satisfied for unpolarized thermal neutron scattering from nuclei. Additionally, for most nuclei, the thermal neutron coherent scattering lengths are almost purely real and have the same sign. The real part of the coherent scattering length distribution can be directly reconstructed from the measured hologram modulation function using the following relations.

$$\tilde{b}_{\text{even}}(\mathbf{r}) = \frac{b_{\text{real}}(\mathbf{r}) + b_{\text{real}}(-\mathbf{r})}{kr} \cos(kr) \propto \int_{\text{constant}|\mathbf{k}|} \chi(\mathbf{k}) \cos(\mathbf{k}\cdot\mathbf{r}) d\mathbf{k} . \tag{9.8}$$

$$\tilde{b}_{\text{odd}}(\mathbf{r}) = \frac{b_{\text{real}}(\mathbf{r}) - b_{\text{real}}(-\mathbf{r})}{kr} \sin(kr) \propto \int_{\text{constant}|\mathbf{k}|} \chi(\mathbf{k}) \sin(\mathbf{k}\cdot\mathbf{r}) d\mathbf{k} . \tag{9.9}$$

Strictly speaking, for a single wavelength or constant k, the even and odd parts are reconstructed at nonoverlapping points in space (i.e., the zeros of the even part reconstruct at turning points of the odd part and vice versa). Therefore, it is possible to eliminate the conjugate image $b_{\text{real}}(-\mathbf{r})$ from the reconstruction of a single hologram data set by a suitable combination of $\tilde{b}_{\text{odd}}(\mathbf{r})$ and $\tilde{b}_{\text{even}}(\mathbf{r})$. For instance, $\tilde{b}_{\text{odd}}(\mathbf{r}) \sin(kr)$ and $\tilde{b}_{\text{even}}(\mathbf{r}) \cos(kr)$ can be summed with a moving box average of dimensions $\lambda/2$ to give $b(\mathbf{r})$. A consequence of this procedure is that the resolution of the reconstructed image is broadened by $\lambda/2$. Alternatively, the quality of the reconstructed image can be improved by combining holograms obtained at several different wavelengths.

For a single-point incoherent scatterer per unit cell, the above formulation will reconstruct the coherent scattering length with the incoherent scatterer located at the origin. For a uniformly random distribution of point incoherent scatterers, the reconstruction origin will be the "center of illumination," which by definition occupies exactly the same volume as the sample.

9.7 Inside Source Neutron Holography

The first inside source neutron holography experiments were carried out at the N5 triple-axis spectrometer (Fig. 9.3A) located at the NRU reactor at Chalk River Laboratories, Ontario. Those experiments used 1.3-Å neutrons with a $\Delta\lambda/\lambda \sim 1.5\%$ using the (113) reflection from a germanium (Ge) single-crystal monochromator. The sample-to-detector angular resolution was limited by distance collimation, defined by a combination of an aperture in neutron-absorbing cadmium masks and the size of the sample. Sample orientation (ϕ)

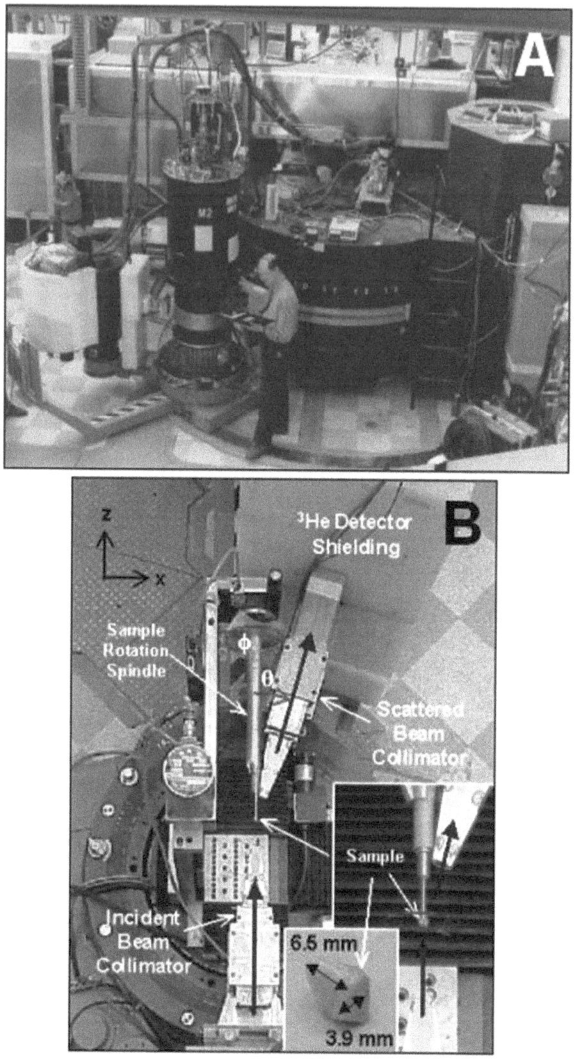

Fig. 9.3 (**A**) The N5 triple-axis spectrometer on which the first neutron holography experiments were carried out, located at the National Research Universal reactor, Chalk River Laboratories. (**B**) Plan view photograph and schematic of the neutron holography experimental setup at the N5 triple-axis spectrometer. The incident 1.3-Å neutron beam (originating from the bottom of the photograph) was parallel to the axis of rotation, ϕ. The angle between the detector and the ϕ axis is denoted by θ. In order to efficiently obtain the hemisphere of scattering, the crystal was rotated, in optimal ϕ steps of $2^\circ/\sin\theta$, from 0 to 2π for a given θ ($17^\circ \leq \theta \leq 83^\circ$). The lower value for θ (17°) was dictated as a result of the physical interference between the scattered beam collimator and the sample rotation spindle. The scattered neutrons were recorded using a ^{3}He detector. Figure adapted from Sur et al. [14]

and detector angle (θ) were manipulated in such a way that nearly a hemisphere of scattering was recorded (Fig. 9.3B). However, as a result of poor scattered intensity and a single-point detector, collecting good statistical quality data (total scattering and background) over 1.7π radians in scattering wavevector ($\mathbf{k}_{out}$) took ~10 days (Fig. 9.4A).

The structure of simpsonite contains two layers of oxygen atoms perpendicular to the c axis of the crystal lattice. The first layer, which is closest to the ^1H atom, has an oxygen atom located directly above (positive direction) the ^1H

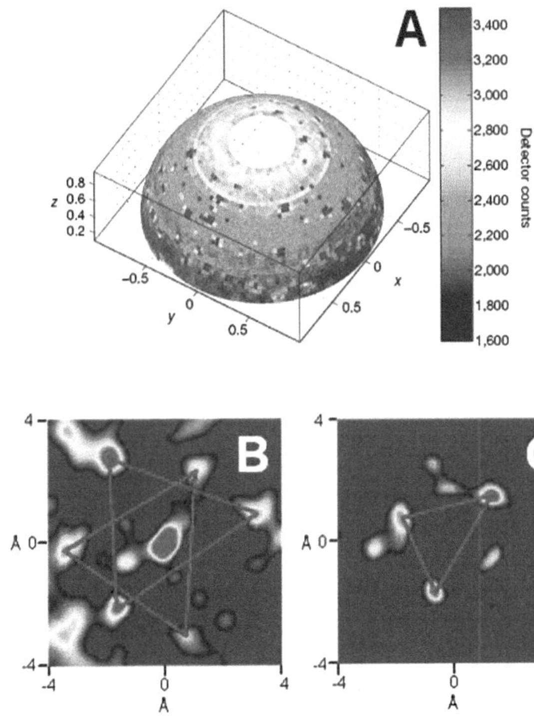

Fig. 9.4 (**A**) Raw hologram data containing 4,334 2° by 2° pixels plotted on a surface of constant scattered wave vector magnitude ($|\mathbf{k}_{out}| = 4.8$ Å^{-1}). Because of the experimental geometry, data collection was limited between 17 and 83° in θ, and about 13% of the pixels containing Bragg reflections were excluded from the reconstruction. The remaining data contained the hologram and θ-dependent background from the vanadium sample holder and aluminum wires (e.g., aluminum powder lines at θ = 32, 37, 54, and 64°) used to secure the simpsonite crystal to the vanadium pedestal. (**B**) Reconstructed plane located approximately + 0.9 Å from the hydrogen atom at the origin showing the seven oxygen atoms. The plane is roughly coplanar to the basal plane of the crystal. The central spot is the oxygen atom located directly above the origin. The triangles are used to indicate the two triplets of oxygen atoms located slightly below this central oxygen atom. (**C**) Reconstructed plane located approximately −1.4 Å from the hydrogen atom at the origin showing the positions of the three oxygen atoms. The distortion of the triangles was attributed to limitations in the quality of the demonstration data. Figure adapted from Sur et al. [14]

atom along the direction of the c axis. This oxygen atom is surrounded by two triangular sublattices of oxygen atoms slightly below the position of the central oxygen atom. In the second layer, located below (negative direction) the ^1H atom, the oxygen atoms form triangular lattices about the perpendicular projection of the ^1H atom onto this layer.

Figure 9.4B shows a reconstructed plane, which is approximately coplanar to the basal plane, at a distance of +0.9 Å from the ^1H atom, which is at the origin of the reconstruction. There is a systematic distortion of the triangular lattices (Fig. 9.4B) owing to the just adequate counting statistics. Nevertheless, all of the expected oxygen atoms, six in the two triangles and the central atom, are present in the reconstruction. Figure 9.4C shows the next closest plane of oxygen atoms found at a distance of –1.4 Å from the origin. From the reconstruction, it is easily seen that these oxygen atoms form a triangle of nearest neighbors, albeit distorted.

9.8 Inside Detector Neutron Holography

The example discussed used a large incoherent neutron scatterer (i.e., hydrogen) to produce nearly S waves, "imaging" its surrounding atoms. By using the principle of optical reciprocity, the inside detector concept interchanges the positions of source and detector. In 2002, Cser et al. [16] demonstrated the feasibility of the inside detector technique to image lead (Pb) nuclei in a single crystal of Pb doped with the strong neutron absorber cadmium (Cd). In this case, Cd atoms act as point-like detectors, with a neutron wave reaching these detector nuclei either directly (reference beam) or after scattering from the Pb atoms in the crystal (object beam). The capture of neutrons by Cd results in the emission of γ radiation. The intensity of the γ radiation is directly proportional to the neutron intensity experienced by the detector nuclei. When recorded as a function of sample orientation with respect to the incident neutron beam, this fluctuating γ radiation intensity is effectively the hologram.

To demonstrate inside detector neutron holography, Cser et al. [16] used a 7-mm-diameter spherically shaped single crystal of $Pb_{0.9974}Cd_{0.0026}$ (the use of spherical samples greatly simplifies the data analysis). Cadmium atoms are strong neutron absorbers with a thermal absorption cross section over four orders of magnitude greater than that of Pb atoms [40]; this accounts for their ability to act as efficient detectors. The crystal structure of the $Pb_{0.9974}Cd_{0.0026}$ alloy is the same (face-centered cubic, fcc) as for pure lead [16], with the Cd atoms randomly distributed. As Cd was used in low concentrations, on average, every Cd atom was surrounded by 12 nearest-neighbor Pb atoms.

The (220) reflection from a Cu single crystal was used to select 0.8397-Å neutrons. Prompt γ radiation from the Cd nuclei was recorded using two scintillation counters shielded by ~20 cm of lead. The sample was rotated about its axis (ϕ) in 3° steps through ~350° and tilted (χ), also in 3° steps, through 120°. Figure 9.5A shows the reconstruction from the hologram of the 12 Pb atoms surrounding the detector Cd atom plotted in χ and ϕ coordinates, and Fig. 9.5B shows the Cd atom and its 12 neighboring Pb atoms plotted on the surface of a sphere of radius 3.49 Å. The x-axis is the direction of the incident neutron beam.

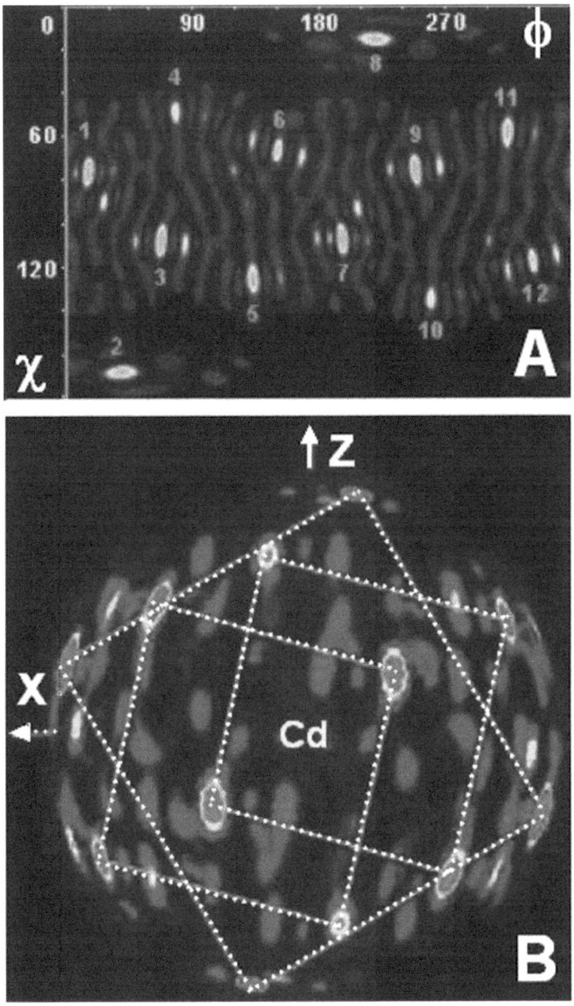

Fig. 9.5 (**A**) Reconstruction showing the positions of the 12 Pb atoms surrounding the detector Cd atom plotted in χ and ϕ coordinates. (**B**) The 12 Pb atoms surrounding the Cd detector atom plotted on the surface of a sphere of radius 3.49 Å. The x-axis denotes the direction of the incident neutron beam. Data were collected using the D9 diffractometer located at the Institute Laue-Langevin (Grenoble, France). Figure adapted from Cser et al. [16]

The most recent inside detector neutron holography study by Cser et al. [41] demonstrated for the first time the atomic positions with precision down to 1 picometer. To date, this is the only work based on a neutron holography experiment that contains information not previously made available by other experimental techniques.

9.9 Holographic Reconstruction from Multiple Incoherent Scatterers

The mathematical formulations for atomic structure holography have, until recently, been limited to samples with one S wave scatterer per unit cell. However, such samples are extremely rare. On the other hand, materials rich in hydrogen (e.g., biological or polymeric) are ubiquitous. Recently, Sur et al. [17] derived a kinematical formulation for the diffraction pattern of monochromatic plane waves scattering from a mixed coherent and incoherent scattering length distribution. From this second-order kinematical formulation, it became evident that samples with a uniformly random distribution of incoherent scatterers can be reconstructed to atomic resolution. Also, Sur et al. [17] provided a formulation for holographic reconstruction eliminating the so-called conjugate, or twin image problem, inherent in the holographic inversion of single-wavelength data.

9.10 Holography and Poorly Crystallized Proteins

Proteins are linear heteropolymers that play crucial roles in virtually all biological processes. The linear chain folds into an intricate 3D structure, which is determined by the sequence of amino acids, and is unique to each protein. It is this 3D structure that is the critical determinant of a protein's biological function. In principle, if one has knowledge of the amino acid sequence, it should be possible to predict the protein's 3D structure. However, a number of problems make this *ab initio* approach very difficult.

Traditionally, protein structures have been determined by X-ray diffraction (and to a much lesser extent by neutron diffraction) studies of crystallized samples and more recently, in the case of small (< 25 kDa) proteins in solution, by 2D and 3D nuclear magnetic resonance spectroscopy. In order to decipher the 3D structure using X-ray diffraction, well-ordered protein crystals must be available. However, the availability of quality crystals constitutes a major obstacle for traditional diffraction methods, particularly for those proteins integral to the cell's membrane. As a result, practically all of the approximately 34,000 known 3D protein structures are water soluble as opposed to membrane proteins. Fewer than 100 membrane proteins have been registered in the RCSB (Research Collaboratory for Structural Bioinformatics) protein data bank.

This is a critical shortfall since membrane-associated proteins are known to constitute approximately one-third of all known proteins [42], and approximately 40% of the human genome encodes for membrane proteins.

The first membrane-associated protein structure determined to near-atomic resolution (3 Å) was that of the photosynthetic reaction center of *Rhodopseudomonas viridis*, a purple sulfur photosynthetic bacterium whose 3D structure was solved by Deisenhofer et al. [43, 44]. In the case of the photoreaction center and other membrane-associated proteins, [45–47] detergents were essential to producing high-quality crystals suitable for Bragg and Laue diffraction studies. At present there does not appear to be a systematic method for obtaining high-quality crystals for the majority of membrane proteins. However, low-quality membrane protein crystals, which do not diffract to high resolution, may be more readily obtainable using the present crystal growing techniques. The concept in holography is to reconstruct the local atomic structure in the vicinity of the source or the detector atoms. Thus long-range translational order (i.e., periodicity) of the protein unit cell structure is not required for a solvable hologram. The only requirement is that the unit cells be oriented in the same direction (within an acceptable mosaic spread, consistent with the unit cell dimensions). Thus poor crystals or quasicrystals are especially useful candidates for the determination of atomic-resolution 3D structure via holography.

Depending on the protein's chemistry, thermal neutron and X-ray holography techniques may be used to resolve, to atomic resolution, the 3D structure of proteins forming either poor crystals or quasicrystals. One strategy may be to start with a low-resolution ($\sim$ 6 Å data) 3D X-ray structure that can be refined to atomic resolution using the appropriate X-ray or neutron hologram of the protein. Assuming that the quasi-protein crystal contains only positional and not orientational disorder, this strategy, if successful, will enable atomic-resolution studies of an entire class of macromolecules with structures that are inaccessible presently and for the foreseeable future. However, there are potential difficulties.

Until now the mathematical formulations for atomic structure holography have been limited to samples with one source atom per unit cell (not common). As mentioned, Sur et al. [17] have mathematically formulated the reconstruction of samples with multiple source atoms per unit cell. In practice, for thermal neutron holography, this condition can be realized when the samples used have a uniformly random distribution of hydrogen atoms, an approach very much analogous to that of the commonly used diffuser in optical holography. However, the experimental feasibility of this proposed technique has yet to be demonstrated. On the other hand, if this approach proves to be feasible, holographic techniques will drive a worldwide expansion of knowledge about the 3D structure of membrane-associated proteins and other macromolecules, leading to better insights into biological processes and improved designs of existing drugs.

9.11 The Future

The technique of atomic-resolution holography is not yet established for routinely determining atomic structure. It is, however, a promising and technically feasible method for solving problems that cannot be resolved with traditional diffraction methodologies. Presently, there has yet to be an atomic-resolution holography study using a sample that could not be successfully studied using traditional crystallography (i.e., a poor crystal or quasicrystal).

A recent technological advance challenging the holographic techniques described is the development of hard X-ray free electron lasers (XFELs), making coherent X-ray diffraction imaging a possibility. The first XFEL, the Linac Coherent Light Source at the Stanford Linear Accelerator Center, is slated to begin experiments in 2009. The European XFEL Facility [48] (Germany) will follow in 2013. Unlike third-generation synchrotron sources, which use lenses to produce coherent X rays, the inherently coherent X-ray beams produced by XFELs have the potential to image materials with 10 times better spatial resolution. XFEL sources thus offer the possibility to resolve structural details of materials to atomic resolution with high-brilliance femtosecond coherent X-ray pulses. In a study headed by scientists from Lawrence Livermore National Laboratory, a soft XFEL (λ = 320 Å) was used to demonstrate that a coherent diffraction pattern from a nonperiodic object could be obtained prior to destroying the sample at 60,000 K [49].

9.12 Concluding Remarks

This brief review has presented the current state of neutron holography. Neutron holography is a nondestructive, *in situ* technique capable of atomic-resolution measurements of biologically relevant materials and appears to be an excellent option for the solution of poorly crystallized or quasicrystal proteins. However, an experimental demonstration of "diffuse source" atomic-resolution holography, utilizing incoherent scattering from the large numbers of hydrogen atoms in biological materials, is key to future developments. Future neutron holography experiments will benefit greatly from spallation source instruments capable of acquiring neutron scattering data with simultaneous large solid angle coverage and at multiple wavelengths, capabilities that will become available at the Spallation Neutron Source.

References

1. W.L. Bragg, The X-ray microscope. *Nature* **149,** 470 (1942).
2. D. Gabor, A new microscopic principle. *Nature* **161,** 777–778 (1948).
3. G.R. Harp, D.K. Saldin, and B.P. Tonner, Atomic resolution holography in solids with localized sources. *Phys. Rev. Lett.* **65,** 1012–1015 (1990).

4. M. Zharnikov, M. Weinelt, P. Zebish, S M. tichler, and H. -P. Steinrück, First experimental determination of an adsorption site using multiple wave number photoelectron diffraction patterns. *Phys. Rev. Lett.* **73,** 3548–3551 (1994).
5. M.T. Sieger, J.M. Roesler, D. -S. Lin, T. Miller, and T. -C. Chiang, Holography of Ge(111)-c(2×8) by surface core-level photoemission. *Phys. Rev. Lett.* **73,** 3117–3120 (1994).
6. A. Orchowski, W. D. Rau, and H. Lichte, Electron holography surmounts resolution limit of electron microscopy. *Phys. Rev. Lett.* **74,** 399–402 (1995).
7. T. Matsushita, F.Z. Guo, F. Matsui., Y. Kato., and H. Daimon, Three-dimensional atomic-arrangement reconstruction from an Auger-electron hologram. *Phys. Rev. B* **75,** 085419(1)–085419(5) (2007).
8. M. Tegze and G. Faigel, Atomic resolution X-ray holography. *Europhys. Lett.* **16,** 41–46 (1991).
9. T. Gog., P.M. Len, G. Materlik., D. Bahr., C.S. Fadley, C. Sanchez-Hanke, Multiple-energy holography: Atomic images of hematite (Fe_2O_3). *Phys. Rev. Lett.* **76,** 3132–3135 (1996).
10. M. Tegze and G. Faigel. X-ray holography with atomic resolution. *Nature* **380,** 49–51 (1996).
11. M. Tegze, G. Faigel, S. Marchesini, M. Belakhovski, and A.I. Chumakov, Three dimensional imaging of atoms with isotropic 0.5 Å resolution. *Phys. Rev. Lett.* **82,** 4847–4850 (1999).
12. S. Eisebitt, J. Lüning, W.F. Schlotter, M. Lörgen, O. Hellwig, W. Eberhardt, and J. Stöhr, Lensless imaging of magnetic nanostructures by X-ray spectro-holography. *Nature* **432,** 885–888 (2004).
13. H.N. Chapman, et al. Femtosecond time-delay X-ray holography. *Nature* **448,** 676–670 (2007).
14. B. Sur, R.B. Rogge, R.P. Hammond, V.N.P. Anghel, and J. Katsaras, Atomic structure holography using thermal neutrons. *Nature* **414,** 525–527 (2001).
15. B. Sur, R.B. Rogge, R.P. Hammond, V.N.P. Anghel, and J. Katsaras Observation of Kossel and Kikuchi lines in thermal neutron incoherent scattering. *Phys. Rev. Lett.* **88,** 065505(1)–065505(4) (2002).
16. L. Cser, Gy. Török, G. Krexner, I. Sharkov, and B. Faragó, Holographic imaging of atoms using thermal neutrons. *Phys. Rev. Lett.* **89,** 175504(1)–175504(4) (2002).
17. B. Sur, V.N.P. Anghel, R.B. Rogge, and J. Katsaras, Diffraction pattern from thermal neutron incoherent elastic scattering and the holographic reconstruction of the coherent scattering length distribution. *Phys. Rev. B* **71,** 014105(1)–014105(12) (2005).
18. P.G. Tanner, and T.E. Allibone, The patent literature of Nobel laureate Dennis Gabor (1900 – 1979). *Notes Rec. R. Soc. Lond.* **51,** 105–120 (1979).
19. A.L. Schawlow, and Townes C.H. Infrared and optical masers. *Phys. Rev.* **112,** 1940–1949 (1958).
20. A.L. Schawlow, and Townes C.H. U.S. patent no. 2,929,222 (March 22, 1960).
21. A. Einstein, *Zur quantentheorie der strahlung. Physik. Zeitschr.* **18,** 121–128 (1917).
22. Y.N. Denisyuk, On the reflection of optical properties of an object wave field scattered by it. *Doklady Akademii Nauk SSSR* **144,** 1275–1278 (1962).
23. E.N. Leith, and J. Upatnieks, Reconstructed wavefronts and communication theory. *J. Opt. Soc. Am.* **52,** 1123–1130 (1962).
24. A. Szöke, X-ray and electron holography using a local reference beam, in Short Wavelength Coherent Radiation: Generation and Applications, eds. D.J. Attwood, and J. Boker AIP Conf. Proc. No. 147, pp. 361–467 (American Institute of Physics, 1986).
25. S. Kikuchi, Diffraction of cathode rays by mica. *Jpn. J. Phys.* **5**, 83–96 (1928).
26. W. Kossel, V. Loeck, and H. Voges, Die richtungsverteilung der in einem kristall entstandenen charakteristichen Röntgenstrahlung. *Z. Phys.* **94,** 139–142 (1935).
27. T. Gog, D. Bahr, and G. Materlik, Kossel diffraction in perfect crystals: X-ray standing waves in reverse. *Phys. Rev. B* **51,** 6761–6764 (1995).

28. G. Faigel, and M. Tegze, X-ray holography. *Rep. Prog. Phys.* **62,** 355–393 (1999).
29. J. Chadwick, Possible existence of a neutron. *Nature* **129,** 312 (1932).
30. J. Chadwick, The existence of a neutron. *Proc. Roy. Soc. (London)* **A136,** 692–708 (1932).
31. E. Rutherford, Nuclear constitution of atoms. *Proc. Roy. Soc.* **A97,** 374–400 (1920).
32. W. Mampe, P. Ageron, C. Bates, J.M. Pendelbury, and A. Steyerl, Neutron lifetime measured with stored ultracold neutrons. *Phys. Rev. Lett.* **63,** 593–596 (1989).
33. B. Jacrot, The study of biological structures by neutron scattering from solution. Rep (sp. nov.) from Tabba Tabba, western Australia. *J. R. Soc. W. Aust. Prog. Phys.* **39,** 911–953 (1976).
34. L. Cser, G. Krexner, and Gy. Török, Atomic resolution neutron holography. *Europhys. Lett.* **54,** 747–752 (2001).
35. H. Bowley, Simpsonite (sp. nov) from Tabba Tabba, western Australia. *J. R. Soc. W. Aust.* **25,** 89–92 (1939).
36. L.E.R. Taylor, X-ray studies of Simpsonite. *J. R. Soc. W. Aust.* **25,** 93–97 (1939).
37. T.S. Ecrit, Černý, P., and F.C. Hawthorne, The crystal chemistry of Simpsonite. *Can. Mineral.* **30,** 663–671 (1992).
38. G.L. Clark, and W. Duane, A new method of using x-rays in crystal analysis. *Proc. Natl. Acad. Sci. U.S.A.* **8,** 90–96 (1922).
39. M. v. Laue, Die fluoreszensröntgenstrahlung von einkristallen. *Ann. Physik* **23,** 705–746 (1935).
40. V.F. Sears Neutron scattering lengths and cross sections. *Neutron News* **3,** 26–37 (1992).
41. L. Cser, G. Krexner, Markó, M., I. Sharkov, and Gy. Török, Direct observatotion of local distortion of a crystal lattice at picometer accuracy using atomic resolution neutron holography. *Phys. Rev. Lett.* **97,** 255501(1)–255501(4) (2006).
42. A. Goffeau, Life with 482 genes. *Science* **270,** 445–446 (1995).
43. J. Deisenhofer, O. Epp, K. Miki, R. Huber, and H. Michel, X-ray structure analysis of a membrane protein complex: electron density map at 3 Å resolution and a model of the chromophores of the photosynthetic reaction center from *Rhodopseudomonas viridis. J. Mol. Biol.* **180,** 385–398 (1984).
44. J. Deisenhofer, O. Epp, K. Miki, R. Huber and H. Michel, Structure of the protein subunits in the photosynthetic reaction center of *Rhodopseudomonas viridis* at 3 Å resolution. *Nature* **318,** 618–624 (1985).
45. M. Garavito and J.P. Rosenbusch, Isolation and crystallization of bacterial porin. *Meth. Enzymol.* **125,** 309–328 (1986).
46. S.W. Cowan, et al., The structure of OmpF porin in a tetragonal crystal form. *Structure* **3,** 1041–1050 (1995).
47. R. Dutzler, et al., Crystal structure and functional characterization of OmpK36, the osmoporin of *Klebsiella pneumoniae. Structure* **7,** 425–434 (1999).
48. I.A. Vartanyants, I.K. Robinson, I. McNulty, C. David, P. Wochner, and Th. Tschentscher, Coherent X-ray scattering and lensless imaging at the European XFEL Facility. *J. Synch. Rad.* **14,** 453–470 (2007).
49. H.N. Chapman, et al., Femtosecond diffractive imaging with a soft-X-ray free-electron laser. *Nat. Phys.* **2,** 839–843 (2006).

Chapter 10
Novel Imaging Techniques: Polarized Neutrons and Neutron-Based Magnetic Resonance Imaging

N. Kardjilov, W.T.H. Lee, and G.E. Granroth

Abstract This chapter describes three rather novel neutron-based imaging methods all of which involve magnetic fields. The first two take advantage of the neutron magnetic moment which makes it sensitive to the presence of magnetic fields. In the first application the Larmor precession of the neutron in a magnetic field is used to produce two- and three-dimensional (3D) visualization of magnetic field distributions in free space and in bulk materials.

In the second potential application, the Larmor precession is used as a measure of the material density through which the beam passes. Although this technique of spin contrast imaging is still very much in the concept stage, initial estimates show that it is feasible with new sources and may open up the possibility of obtaining 3D images without moving the sample.

Polarization of the nuclei in a material can also provide unique ways of imaging a system. By partially polarizing the nuclei with a sufficiently large magnetic field, the Zeeman splitting of the nuclei can be detected by high-resolution neutron spectroscopy. If the field is then reduced, the time for the Zeeman splitting to relax to the equilibrium value is a measure of the nuclear relaxation time; this is the physical quantity measured in magnetic resonance imaging. A full image of the relaxation times with respect to position could then be obtained by rastering the sample in the neutron beam.

Keywords Polarized neutrons · Magnetic fields · Magnetic resonance imaging · Spin-echo · Time-of-flight · three-dimensional imaging

10.1 Neutron Spin Polarized Imaging of Magnetic Field

Neutrons are sensitive to magnetic fields due to their magnetic moment, i.e., spin. Therefore, similar to the conventional attenuation contrast image, the magnetic field inside and around a sample can be visualized independently by detection of

N. Kardjilov (✉)
Helmholtz Centre Berlin for Materials and Energy, Berlin, Germany
e-mail: kardjilov@helmholtz-berlin.de

I.S. Anderson et al. (eds.), *Neutron Imaging and Applications*,
Neutron Scattering Applications and Techniques, DOI 10.1007/978-0-387-78693-3_10,

the polarization changes in the transmitted beam. Polarized neutron radiography is based on the spatially resolved measurement of the final precession angles of a collimated and polarized monochromatic neutron beam that is transmitted through the magnetic field, which is present inside and outside of a sample.

This new technique can be described using theoretical considerations related to the interaction of the magnetic moment of the neutron with the magnetic field. In the presence of a magnetic field $\vec{B}$, the magnetic moment (i.e., the spin $\vec{S}$) of the neutron will perform Larmor precession with a frequency ω_L. For a monochromatic neutron beam having a uniform neutron velocity v, the precession angle φ is proportional to the integral of the magnetic field $\vec{B}$ along a certain path:

$$\varphi = \omega_L t = \frac{\gamma_L}{v} \int_{path} B ds \tag{10.1}$$

where γ_L is the neutron gyromagnetic ratio. φ can be measured experimentally using a spin polarizer-analyzer arrangement as shown in Fig. 10.1. Here the neutrons are first polarized (Chapter 3) and, after the interaction with the magnetic field, the final spin orientation is analyzed with respect to the initial state. The signal transmitted through the spin analyzer depends on the spin rotation angle (Eq. 10.1) and has a maximum for parallel and a minimum for anti-parallel spin orientation. In this case the detected image behind the analyzer is determined by a superposition of conventional attenuation contrast $I_a(x,y)$ and the contrast variations due to spin rotation $I_m(x,y)$:

$$I(x,y) = \underbrace{I_0(x,y) \cdot \exp\left(-\int_{path} \Sigma(s) ds\right)}_{I_a(x,y)} \cdot \underbrace{\frac{1}{2}(1 + \cos\varphi(x,y))}_{I_m(x,y)} \tag{10.2}$$

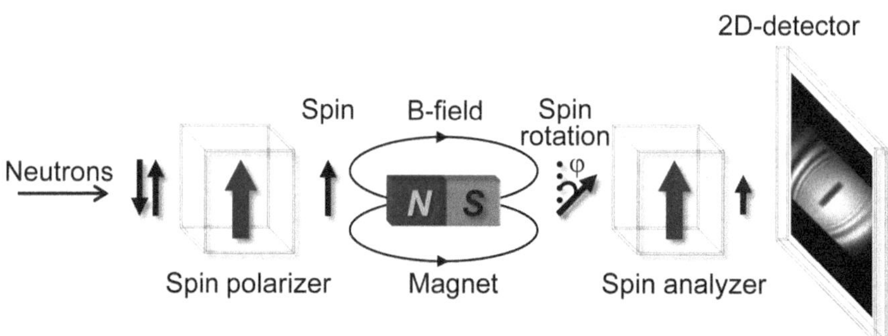

Fig. 10.1 Schematic of the set-up used for imaging magnetic materials on CONRAD. The neutron beam is first polarized, then precesses around a magnetic field is analyzed and finally detected. Note how the intensity (represented by the height of the *dark arrow*) behind the analyzer is smaller than the intensity behind the polarizer [1]

where $I_0(x,y)$ is the incident beam intensity, $\Sigma(x,y)$ is the linear attenuation coefficient of the sample and (x,y) are the coordinates in the detector plane. The cosine implies a periodic transmission function for the analyzed precession angles and complicates a straightforward quantification with respect to the traversed magnetic fields. By use of a calculation model based on the Biot-Savart law, it was shown that quantitative analysis is possible [1].

Imaging experiments using the setup shown in Fig. 10.1 were carried out at the neutron tomography facility CONRAD at HMI [1]. For this purpose the instrument was equipped with solid-state polarizing benders [2] providing a beam with a cross-section of 15 mm width and 45 mm height. For investigation of larger samples, up to 20 cm width, a scanning arrangement was adapted. A double crystal monochromator [3] was used to select a defined wavelength from the cold neutron spectrum. The spatial resolution achieved in the radiography images was around 500 μm for the given experimental geometry.

The potential of the method was demonstrated by visualization of the magnetic field around a simple dipole magnet (Fig. 10.2). This shows the

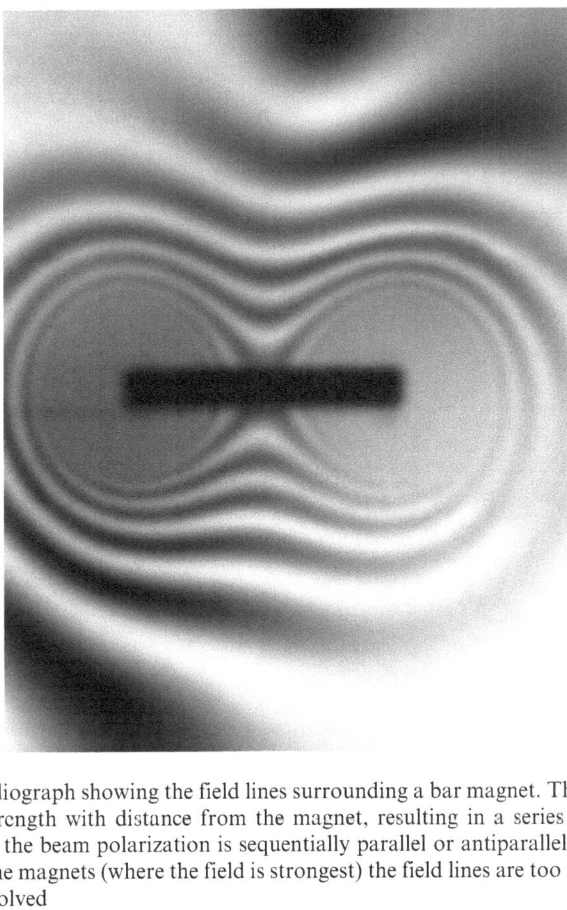

Fig. 10.2 A radiograph showing the field lines surrounding a bar magnet. The magnetic field decreases in strength with distance from the magnet, resulting in a series of maxima and minima, where the beam polarization is sequentially parallel or antiparallel to the analyzer. Very close to the magnets (where the field is strongest) the field lines are too close together to be spatially resolved

decay of the magnetic field strength with increasing distance from the magnet, resulting in an annular structure around the sample with an increasing period due to the changing precession angles of the neutron spins on their path through the strongly decaying field. The gray level scale used in the images is related to the intensity variations induced by the sample and the magnetic field (from black = minimum to white = maximum, representing the periodic 2π rotation of the neutron spin). The structure in the center part cannot be seen due to the limited spatial resolution of approximately 500 μm [4].

This imaging method can in some instances be extended into three dimensions by a standard tomographic technique. Figure 10.3 shows the distribution of a magnetic field trapped inside a polycrystalline lead cylinder, with a diameter of 1 cm and length of 3 cm, that becomes superconducting when cooled below the critical temperature, $T_c = 7.2$ K. The sample was cooled down to 6.8 K in a homogenous magnetic field of 10 mT. After this the magnetic field was switched off, resulting in partially trapped magnetic fields in the

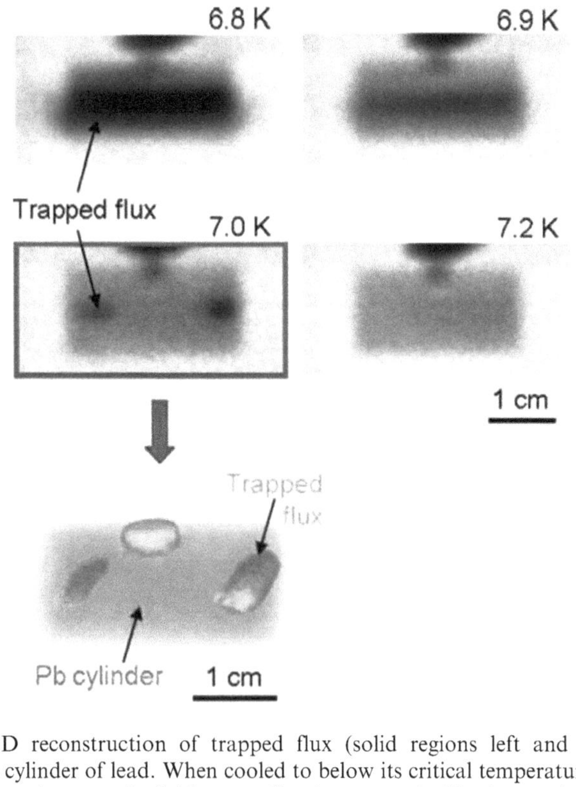

Fig. 10.3 A 3D reconstruction of trapped flux (solid regions left and right) inside a polycrystalline cylinder of lead. When cooled to below its critical temperature (7.2 K) in the presence of a weak magnetic field, some flux is present inside due to defects and grain boundaries and this remains trapped even after the field is switched off [1]

superconductor due to grain boundaries and other defects [5, 6]. The temperature dependence of the residual field distribution inside the sample was visualized by recording radiographic images during the heating process. The images (Fig. 10.3) show an inhomogeneous residual field which decreases during heating and vanishes completely when the critical temperature T_c is reached. For the weak trapped residual field at 7.0 K, a tomographic investigation was performed by rotating the sample around its vertical axis. The beam attenuation for each pixel can be related to absorption and magnetic contrast (Eq. 10.2) assuming that the trapped magnetic field conserves its main orientation perpendicular to the beam polarization and is weak enough to cause spin rotations smaller than π for all recorded projections. The volumetric data set was reconstructed from 60 2D images, collected through rotation over 180°, using a filtered back projection algorithm (Chapter 6). The results show the three-dimensional (3D) representation of the flux trapped in the sample. Flux concentrations could be found close to the end surfaces of the cylinder and at the position where the sample was held with a screw.

This imaging method has many potential applications. The presence and controlled application of magnetic fields are essential in many fields of science and technology as well as in fundamental physics. For example, the flux distribution and flux pinning in large superconducting samples, the skin effect [7] in conductors, or magnetic domain distributions in bulk ferromagnets could be visualized and studied in detail.

10.2 Spin Echo Imaging

The Larmor precession described in the previous section forms the base of the neutron spin-echo technique which is typically used to determine small energy or momentum changes upon scattering [7], see also Neutron Scattering Instrumentation [8]. In a typical spin-echo setup (Fig. 10.4), two identical regions of magnetic field are applied before and after a "π-flipper." A neutron with polarization perpendicular to the field undergoes Larmor precession in the first field region and its spin rotates by an angle of ϕ radians. Due to the energy spread of the neutron beam, the spins of different energy neutrons acquire different precession angles and the neutron beam appears to be depolarized after the first field region. The π-flipper between the two magnetic field regions rotates the neutron spins about an axis perpendicular to the applied field by 180°. This has the effect of reversing the angle of the neutron spin from ϕ to $-\phi$. Since the two magnetic field regions are identical, the neutron spin rotates an additional angle of ϕ radians in the second field region. The spin precession in the two field regions cancels and the beam "recovers" its polarization after going through the field regions. This setup constitutes a neutron spin-echo instrument, and the process through which the neutron beam recovers its polarization is called "spin echo."

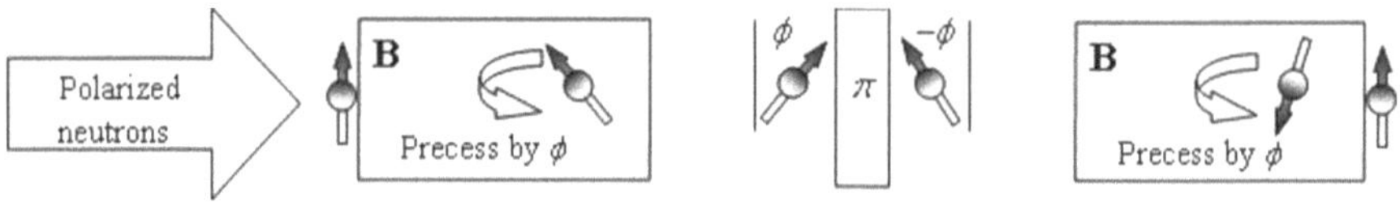

Fig. 10.4 A basic neutron spin-echo instrument setup. The spin of a neutron undergoes Larmor precession in the upstream magnetic field. A π-flipper between the magnetic field regions inverted the spin rotation angle from θ to $-\theta$. The neutron spin rotates an additional angle equal to θ in the identical downstream field region and returns to its starting direction

In neutron spin-echo spectrometers, a sample is placed between the two magnetic field regions in the neutron flight path. Inelastic scattering of the neutron beam changes the energy (velocity) of the neutrons, so that the times that a neutron spends in the two field regions are no longer equal. Since the spin precession is proportional to the time spent in the magnetic field, this prevents the beam from completely recovering its polarization. The rate of depolarization is proportional to the difference in the times spent in the two field regions; hence the neutron energy change due to inelastic scattering by the sample can be determined.

In a previous development [9], experiments were carried out to measure the neutron optical potential in nonmagnetic materials by applying the neutron spin-echo technique. When a neutron beam goes through a material without being scattered, it experiences an optical potential and changes its momentum. In neutron optics, the optical potential manifests as the neutron index of refraction n,

$$n = 1 - \frac{Nb\lambda^2}{2\pi}, \qquad (10.3)$$

where N is the number density of the material, b is the coherent scattering length, and λ is the neutron wavelength. The velocity of a neutron in the material is related to its velocity in vacuum v_0 by $v = nv_0$. Due to the change in the neutron velocity, neutrons that go through a material along a direction x over a distance X acquire an additional spin angle ϕ, given by

$$\phi = \int_0^X \gamma_L B \frac{1}{n(x)v_0} dx - \int_0^X \gamma_L B \frac{1}{v_0} dx = \frac{m_n \gamma_L}{2\pi h} B\lambda^3 \int_0^X N(x)b(x)dx, \qquad (10.4)$$

where m_n is the neutron mass, h is the Planck's constant, and b is the coherent scattering length. The quantity $N(x)b(x)$ is more commonly known as the neutron scattering length density. The magnitude of the neutron scattering length density ranges from 10^{-7} Å^{-2} to 10^{-5} Å^{-2}. Hence the spin angle change ϕ measures the path integral of the scattering length density.

In a modification of the neutron spin-echo instrument discussed above, placing a material in one arm of the pair of magnetic field regions results in a net neutron spin angle that corresponds to the path integral of the scattering length density. With a proper neutron beam arrangement, this effect can be applied to measure the optical potential as a function of position in a material – an imaging technique. A typical benchmark for an imaging technique is a 1 mm spatial resolution. A common neutron polarization analyzer such as a polarized ^3He analyzer (Chapter 3 in this book) may resolve a spin angle of 10°. If the magnetic field strength is strong enough to rotate the neutron spin by 10° over 1 mm of its flight path, a 1 mm imaging resolution in the direction along the beam path can be achieved. Using a typical scattering length density of $Nb = 10^{-6}$ Å^{-2}, Fig. 10.5 shows the magnetic field required to reach $\phi = 10°$ over a 1 mm flight path as a function of the wavelength. A potential imaging device using this spin-echo technique would therefore use cold neutrons with wavelengths ranging from 20 to 100Å and an applied field from several gauss for 100Å neutrons to 1 Tesla for 30Å neutrons. The magnetic potential energy $U = -\vec{\mu} \cdot \vec{B}$, where $\bar{\mu}$ is the neutron magnetic moment, is equal to 0.06 μeV in a 1 Tesla field. Hence, compared with a kinetic energy of 8.181 μeV for 100Å neutrons, the optical effect of the magnetic potential can be ignored.

One main difference between the imaging application and the neutron spin-echo spectrometer is that in imaging, the additional change in the spin angle occurs only when the neutron is inside the material, whereas the change accumulates over the entire field region after the sample in the case of spin echo spectrometer. This makes the spin angle change insensitive to the beam divergence. The beam divergence, however, is still important in that neutrons through different parts of a sample may end up striking the same detector pixel. This blurring of the image is, however, a common concern for all imaging

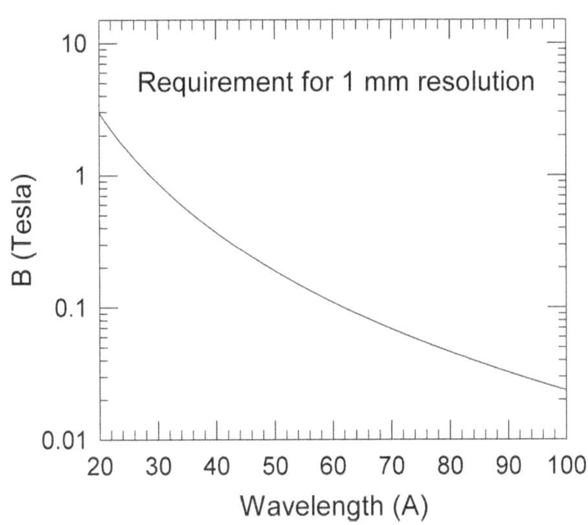

Fig. 10.5 Applied field strength necessary to achieve a 1 mm resolution

techniques rather than being specific to the spin contrast imaging. In neutron imaging, collimation is used to reduce the divergence and anti-grids can be used between the sample and the detector system. The same setup can be applied in the techniques discussed here.

To acquire a three-dimensional image of an object using imaging techniques based on direct neutron absorption, measurements with the neutron beam passing the object at different angles are first carried out and tomography reconstruction is then applied to render the 3D image. The advantage of the imaging technique discussed here is that it can be used with a time-of-flight (TOF) technique to achieve 3D imaging without the need to measure the object at many different angles and without using tomographic reconstruction.

The idea behind the technique is illustrated in Fig. 10.6. Initially the neutron spins are aligned parallel to the applied field. At time $t=0$, a $\pi/2$ pulse is applied over the length of the sample to rotate the neutron spins to a perpendicular direction with respect to the field, which starts the precession of the neutron spins. Along the neutron flight path, neutrons at different position along a flight path go through a different path length through the material and therefore acquire a different spin angle change ϕ. For instance, the spin change angle is ϕ_1 for neutrons starting at point 1 and ϕ_2 for neutrons starting at point 2, respectively. The spin angle change is converted to intensity change by a polarization analyzer before the detector. Since neutrons at point 1 at $t=0$ reach the detector later than neutrons at point 2 at $t=0$, by recording the time-of-flight of each neutron at the detector, the position of each neutron at $t=0$ can be determined. Subtracting ϕ_2 from ϕ_2 gives the spin angle change that corresponds to the scattering length density from point 1 to point 2. Applying this analysis to a series of images taken at consecutive times-of-flight, a slice-by-slice 3D distribution of the scattering length density of an object can be obtained. The speed of 30Å neutrons is 131.87 m/s. It takes 7.6 μs to travel a distance of 1 mm. A timing resolution of 1 μs would be sufficient to resolve the initial location of the neutrons at the time the precession is started.

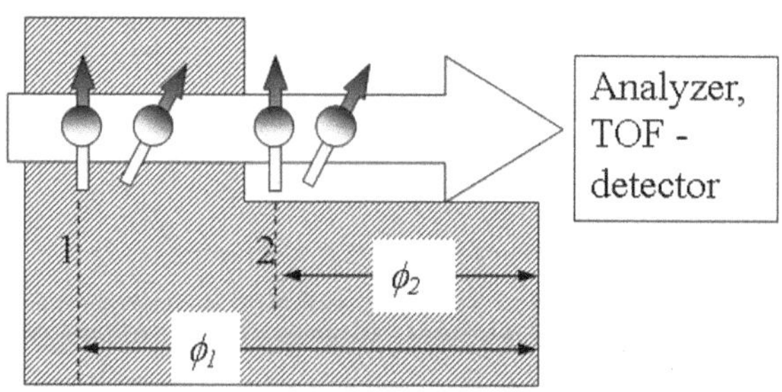

Fig. 10.6 Technique for direct 3D imaging. At time $t=0$, a $\pi/2$ flip starts the neutron precession; neutrons at point 1 precess through a longer path in the sample than neutrons at point 2. Subtracting the two gives the precession angle from point 1 to point 2

In order to demonstrate the technique, a Monte Carlo simulation was carried out that takes into account bandwidth, beam divergence, and placements of the neutron source and neutron detectors. The sample model is a 1 × 1 cm plate. It is 2 mm thick in the beam direction with one-quarter of the sample cross-section only 1 mm thick. A nominal scattering length density $Nb = 10^{-6}$ Å^{-2}was used for the calculations. The field strength at the sample was 1 Tesla. For the purposes of the simulation a 5 mm diameter, isotropic neutron source located at 4 m from the sample was assumed to produce a 30Å neutron beam with a typical wavelength spread $\Delta\lambda/\lambda = 10\%$. This wavelength spread allows for both 1 mm resolution and sufficient neutron intensity. A detector was located 1m away from the sample. For simplicity, only the additional amount of spin precession in the material was computed instead of the spin precessions in the two magnetic field regions. Also for simplicity, the incident neutron polarization efficiency was assumed to be 1 and the efficiency of the polarization analyzer was also considered to be ideal. In a real instrument, the polarization and the efficiency of the analyzer is typically 95–98%. The neutron intensity after the analyzer is given by $I = \cos^2(\phi)$. In this setup, $\phi \sim 11.5^\circ$ for a passage of 1 mm through the material.

The intensity profile at the image plate detector for different TOF neutrons is shown in Fig. 10.7. The upper-right corner is the region where the material is thinner (1 mm thick compared with 2 mm for the rest of the sample). In Fig. 10.7(a), the neutrons began to precess before entering the sample. In Fig 10.7(b), the neutrons are half-way through the sample when the precession begins. Due to the finite size of the source and the beam divergence, there is blurring of the edges of the modeled sample. The diverging beam also gives rise to an amplified image. To further evaluate the simulation results, the intensity across the x-direction of the image plate at $y = +3$ mm to $+3.2$ mm was extracted to calculate the spin angle change ϕ. The intensity results are shown in Fig. 10.7(c) and the spin angle change results are shown in Fig. 10.7(d). The simulation verifies that $\phi \sim 23^\circ$ for neutrons passing the 2 mm thick area and $\phi \sim 11.5^\circ$ for those passing the 10 mm thick area. Subtracting the two spin angle changes gives the spin angle change through the first 1 mm thick area and it gives the expected $\phi \sim 11.5^\circ$ as shown in Fig. 10.7(d). This demonstrates that, by extracting the spin angle change ϕ from the intensity map at different TOFs, the distribution of the scattering length density in an object can be mapped in a slice-by-slice fashion to give a 3D rendering of the composition.

There are several technological challenges to reaching 1 mm resolution with this method. The first is the requirement to flip the neutron spins by $\pi/2$, i.e., from being parallel to being perpendicular to the 1 Tesla field. The Larmor precession frequency for a 1T field is 29.164 MHz. The neutron spin will rotate 90° in 8.6 ns. To avoid a broadening of the radio-frequency pulse spectrum, a pulse of $2\,n\pi + \pi/2$ can be used. However, a setup accurate to 0.5 ns in the application of the pulse is needed to reach accuracy better than 5° in the $\pi/2$-flipping. A second challenge is the precision required to cancel the spin precession (the spin-echo point) using

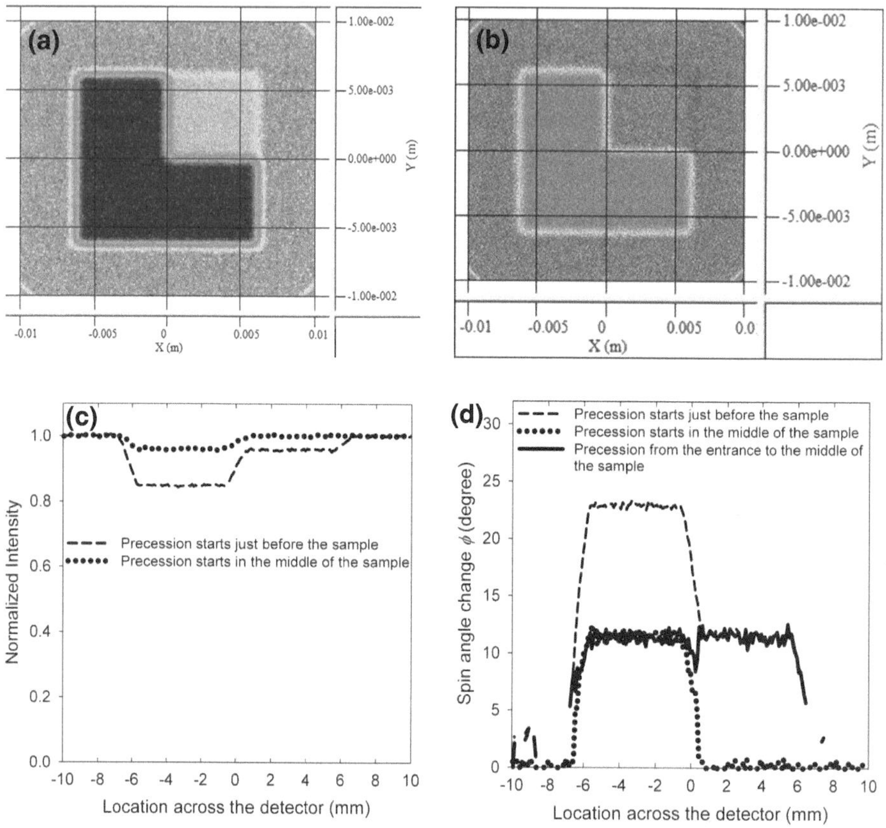

Fig. 10.7 Neutron intensities at the detector if spin precession starts (**a**) when the neutrons are right before the sample, and (**b**) when the neutrons are passing the middle of the sample. The color scales are relative; (**c**) The normalized integrated intensity at the detector along the x-direction of the detector from $y = +2$ mm to $y = +3$ mm from the two detector images shown in (a) and (b); (**d**) The spin angle change ϕ calculated from the two time-of-flight intensities

both high magnetic field and very cold neutrons: From Eq. (10.1) above, 30Å neutrons in a 1 T field precess 221 times every 1 mm flight path. If identical upstream and downstream field regions are used, the location where the π-flipping occurs will need to be accurate to 0.1 μm for a 10° accuracy. This would be a formidable task using currently available technology. However, only a small region that covers the sample will need the 1 T field. The rest of the field regions can use a field that can be substantially lower. This is because the spin precession angle depends upon both the field strength and the path length along which the field is imposed. A smaller field over a long distance can result in the same spin rotation as a larger field over a shorter distance. If a nominal 1 T field is applied over a 1 cm length where the sample is located and the rest of the setup is in a 20 gauss field, the second field region can extend 5 m and require an accuracy of 0.06 mm in the π-flipping location for a 10° accuracy. The third challenge results from the fact that the precession of the neutrons starts at different locations along the beam. Hence either the downstream $\pi/2$ pulse or the π-flipping will need to be applied to different locations along the flight path. This would require a more sophisticated π-flipper setup than those currently used at neutron beam lines, which typically apply π-flipping over a small distance along the beam.

Hence although we have shown that spin contrast tomography is theoretically feasible, there are some interesting technical challenges to be overcome before the method can be applied in practice.

10.3 Prospects for Neutron-Probed Magnetic Resonance Imaging

Magnetic resonance imaging (MRI) is a powerful tool used primarily for internal imaging in biological systems. It is basically a position-dependent map of the nuclear spin relaxation time. There are many other systems that one would like to image using MRI, but they have a negligible penetration depth for radio frequency (RF) radiation. Therefore traditional MRI techniques cannot be used, and an alternative method to perform the nuclear magnetic resonance (NMR) measurement that is the core technology behind MRI must be found. The main classes of samples that are of interest are metallurgical samples, hydrogen storage materials, and biological samples in ionic salt solutions. As discussed elsewhere in this book, many of these systems can be addressed with conventional neutron radiography. Where MRI brings additional information is in systems where there are magnetic interactions of interest. For example in a magnetic alloy, MRI would provide a position-dependent measure of the interaction between one magnetic region and its environment, where conventional imaging only provides the positional dependence of the constituents of the alloy.

Neutrons are a promising probe to overcome the aforementioned difficulty. They penetrate deeply into materials. However, neutrons present their own challenges. Several of these are discussed. First and foremost is performing the NMR measurement.

Fundamentally, NMR and neutron spectroscopy are MHz and THz techniques, respectively. Therefore, although the Zeeman splitting of nuclear magnetic moments can be observed with conventional spectroscopy, the additional fine splittings of interest in NMR measurements cannot. Alternatively, a time-dependent measurement can be used to measure the decay of a nuclear spin (τ) and has been demonstrated in some limited cases. These studies used low temperature (~mK) [10], or microwaves [11], to saturate the nuclear spins. These oriented spins changed the intensity of Bragg peaks. Then studies the temperature was raised, or the microwaves turned off, and the intensity decay of the Bragg peak was recorded as a function of time. The resultant decay is proportional to τ. However, mK temperatures are difficult to attain, would freeze liquid samples, and the microwaves suffer from the same penetration depth problems as the lower-frequency RF. An alternative approach to measuring the decay time must be found to make this technique accessible to more types of samples. A potential solution is to use high magnetic fields.

With the availability of high magnetic fields, a simple extension of the aforementioned techniques would be to use the field to raise the temperature necessary for the saturation of the nuclear spins. Figure 10.8 examines this possibility by looking at the nuclear polarization as a function of temperature for different fields. As can be seen from the figure, $T \sim 10$ mK is required for the current state of the art field for neutron-scattering experiments of 15 *T*. If the field was increased to 55 *T*, the required temperature is in the hundreds of mK range. Such an increase makes the measurement simpler but does not relieve the freezing problem. To use this technique at room temperature requires $\sim 10^5$ *T*, an unrealizable static field.

So the main challenge is to find another way to measure τ. The approach described below eliminates the need for microwaves and/or mK temperatures. The general idea is to use spectroscopy to measure the Zeeman splitting of relaxing nuclear spins at a high field, then quickly drop the field to a lower value and watch the Zeeman splitting relax to the equilibrium value at the lower field. A large static field is required to sufficiently separate the peak associated with the Zeeman splitting from other incoherent processes. Protons are the nuclei of choice for this study, as they are the most familiar NMR active nuclei. The protons are assumed to be provided by a water sample at room temperature. Therefore the nuclear relaxation time is assumed to be 200 ms. A static field of

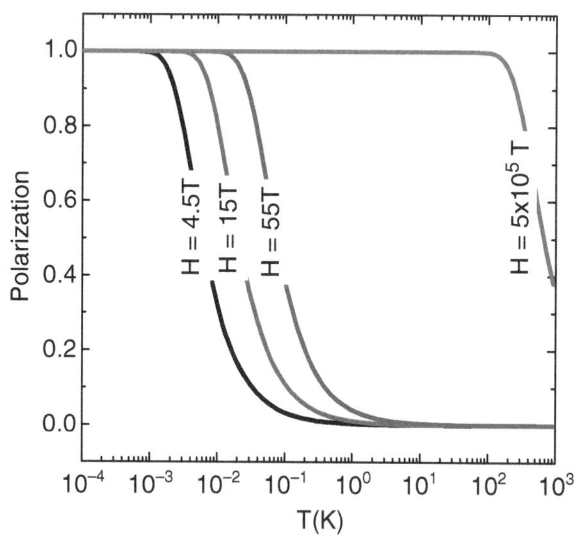

Fig. 10.8 The normalized proton polarization as a function of temperature for different values of magnetic field

15 *T* is sufficient for this system. Extending this technique to significantly more nuclei would require 2–4 times the static field, e.g., ~ 40 *T* [12]. The simulations detailed below are for a 10 *T* pulsed field on top of the 15 *T* static field.

To demonstrate the feasibility of this technique, one must show that, with current magnet technology, the field can cycle fast enough to allow for measurement of the nuclear relaxation time. The current state of the art in pulsed magnetic fields is the long-pulsed magnet at the National High Magnetic Field Laboratory in the United States [13]. A calculation was performed to observe the response of the resonance energy of the aforementioned proton system to a 10 *T* pulse (solid curve in Fig. 10.9) having the shape of the peak of the long pulse magnet [14]. The dashed curve is the resultant solution to the Bloch equation [15]. The equation was solved numerically with MathCAD [16]. This resonance energy will be the center value for the splitting of the scattering away from zero energy transfer. Given the dashed curve, a pulsed neutron source, operating at 60 Hz, will provide 80 pulses over the observed nuclear decay time.

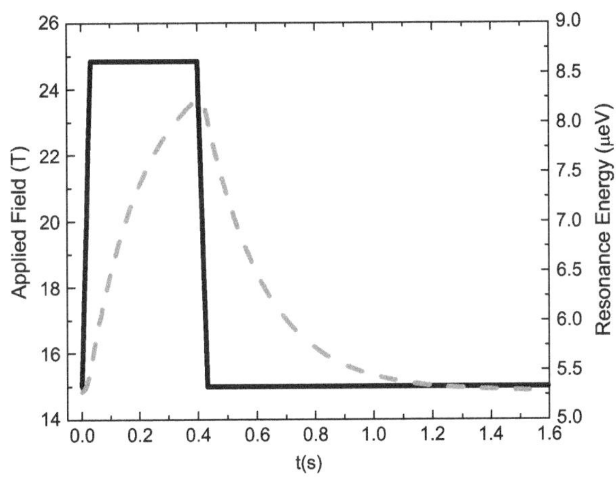

Fig. 10.9 The resonance energy for a system of hydrogen nuclei (*dashed curve*) for a given applied field (*solid curve*)

An estimate of the neutron count rate is also necessary to demonstrate the feasibility of this type of experiment. Approximations of the cross-sections for scattering between spin states were estimated elsewhere [17]. The assumed flux on a 40 mm^3 sample is of the order 10^5 n cm^{-2} s^{-1}. The scattering from the sample will illuminate 0.379 sr of detector area. The energy resolution of the instrument close to the elastic line is 3 μeV. (Detailed balance and absorption are ignored in these calculations.) Given these assumptions, an example trace is simulated and shown in the inset of Fig. 10.10. A similar trace is calculated every 16.7 ms out to 1.7 s after the field is applied. The uncertainty in the curve is simulated Poisson statistics. To obtain sufficient counting statistics, each point

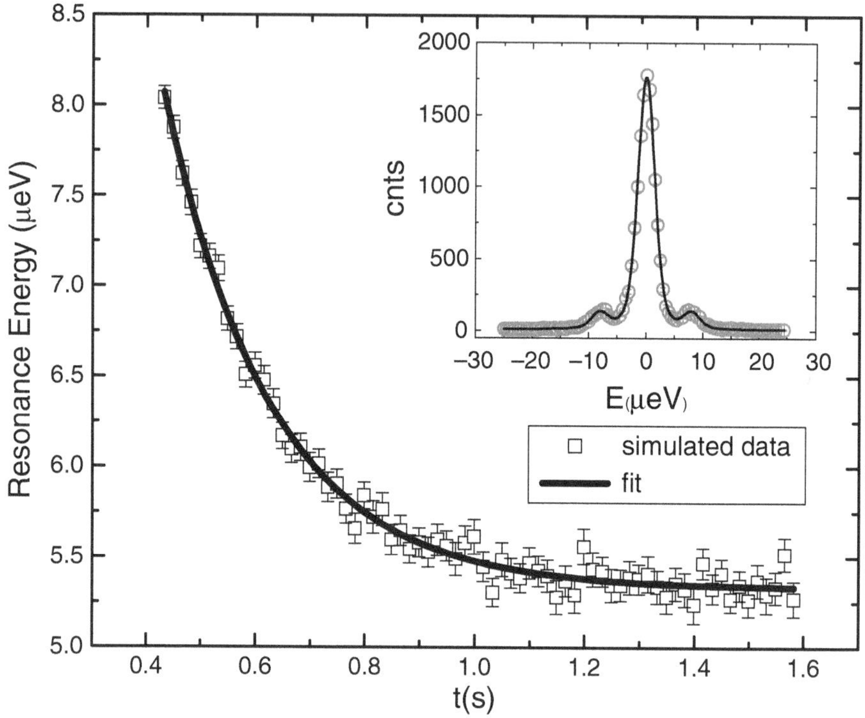

Fig. 10.10 The resonance energy determined from the simulated spectra. The resonance energy is fitted to an exponential decay to determine the relaxation time. The inset shows an example simulated spectrum (*circles*) and a fit (*solid*)

on the decay curve should be counted for a total of 10 s. Therefore the measurement would be repeated many times and the aggregated results analyzed. In practice this means a $\sim$ 0.6 Hz pulsed magnet operated for 600 pulses. A full relaxation time measurement would require $\sim$ 20 min.

Each of these resultant spectra are analyzed by fitting 3 Lorentzians of 1 μeV width, convoluted with the instrumental energy resolution of 3 μeV, at each time interval. The parameter to be tracked, as a function of time, to determine the relaxation time is the splitting of the finite energy transfer peaks from the center. The results from the simulated data for this parameter and its uncertainties are plotted as points in Fig. 10.10. An exponential decay is then fit to the points to determine the relaxation rate. The specific function used for this fit is $E = Ae^{-t/\tau} + E_0$ with E_0 the field splitting at the lower field value and A a normalization constant. The results are as follows: $\tau = 196 \pm 5$ ms, $A = 25 \pm 2$ μeV and $E_o = 5.33 \pm 0.02$ μeV. The fitted relaxation or Zeeman splitting time, τ, is in excellent agreement with the input time of 200 ms. Therefore, simulations show that neutron measurements of the relaxation time could be performed by watching the timed decay of the Zeeman splitting.

To test the limits of this method, similar calculations were performed with a 10 *T* static field. However, for this configuration the Zeeman splitting was too small at the longest times to be resolved from the central incoherent peak. This resulted in an artificially large value of τ from the fit. In principle the τ measurement will be influenced at longer times by the spins already flipped at shorter times. However, less than 10^6 spins are flipped before the next field pulse is applied, which is a negligible quantity when compared with the total number of spins in the system.

In order to do this measurement, a high-resolution spectrometer such as BASIS at the U.S. Spallation Neutron Source (SNS), a high (static) field magnet in excess of 30 *T* such as that proposed for Zeemans at the SNS, and a 10 *T* pulsed magnetic field are required [18, 19].

Though the core technology behind neutron-probed MRI has been described, other significant challenges remain. Position encoding is one of them. In traditional MRI a field gradient is placed across the sample [20, 21]. Then the resonance condition is only met at one spot in the sample. The field gradient is then varied to change this location. A field gradient would only smear out the neutron-probed measurement as it measures all energies. Instead, we would raster the sample through the beam so only a single-point is measured at a time. We assume the state of the art pixel size of 1×1 mm to 0.1×0.1 mm, as defined for the Vulcan instrument at SNS [22]. An additional restriction is that the bore in the proposed magnet [19] is only 50 mm. Therefore rastering a sample within this bore requires a small sample $\sim 1 \times 1$ cm and would provide between 100 and 10000 unique points depending on the beam size. Extending this technique to larger samples is a motivation for increasing the bore size in the next generation of superconducting hybrid magnets. Another approach would be to use small pulsed magnets that can achieve 30 *T*, but have a repetition rate of $\sim$1 pulse per 7–8 minutes [23]. A sample sandwiched between two of these magnets and rastered independently of the magnets provides a way to increase the sample size. For this approach one would have to stop measuring the relaxation after the peaks associated with the Zeeman splitting coalesce with the main incoherent peak. The main restriction of this approach is the long time between pulses. This time is controlled by internal heating of the magnet. The time between pulses could be reduced to the time limit of the capacitor bank power supply by rastering additional magnets while the first pair cools.

For both styles of magnets, field gradients are important. Though for traditional MRI homogeneity on a few PPM [20] is expected, $\sim$100 ppm is acceptable. The pulsed magnets are quite configurable so field gradients of this order are reasonable. For the Zeemans style of magnet, a high-homogeneity (100 ppm) resistive insert would have to be designed. This insert design would drop the peak field to $\sim$28 *T* (M. D. Bird, private communications). Nevertheless, one could envision running the magnet for a series of cycles in a high field mode and then changing for another series of cycles to a high homogeneity mode.

Finally, counting times for the MRI measurement must be considered. Assuming the counting described for the NMR measurement, but reducing the sample to a 1 × 1 mm cross-section, means that beam focusing of ~15,000 is required so that a measurement can be carried out in 1 week. This is assuming that the static and pulsed field method is used. An order of magnitude may be gained by increasing the pulse rate of the pulsed magnet. If the fully pulsed field approach is used, this number would increase, as the repetition rate is even slower. Therefore beam flux is the biggest challenge to realizing neutron-probed MRI.

In summary the major challenges to realizing neutron-probed MRI are technical and not fundamental to the science. The core technique of neutron NMR can be demonstrated with the next generation of high magnetic field beamlines. Nevertheless, flux and the limitations on neutron focusing are the major challenges to using the neutron NMR technique for neutron-probed MRI.

Acknowledgments Garrett Granroth is grateful for useful discussions with J.-K. Zhao, G. Greene, E. Iverson, M. W. Meisel, N. Sullivan, and B. Halperin. His work was performed at Oak Ridge National Laboratory, managed for the U.S. Department of Energy by UT-Battelle, LLC, under Contract No. DE-AC05-00OR22725.

References

1. N. Kardjilov, I. Manke, M. Strobl, A. Hilger, W. Treimer, M. Meissner, T. Krist, J. Banhart, Three-dimensional imaging of magnetic fields with polarised neutrons, *Nat. Phys.* **4**, 399–403 (2008).
2. Th. Krist, S. J. Kennedy, T.J. Hick, F. Mezei, New compact neutron polarizer. *Physica B* **241–243**, 82–85 (1998).
3. W. Treimer, M. Strobl, N. Kardjilov, A. Hilgar, I. Manke, Wavelength turable device for neutron radiography and temography, *Appl. Phys. Lett*, **89**, 203504 (2006).
4. A. Hilger, N. Kardjilov, M. Strobl, W. Treimer, J. Banhart, The new cold neutron radiography and tomography instrument CONRAD at HMI Berlin, *Physica B* **385–386**, 1213 (2006).
5. P. Gammel, D. Bishop, Fingerprinting vortices with smoke, *Science* **279**, 410–411 (1998).
6. Ch. Jooss, J. Albrecht, H. Kuhn, S. Leonhardt, H. Kronmüller, Magneto-optical studies of current distributions in high-T_c superconductors. *Rep. Prog. Phys.* **65**, 651–788 (2002).
7. I. Manke, N. Kardjilov, M. Strobl, A. Hilger, J. Banhart, Investigation of the skin effect in the bulk of electrical conductors with spin-polarized neutron radiography, *J. Appl. Phys*, 104, 076109 (2008).
8. http://www.springer.com/series/8141, this series of books
9. A.I. Frank, I. Anderson, I.V. Bondarenko, A.V. Kozlov, P. Hoghoj, and G. Ehlers, *Phys. Atom. Nucl.* **65**, 2009 (2002).
10. T.A. Jyrkkiö, M.T. Huiku, K.N. Clausen, K. Siemensmeyer, K. Kakurai, and M. Steiner, Calibration and applications of polarized neutron thermometry at milli- and microkelvin temperatures, *Z. Phys. B* **71**, 139 (1988).
11. J.B. Hayter, G.T. Jenkin, and J.W. White, Polarized-Neutron Diffration from Spin-Polarized Protons: A Tool in Structure Determination? *Phys. Rev. Lett.* **33**, 696 (1974).
12. M. Dobhal, *NMR Data for Carbon-13. Part 4: Natural Products*, Springer, Heidelberg (2006).

13. http://www.lanl.gov/orgs/mpa/nhmfl/60TLP.shtml
14. http://www.lanl.gov/orgs/mpa/nhmfl/images/60TLPpulse.jpg
15. C.P. Slichter, *Principles of Magnetic Resonance*, Springer-Verlag, Heidelberg (1990).
16. MathCad, Parametric Technology Corporation.
17. G.E. Granroth, *Prospects for Neutron probed NMR* in The proceedings of ICANS–XVII, J. Wei, S. Wang, W.L. Huang, J.S. Zhao, ed. http://www.icans-xviii.ac.cn/proceedings/indexed/copyr/15. pdf.
18. http://www.sns.gov/instrument_systems/beamline_02_basis/index.shtml, (2007).
19. G.E. Granroth, C. Broholm, F. Klöse, G. Srajer, Z. Islam, Y. Lee, and J. Lang, *Report from Probing Matter at High Magnetic Fields with X-Rays and Neutrons*, SNS Report SNS107000000TR-005-R00, (2005).
20. J. Jin, *Electromagnetic Analysis and Design in Magnetic Resonance Imaging*, CRC Press, Boca Raton (1999).
21. P. Sprawls, *Magnetic Resonance Imaging*, Medical Physics Publishing, Madison (2000).
22. X.-L. Wang, T.M. Holden, G.Q. Rennich, A.D. Stoica, P.K. Liaw, H. Choo, and C.R. Hubbard, *Physica B* **385–386**, 673 (2006).
23. K. Ohoyama, N. Katoh, H. Nojiri, Y.H. Matsuda, H. Hiraka, K. Ikeda, H.M. Shimizu, Neutron diffraction under 30 T pulsed magnetic fields, *J. Mag. Mag. Mat.* **310**, e974 (2007).

Section C

Chapter 11
Neutron Imaging for the Hydrogen Economy

M. Arif, D.S. Hussey, and D.L. Jacobson

Abstract The development of fuel cells and hydrogen storage materials will be one of the highest global research and development priorities for the foreseeable future. The particular abilities of neutrons to penetrate materials and to image hydrogen will make neutron imaging a key technique, allowing the in-situ study of real operational devices. This chapter describes the current state of the art in this field.

Keywords Fuel cell · PEM · Hydrogen storage · Water dynamics

11.1 Introduction

Compared to most other forms of radiation, neutrons are highly efficient means of probing complex structures because of their tremendous of penetrating capability many common materials and their unique capability of distinguishing different materials with similar chemical and physical properties. Consequently, neutron imaging has always been regarded as one of the most potentially important tools for many areas of industrial research. However, these promises have remained largely unfulfilled, and neutron imaging techniques have seen limited applications, primarily due to traditionally poor spatial resolution and slow processing time as well as lack of identification of critical, high-impact areas of research. Most industrial and academic research that can significantly benefit from this technique involves dimensions that are a few tens of micrometers in length and interaction times that last at longest a few seconds. The current practical limitations for neutrons are about 100 μm in spatial resolution and tens of seconds in time resolution. These limitations are mainly due to the relatively weak neutron sources (nuclear reactors that are many orders of magnitude weaker than even table-top photon sources), large mean free paths of neutrons and secondary radiation in

M. Arif (✉)
National Institute of Standards and Technology, 100 Bureau Drive, MS 6100, Gaithersburg, MD 20899-6100, USA
e-mail: muhammed.arif@nist.gov

I.S. Anderson et al. (eds.), *Neutron Imaging and Applications*,
Neutron Scattering Applications and Techniques, DOI 10.1007/978-0-387-78693-3_11,

materials (limits spatial resolution), and the inability to process data in real-time while being statistically significant (limited by weak source).

However, in the past decade enough gains have been made to enable neutron imaging to be applied in different fields of science and technology. One of the most important prospects is the understanding of the interface physics, chemical kinetics, and proper modeling of the processes that govern the operation of the future generation of energy production and storage devices based on hydrogen. The sensitivity of neutrons to hydrogen, together with other unique properties, makes them an ideal probe for characterizing the structure, morphology, and hydrogen dynamics that are key to the future use of these promising devices. The neutron imaging technique, in particular, has emerged as one of the most important tools in the development of robust and efficient fuel cells accompanying hydrogen storage materials and devices. This article will focus on the application of neutron imaging research in the development of low-temperature fuel cells that may power homes, consumer electronics, and transport vehicles in the emerging hydrogen economy.

11.2 Neutron Imaging of Fuel Cells and Hydrogen Storage Devices

11.2.1 Neutron Imaging of Fuel Cells

Fuel cells are efficient, silent, and virtually non-polluting electrochemical devices that have immense potential for powering homes, businesses, and transport vehicles of the future. Major energy-related and automobile companies worldwide are currently engaged in extensive research to perfect fuel cells for everyday, cost-efficient use.

Fuel cells containing solid proton exchange membranes (PEM) are of special interest, as they enable fuel cell operation at lower temperatures ($< 100^\circ C$) than conventional solid oxide fuel cells ($> 500^\circ C$). PEM fuel cells convert the chemical energy of hydrogen-containing fuels directly into electricity in the presence of a catalyst, without combustion and without any moving parts. Fuel cells provide a quantum leap in fuel efficiencies compared to traditional heat engines because conversion to electricity does not require the intermediate generation of heat. Similar to batteries, fuel cells generate power electrochemically and thereby produce extremely low to zero emissions, depending on the fuel supply and operating conditions.

Current PEM fuel cells use perfluorosulfonic acid-based membrane technology such as Nafion® and Gore-Select® [1]. Platinum catalyst is applied to both sides of the membrane to form the membrane electrode assembly (MEA). During operation, hydrogen gas is brought to the anode side of the MEA while air is brought to the cathode side. H_2 is oxidized to protons and electrons. The protons migrate through the membrane (Fig. 11.1) to the cathode side and react with O_2 in the air to form water. The electrons, on the other hand, conduct externally to provide electricity. In this process, water is "dragged" from the

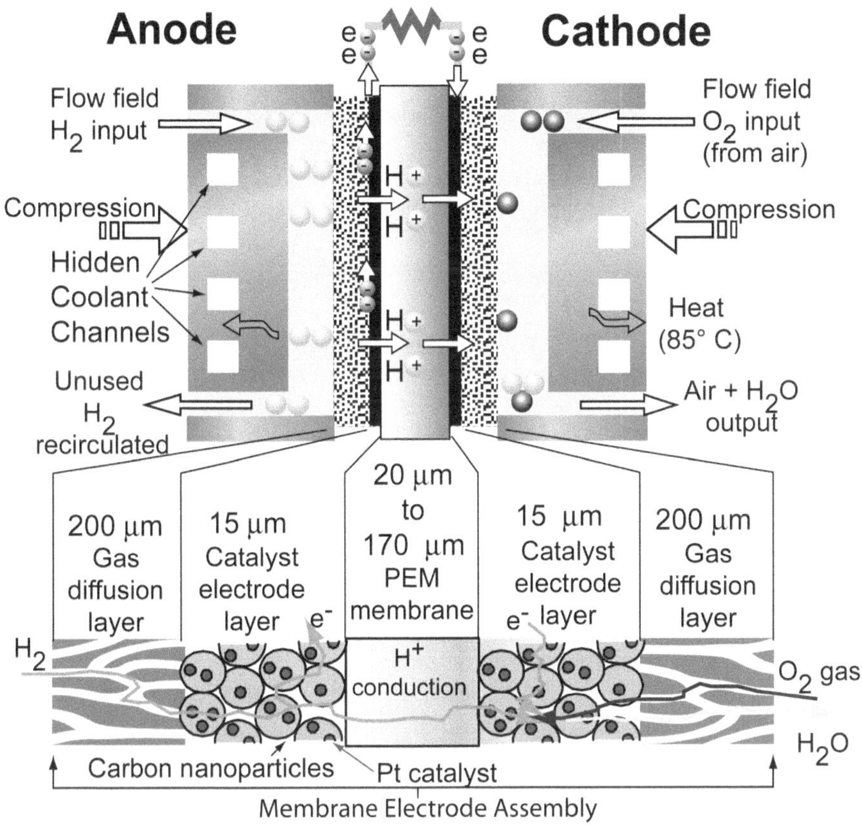

Fig. 11.1 Schematic of a fuel cell. The platinum (Pt) catalyst is generally applied to both sides of the membrane to form the membrane electrode assembly. Formation of water at the membrane can cause the cell to perform poorly

anode side of the MEA to the cathode side with the proton transport. In the meantime, water is generated on the cathode side of the MEA. A water gradient thus exists within the membrane. The PEM water content is extremely critical as it impacts the ionic conductivity and mechanical properties such as strength and gas permeability. Depending on the water concentration, the membrane performance and lifetime can be significantly reduced. In order to maintain the proton conductivity and preserve membrane life, PEM fuel cells typically require humidified reactant gases. This vapor, together with the water generated during operation, poses significant challenges for water management. A hydrophobic porous gas diffusion layer (GDL) is typically used to help manage water distribution. Furthermore, the electrode composition may be hydrophobic or hydrophilic, which is another area of water management difficulty.

PEM fuel cells have gone through significant evolution in recent years, but one of the remaining key challenges is efficient water management. The control of incoming humidity and the product water within the MEA and GDL is generally acknowledged as one of the most crucial aspects of operating a PEM fuel cell. Too much water within the MEA or GDL will result in flooding conditions that impede gas diffusion; too little water will reduce the membrane proton conductivity, thereby decreasing the cell performance; and cycling between water levels will reduce membrane life. Thus, proper water management is the key to a stable and long-life PEM fuel cell and must be achieved by properly designing the flowfield, the GDL, the MEA, and their interfaces. This requires a fundamental understanding of the *in situ* water distribution in an operating fuel cell.

It is critically important for both optimal design and efficient operation of the fuel cells to study and quantify these adverse effects in situ using non-destructive means [2]. At present, neutron imaging is the only technology that has the ability to look inside a standard, commercially viable PEM fuel cell while the cell is operating and characterize the water content non-destructively. Since neutrons are highly penetrating in most materials, they are ideal probes in the investigation of internal structures of bulk objects like the fuel cell. Thermal neutrons (low-energy neutrons) passing through matter experience the following interactions: (a) absorption, (b) scattering (elastic, inelastic, small angle scattering), (c) refraction, (d) change in the phase of the neutron wave, (e) depolarization of an incident spin direction.

These interactions are used for signals in thermal neutron two-dimensional (2D) (radiographic) and 3D (tomographic) imaging (Chapters 5 and 6, in this volume). The most common processes are absorption and scattering. The thermal cross-sections for absorption and scattering events are complex functions of atomic mass, atomic number, nuclear isotope, and neutron energy. The contrast obtainable between materials in radiographic images is a function of their different cross-sections. For thermal neutrons, hydrogen-based compounds (such as water) would be highly visible against aluminum, iron, or carbon, which are commonly used in fuel cell construction (Fig. 6.1). This is due to the fact that the so-called mass attenuation coefficient of water is nearly two orders of magnitude larger than that of the background materials. This unique signature of hydrogen allows for the study of water transport phenomena in fuel cells with sub-microgram precision.

11.2.2 Neutron Imaging of Hydrogen Storage Devices

Vehicles powered by PEM fuel cells require, among other things, the presence of onboard hydrogen storage vessels. The design of metal-hydride beds with suitably low weight and volume to meet the demands of automotive applications is no trivial matter. The characterization of prototype technologies would benefit greatly from an in situ imaging technique such as neutron radiography to aid in understanding the hydrogen concentration gradients across these beds during absorption/desorption cycling. As a demonstration, the hydrogen uptake in Sieverts' Reactor was imaged in situ during a hydrogen charging step. Figure 11.2 shows a photo of the Sievert's Reactor and neutron images at two different points during the hydrogen charge. Since hydrogen has such a large total neutron scattering cross-section, it is straightforward to measure the real-time metal hydride uptake of gaseous hydrogen.

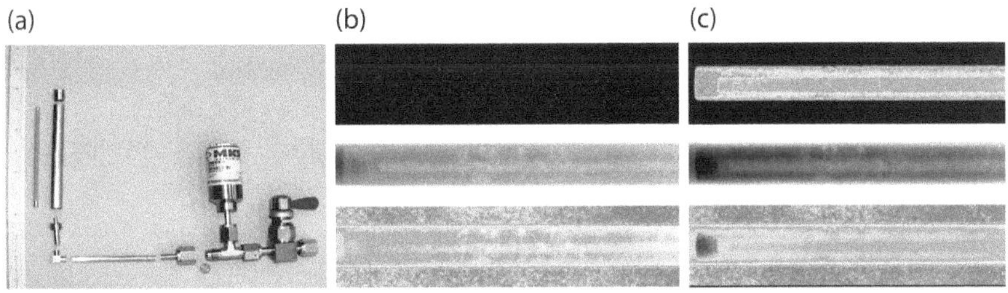

Fig. 11.2 Neutron radiography study of hydrogen storage in $LaNi_{4.78}Sn_{0.22}$. (**a**) Photo of the Sievert's reactor. (**b,c**) Colorized (*bottom*) and raw greyscale (*middle*) images of the reactor and the hydrogen content (*top*) in the uncharged and charged states, respectively

Future research on hydrogen storage will correlate local hydrogen reaction rates and contents with temperature profiles numerically simulated by detailed thermal models based on designs such as those previously developed [3,4] at the Jet Propulsion Laboratory (JPL) for hydride compressor beds to be used in sorption cryocoolers on the spacecraft of the Planck Mission [5,6]. A cross-sectional drawing of such a highly integrated hydride bed is shown in Fig. 11.3.

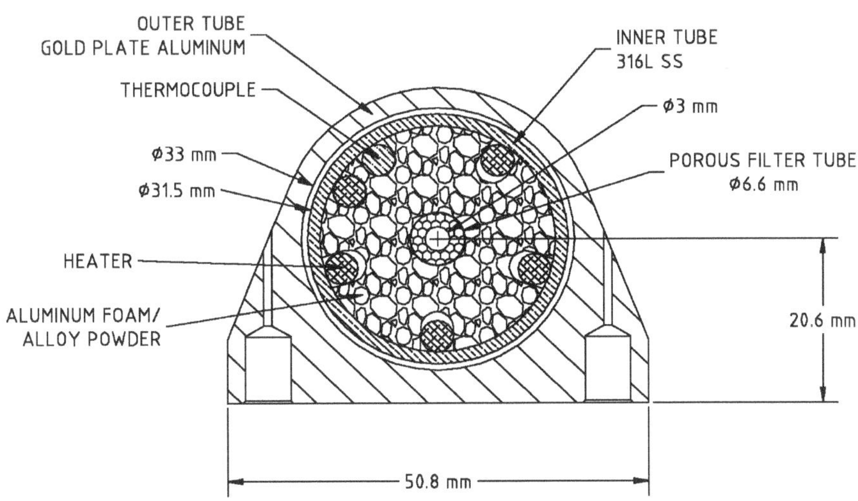

Fig. 11.3. Cross-sectional drawing of a prototype hydrogen storage bed. The porous aluminum is filled with $LaNi_{4.78}Sn_{0.22}$. With neutron tomography, the effects of hydrogen uptake, such as swelling and temperature gradients, can be revealed

This prototype hydrogen storage vessel consists of an outer aluminum shell surrounding an inner bed enclosed by a ~3 cm diameter stainless steel tube (~30 cm long) that contains an annular, open-pore Al foam matrix filled with $LaNi_{4.78}Sn_{0.22}$ hydrogen-absorbing powder. The center of the annulus contains a porous 316L stainless steel (SS) filter tube for hydrogen delivery.

In such a study, real-time images of concentration gradients within the $LaNi_{4.78}Sn_{0.22}$/Al foam hybrid matrix would need to be collected as a function of controlled stepwise hydrogen addition and removal (hydrogen coming from an adjacent control volume). Also, neutron tomography would be performed for each step after steady state had been reached. In this way, it may be possible to characterize the presence and nature of any hydrogen concentration gradients.

11.2.3 Primary Areas of Applications

11.2.3.1 Efficient Flow Field Design

Flow channels are the conduit through which hydrogen and air are drawn into a fuel cell. They also act as conduits to expel excess water. Effective design of the flow channels is essential to achieve the desired proper balance of incoming humidity and outgoing water. The design should ensure proper hydration of the MEA that in turn determines the fuel cell operating efficiency and durability. The flow channels have a typical dimension of a few tenths of a millimeter to a few millimeters and can be arranged in many different geometric patterns. By optimizing the channel geometry, efficient, passive water removal from the fuel cell can be realized, reducing the need for energy-intensive gas purges [7]. Neutron imaging has played a critical part in describing the correlation between current density and total measured water content and its distribution in operating fuel cells, since the spatial resolution of scintillators easily resolves the channel separation, as shown in Fig. 11.4.

11.2.3.2 Water Dynamics Through the PEM

Fuel cell performance depends on the complex interplay of several chemical and mass transport processes that occur at interfaces. A typical MEA consists of electrodes with catalytic particles on larger particles of a support medium, which are embedded on an electrode. Two electrodes sandwich a PEM. The structure of the electrode involves pores that have a typical length scale of 10–50 nm. For a fuel cell to be efficient, these pores must remain open for reactant transport during the operating cycle.

The first reported fuel cell neutron imaging measurement was carried out more than a decade ago [8]. It was a pioneering experiment attempting to improve the understanding of water mass transport and its effects on the electrochemical performance of PEM fuel cells via detailed knowledge of the

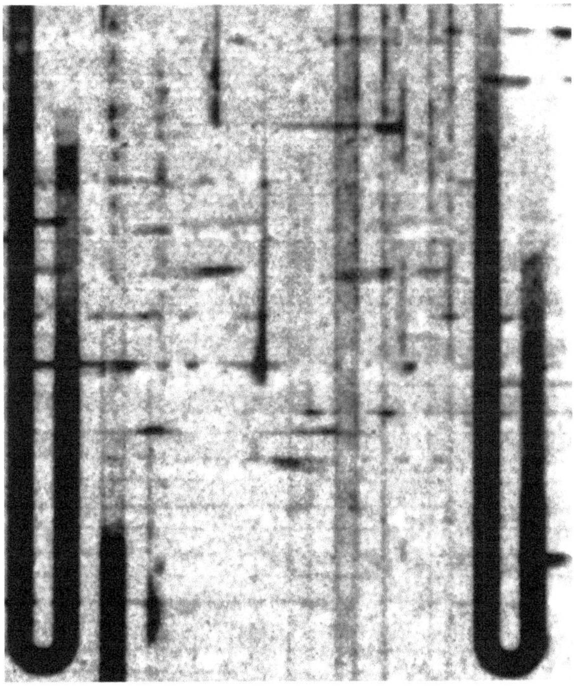

Fig. 11.4 A sub-image of a neutron radiograph of a 50 cm^2 active area fuel cell clearly showing the water content (dark areas) of the separate channels. In this case the anode and cathode channels were oriented perpendicularly to permit discrimination of anode vs. cathode channel water.

through-plane water distribution. The measurements, as shown in Fig. 11.5, involved direct neutron imaging and quantification of the variation of water content normal to the MEA surface cross sections. The resolution and efficiency of the detector used for the experiment were poor and the available neutron flux was low. As a result, the measurement was actually done on a thermally bonded membrane assembly that was nearly ten times thicker than that in modern fuel cells. For commercially competitive materials, a similar measurement requires a detector resolution of 10 μm or better in order to spatially resolve the water distribution. A future generation of neutron imaging techniques and technologies are being developed to achieve this.

11.2.3.3 GDL Characteristics and Water Retention

The structure of the GDL plays a large role in determining the water retention of the fuel cell. Placing a highly hydrophobic microporous layer (MPL) near the cathode catalyst helps to reduce flooding by presenting a barrier to condensed water [9]. Adding PTFE to the bulk GDL substrate can decrease or increase

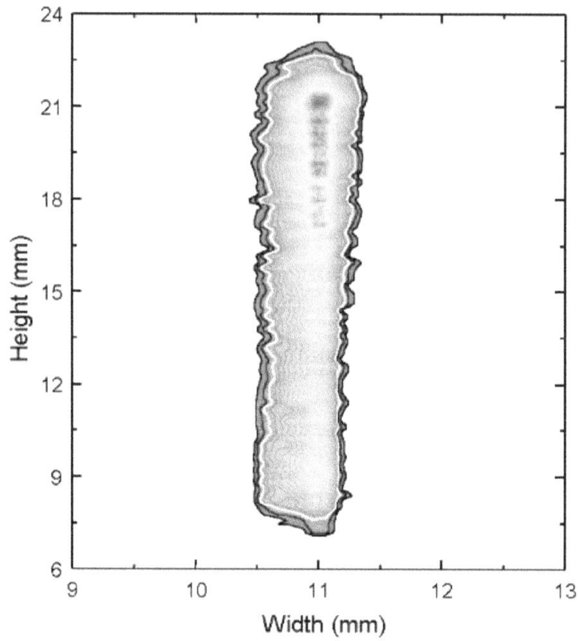

Fig. 11.5 One of the first neutron radiographs of an operating fuel cell, showing the water gradient across a 0.5 mm thick (wide) membrane, comprising several thermally bonded Nafion 117 membranes. The darker region in the membrane center indicates more water

water retention [10]. Having a broad range of pore sizes can ensure that there are sufficient pathways for reactant gas to reach the catalyst layer [11].

11.2.3.4 Characterization of Two-Phase Flow Phenomena in the Flow Field

The water removal from the fuel cell GDL and channels is a difficult process to model since it is a non-isothermal, two-phase flow system. For instance, the waste heat produced in the cathode can drive evaporation so that the fuel cell retains significantly less water at higher current production, despite the corresponding increase in water production [12]. Additionally, the operating pressure and inlet gas flows impact the condensation rate and the convection of water in the flow channels [13].

11.2.3.5 Membrane Durability

One of the failure modes of PEM fuel cells stems from dimensional changes of the polymer layers resulting from repeated drying and wetting cycles or freeze–thaw cycling. Using high-resolution neutron imaging, the effects of freezing on the formation of "frost heaves" between the GDL and MEA can be resolved [14].

Also, it may be possible to study interface defects by neutron phase contrast imaging techniques that are currently being developed (Chapter 8).

11.2.3.6 Materials for Hydrogen Storage

The success of any future hydrogen economy depends, in no small part, on the ability to develop inexpensive, state-of-the-art materials with high hydrogen storage capacity. Such targets require the search for new and improved materials. The utility of a particular material for these applications depends critically on ionic transport. Thus, the fundamental questions that must be answered in order to understand and tailor the performance of these systems are "Where do the ions go?" and "How do they get there?" Many types of materials have been tried or suggested for use as hydrogen storage media. Proper characterization of these materials requires information regarding hydrogen diffusion properties as well as the macroscopic morphologies that govern their useful properties. As shown in Fig. 11.2, neutron imaging can play a key role in assessing device performance.

11.3 Neutron Imaging Facilities and Techniques

11.3.1 Basic Principles

The quality of the image depends on a complex interplay of beam properties, sample material, neutron intensity, and temporal and spatial resolution of the detector system. For hydrogen or water, the primary form of interaction is scattering with a cross section of $0.8\times10^{-22}\,\mathrm{cm}^2$. This is nearly two orders of magnitude larger than that of aluminum, a common material used in fuel cell construction. The total transmission of neutron flux is given by the Lambert-Beer law:

$$\frac{I}{I_0} = \mathrm{e}^{-\sum_i (N\sigma t)_i}, \tag{11.1}$$

where I is the transmitted and I_0 the incident neutron flux, N is the atom number density, t is the sample thickness, and σ is the total scattering cross-section responsible for the neutron loss in the forward direction passing through the sample. Since the primary goal is to image and quantify the water content, all measurements involve the basic steps of taking a dry and wet image of the device. The wet image is normalized with respect to the dry image to determine the profile of the water thickness. The detection system typically consists of a scintillator viewed by either a charge-coupled device CCD or an amorphous silicon flat panel. The spatial resolution, which is a convolution of detector hardware pixel resolution, geometric blur, and scintillator blooming, determines which components of the fuel cell can be identified.

The water volume can be obtained by combining the water thickness from Eq. (1) with knowledge of the area A and the liquid water attenuation coefficient, $\mu_w = N_w \sigma_w$. Since σ_w depends on neutron energy, μ_w must be measured for each beamline or spectrum and detector system used. The total uncertainty in the water thickness, δt_w, comes from both random and systematic contributions. The random uncertainty, from Poisson counting statistics, is the fundamental limit to the measurement accuracy and is estimated by:

$$\delta t_w \approx \frac{1}{\mu_w} \sqrt{\frac{2}{I_0 A T \eta}} \, , \tag{11.2}$$

where I_0 is the incident fluence rate ($cm^{-2}\,s^{-1}$), T is the acquisition time, and η is the detection efficiency [15]. Typical random uncertainties achieved at the national institute for standards and technology (NIST) neutron imaging facility range from 1 to 10 μm.

11.3.2 Imaging Techniques

Both 2D and 3D neutron imaging are used to study water dynamics in fuel cells. Neutron tomography can provide important information about an object's internal construction and about its constituent materials. The neutron tomography technique allows one to reconstruct a 3D image of an object from the 2D neutron radiographs taken for several rotations of the object about a rotation axis (Chapter 6). Using the technique of filtered back projection (FBP), these projection images are Fourier filtered and then back-projected across planes orthogonal to the rotation axis. The result of this process is a stack of 2D reconstructions of the planes of the object, orthogonal to the rotation axis. The result of the FBP is a 3D map of the object's neutron attenuation coefficient. Subtracting the dry map from the wet map yields the 3D water profile of the fuel cell. The spatial resolution of the map is determined by the resolution of the detector system, on the order of 100 μm. Figure 11.6 shows the 3D water content of an operating fuel cell at steady state.

Because of the time required to rotate the fuel cell by 180°, neutron tomography can be used only as a steady state measurement. Often the desired information is the average water gradient going from the anode to the cathode. This "through-plane" water gradient must be imaged with higher resolution than is typical of scintillator-based detectors, since the total thickness of the PEM sandwich is only on the order of 0.5 mm. Various companies are developing new detector technologies aimed at achieving 10 μm or better resolution in neutron detection in the near future [16]. These detectors are based on microchannel plates and have been demonstrated to have resolution better than 25 μm [17] (Chapter 3). With a smaller pixel size, image acquisition time must

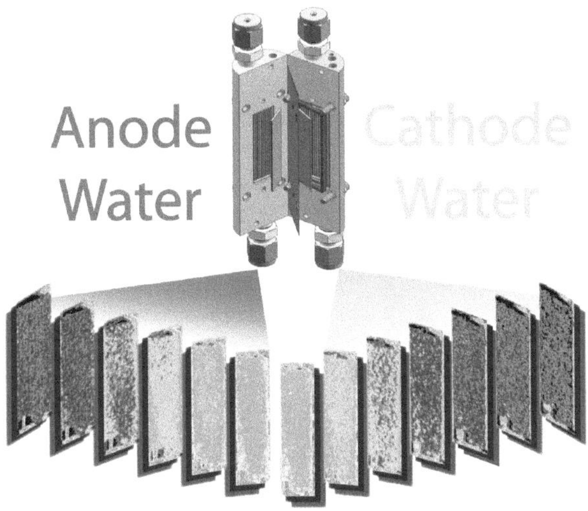

Fig. 11.6 A small fuel cell (4 cm^2) was used to demonstrate neutron tomography. The water distribution in each slice plane from the anode to the cathode shows that anode flooding is occurring in this fuel cell as a result of the thick (1 mm) gas diffusion layers. The data set was acquired within 20 min

be increased in order to maintain signal-to-noise ratios similar to those achieved with scintillators, and a typical time is similar to that of a tomography scan. While a high-resolution through-plane image yields only 2D information, it enables direct measurement of the water gradient in commercial-grade GDLs.

11.3.2.1 Phase Imaging and Other Emerging Techniques

The neutron tomography technique can be complemented by two other related imaging techniques, refraction imaging (Chapter 6) and phase contrast imaging (Chapter 8). The refraction imaging technique uses a mono-energetic neutron beam that is refracted by constituent elements in the sample. The refracted beams from different elements are spatially separated from each other by a very small amount. A single-crystal analyzer redirects this spatially separated beam away from the original direction, and a high-resolution position-sensitive detector detects the redirected beam. Since the refracted beam is now separated from the direct transmitted beam, the inherent background present in normal radiographs is eliminated, and an image with very sharp contrast is formed.

In phase contrast imaging, an image of the locally varying phase shift of a neutron beam caused by the sample is formed. Phase variation may occur because of dimensional and constituent variations, the presence of a magnetic domain structure, the presence of precipitates, or a spatial distribution of stress. The image is recorded with a high-resolution position-sensitive detector. Not only are the bulk properties of the sample measured, but the *elemental distributions* of local inhomogenities are also observable. This technique is typically a few hundred times more sensitive than normal radiography. It is ideally suited for detecting laminate separation and minute thickness variations of polymer electrolyte layers. NIST is developing a cold neutron imaging facility that will enable research to measure the phase gradient, using a variety of such techniques, including a grating method based on the Talbot effect [18].

Ultimately, these technologies can be used as a quality control tool for fuel cell manufacturers. Fuel cells can be subjected to routine scans for the verification of proper assembly and operational characteristics. Additional resolution may be achieved using neutron point sources to magnify the neutron image of the sample.

11.4 Neutron Imaging Facilities for Fuel Cell Research

Successful applications of neutron imaging of fuel cells require access to high-intensity neutron sources and a detection system with fast temporal and high spatial resolution. Also, any large-scale comprehensive research with fuel cell and hydrogen storage devices requires an associated hydrogen support infrastructure capable of safely providing large quantities of hydrogen to these devices. The latter requirement is especially difficult for conventional steady state reactor-based neutron sources. There are facilities in the United States (NIST, PSU), Japan (JAERI, KURRI), and Europe (PSI, HMI, FRM2) that are carrying out fuel cell related research. Among these, the facility at NIST has the most developed and focused infrastructure for hydrogen economy related research. In order to give an idea of the practical considerations, a brief description of this facility follows.

The NIST neutron imaging facility is located at beam tube 2, with a direct view of the reactor core at the NIST Center for Neutron Research; a schematic layout is shown in Fig. 11.7. An aperture of diameter D (variable diameter from 0.1 cm to 2 cm) defines the beam collimation, characterized by the ratio L/D, where $L \approx 6$ m is the distance from the aperture to the detector. A 10 cm section of Bi is used to reduce the gamma-ray and fast neutron background, though it also reduces the thermal neutron fluence rate by about a factor of four from that with no filter. The typical fluence rates used range from about 2×10^6 cm^{-2} s^{-1} to 3×10^7 cm^{-2} s^{-1}. The flight path is evacuated to reduce the neutron scattering in air. The sample position sits directly in front of the detector, and there is a 5-axis manipulation table to align the object and perform tomography. There are several penetrations through the radiation shielding to allow gas lines and electrical connections to be made between the fuel cell and fuel cell test stand. The NIST test stand accurately controls the flow of hydrogen, air, nitrogen, oxygen, and helium over a wide range of flow rates [15]. The maximum current that the test stand can control is about 500 A. The inlet gases are humidified in

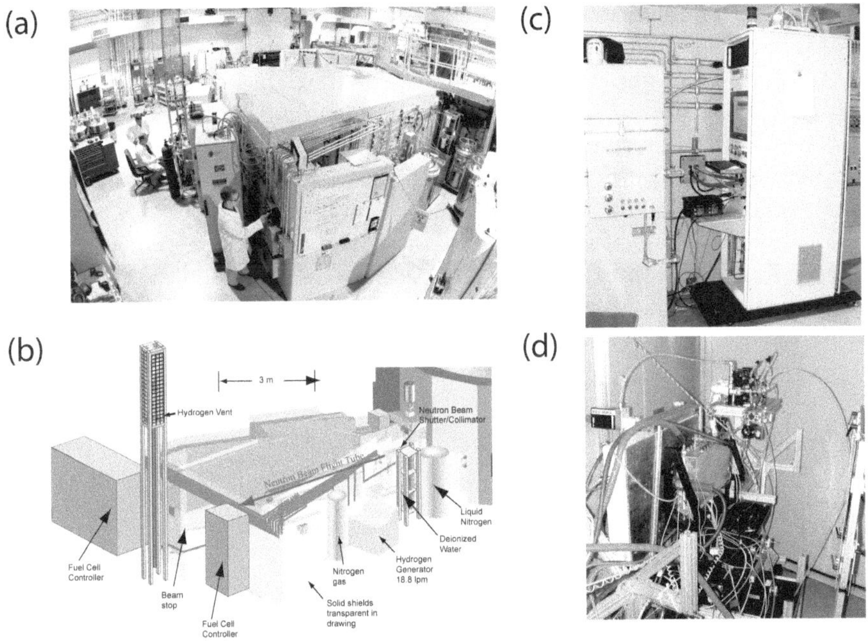

Fig. 11.7 (**a**) Photograph, and (**b**) schematic of the NIST neutron imaging facility. Also shown is (**c**) the fuel cell test stand and (**d**) the sample area with connections to a typical fuel cell

sparger bottles, and heated inlet gas lines prevent condensation. The hydrogen is supplied by a hydrogen generator so there is no stored inventory of hydrogen in the confinement building. There is an extensive set of hydrogen safety features, including several hydrogen sensors, a line pressure sensor, and flame detection in the hydrogen vent that automatically shuts down all hydrogen systems in the event of a detected emergency. Also, there is a freeze chamber, capable of controlling the temperature between –40°C and 50°C for use in freeze/thaw studies.

11.5 Examples of PEM Fuel Cell Neutron Imaging

To demonstrate the capability of 3D neutron imaging, we describe a study of a small PEM fuel cell, with an active area of 4 cm^2 (Fig. 11.6). The tomography was performed with a scintillator-based detector and, in order to measure water gradients, a 1.0 mm thick GDL on the anode and cathode were used, as well as a Nafion 117 membrane. Because of this large thickness, the diffusion through the GDL was significantly slower than in typical fuel cells, which impacts the water content. In particular, the water gradient shows that there is significant flooding in the anode.

Because of the surprising amount of anode flooding, a second experiment using the same small fuel cell was conducted to measure the water diffusion times under different conditions. The fuel cell was operated at 0.4 V with both gas streams humidified to a relative humidity (RH) of 75 % for 2 h to ensure that the GDL saturation was at equilibrium. After this precondition, the fuel cell load was dropped. In the first case, the humidified gas flows were maintained, and in the second case the gas flows were stopped. By so doing, the efficacy of the competing GDL water transport mechanisms of capillary action versus evaporation and advection could be compared. Figure 11.8 shows the first case in which the fuel cell load was stopped but the gas flow was maintained. The water content in the GDL is seen to decline rapidly to the completely dry state within about 15 min. Figure 11.9 shows the second case where

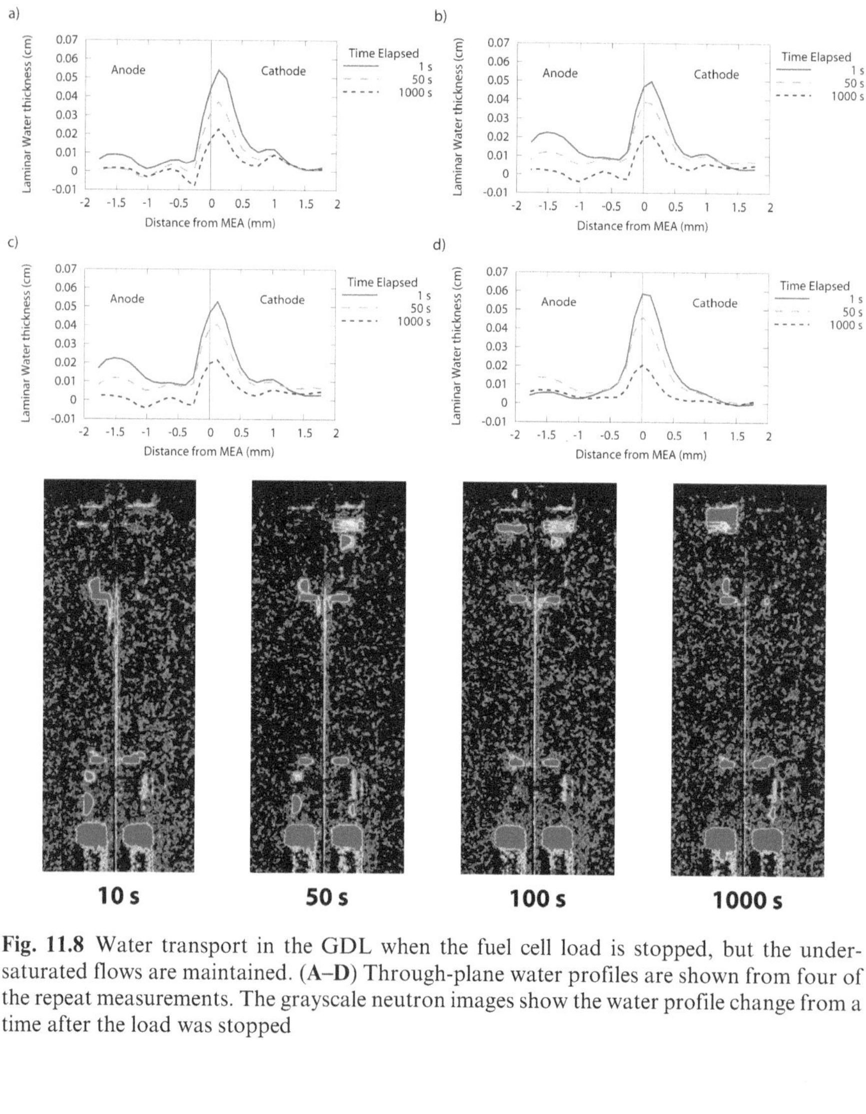

Fig. 11.8 Water transport in the GDL when the fuel cell load is stopped, but the undersaturated flows are maintained. (**A–D**) Through-plane water profiles are shown from four of the repeat measurements. The grayscale neutron images show the water profile change from a time after the load was stopped

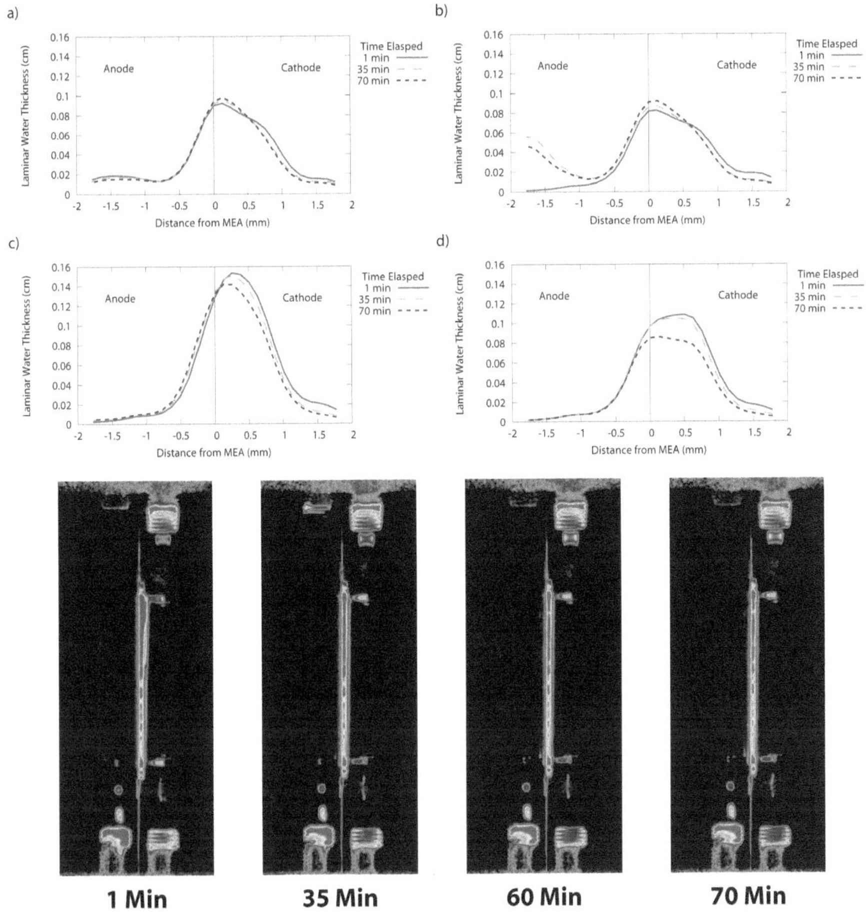

Fig. 11.9 Water transport in the GDL when the fuel cell load and the gas flows are stopped. (**A–D**) Through-plane water profiles are shown from four of the repeat measurements. The grayscale neutron images show the water profile change from a time after the load was stopped. Compared with the evaporation/advection case, the change in water content of the GDL is negligible

both the fuel cell load and gas flows were stopped. The water content of the GDL barely changes, even after 70 min, indicating that capillary action plays a smaller role than evaporation and advection in the water transport through the GDL.

11.6 Conclusion

Neutron imaging has become a powerful tool in the development of hydrogen-based alternate energy devices. It is the only technique that can study the critically important *in situ* water transport in commercial-grade fuel cells.

Today this technique is used by almost all automobile manufacturers and major fuel cell developers. By all accounts, neutron imaging research has shortened the development time of fuel cells and is providing reliable, real-time data important for optimum design and performance that was almost impossible to get only a few years ago. Though the primary focus of neutron imaging research has been on fuel cells during the last few years, it has now started to play an expanded role in the development of hydrogen storage materials and devices. It is expected that this trend will continue in the coming years. In spite of all the success that neutron imaging has enjoyed in recent years, even higher-impact research is limited by the current temporal and spatial resolution of neutron detection devices, the lack of very accurate analytical techniques, and the lack of availability of wide infrastructures for hydrogen research at various neutron facilities around the world. Fortunately, all of these issues are being aggressively pursued by various government laboratories and industrial and academic researchers. It is expected that in the very near future, both temporal and spatial resolution will improve by an order of magnitude. This, coupled with advanced methods such as phase contrast imaging, will provide much needed enhanced capabilities and open new avenues of research that are not possible today. Consequently, in the foreseeable future, neutron imaging will continue to significantly contribute to enhanced efforts to develop and commercialize pollution-free, hydrogen-powered energy devices that are the centerpieces of the emerging hydrogen economy.

References

1. Certain trade names and company products are mentioned in the text or identified in an illustration in order to adequately specify the experimental procedure and equipment used. In no case does such identification imply recommendation or endorsement by NIST, nor does it imply that the products are necessarily the best available for this purpose.
2. J. St-Pierre, *J. Electrochem. Soc.***154**, B724 (2007).
3. P. Bhandari, M. Prina, M. Ahart, R. C. Bowman, and L. A. Wade, Sizing and Dynamic Performance Prediction Tools for 20 K Hydrogen Sorption Cryocoolers in *Cryocoolers 11*, Edited by R. G. Ross, Jr. Kluwer Academic/Plenum Press, New York, pp. 541–549 (2001).
4. M. Prina, P. Bhandari, R.C. Bowman, L. A. Wade, D. P. Pearson, and G. Morgante, Performance Prediction of the Planck Sorption Cooler and initial Validation, *Advances in Cryogenic Engineering*, Vol. 47, edited by S. Breon, et al., Am. Inst. Phys., New York, pp. 1201–1208 (2002).
5. P. Bhandari, M. Prina, R.C. Bowman-Jr, C. Paine, D. Pearson, A. Nash, Sorption Coolers using a Continuous Cycle to Produce 20 K for the Planck Flight Mission, *Cryogenics* **44**, 395–401 (2004).
6. D. Pearson, R. Bowman, M. Prina, P. Wilson, The Planck Sorption Cooler: Using Metal Hydrides to Produce 20 K, *J. Alloys Compounds* **446–447,** 718–722 (2007).
7. J. P. Owejan., T. A. Trabold, D. L. Jacobson, M. Arif, and S. G. Kandlikar, *Int. J. Hydrogen Energy* **32**, 4489 (2007).
8. R. J. Bellows, M. Y. Lin, M. Arif, A. K. Thompson, and D. Jacobson, *J. Electrochem. Soc.* **146**, 1099 (1999).

9. M. Mathias, J. Roth, J. Fleming, and W. Lehnert, 'Diffusion Media for PEM Fuel Cells,' in Handbook of Fuel Cells – Fundamentals, Technology and Applications. *Fuel Cell Technology and Applications*, Vol. 3, W. Vielstich *et al.*(Eds.), John Wiley & Sons, Chapter 46 (2003).
10. R. Mukundan, J. R. Davey, T. Rockward, J. S. Spendelow, B. S. Pivovar, D. S. Hussey, D. L. Jacobson, M. Arif, and R. L. Borup, 'Imaging of Water Profiles in PEM Fuel Cells Using Neutron Radiography: Effect of Operating Conditions and GDL Composition', *ECS Trans.* **11**(1), 411 (2007).
11. K. Yoshizawa, K. Ikezoe, Y. Tasaki, D. Kramer, E. H. Lehmann, and G. G. Scherer, *ECS Trans.* **3**, 397 (2006).
12. M. A. Hickner, N. P. Siegel, K. S. Chen, D. N. McBrayer, D. S. Hussey, D. L. Jacobson, and M. Arif, *J. Electrochem. Soc.***153**, A902 (2006).
13. M. A. Hickner, N. P. Siegel, K. S. Chen, D. N. McBrayer, D. S. Hussey, D. L. Jacobson, and M. Arif, *J. Electrochem. Soc.* **155**(4), B427–B434 (2008).
14. S. Kim, A.K. Heller, M.C. Hatzell, M.M. Mench, D.S. Hussey, D.L. Jacobson, High Resolution Neutron Imaging of temperature-driven flow in polymer electrolyte fuel cells, *American Nuclear Society National Meeting*, Anaheim, CA (2008).
15. D.S. Hussey, D.L. Jacobson, M. Arif, K.J. Coakley, and D.F. Vecchia, *In situ* fuel cell water metrology at the NIST neutron imaging facility, *Proceedings of the ASME Fuel Cell Conference*, New York, June 18–20, 2007.
16. O.H.W. Siegmund, J.V. Vallerga, A. Martin, B. Feller, M. Arif, D.S. Hussey, and D.L. Jacobson, *Nucl. Instrum. Meth. A* **579**, 188 (2007).
17. W.B. Feller, P.L. White, and P.B. White, Gamma Insensitive Highly Borated Microchannel Plates for Neutron Imaging, *Proceedings of the 8th World Conference on Neutron Radiography,* 583–591 (2008).
18. F. Pfeiffer, C. Grünzweig, O. Bunk, G. Frei, E. Lehmann, and C. David, *Phys. Rev. Lett.* **96**, 215505 (2006).

Chapter 12
Material Science and Engineering with Neutron Imaging

D. Penumadu

Abstract This chapter summarizes some of the results related to the use of neutron imaging (radiography and tomography) as applied to the broad area of materials science and engineering research. These include multi-phase flow visualization in metal casting techniques, energy-selective imaging of materials and its use for texture and stress imaging in polycrystalline materials, characterization of discrete particle systems, flow through porous media, and stroboscopic imaging. The importance of spatial resolution and neutron detector type for given engineering applications is also addressed.

Keywords Metals · Composites · Granular · Flow · Pyrolysis · Casting · Numerical · Modeling · Imaging · Neutrons · MultiPhase · Functional · Structural · Materials

12.1 Introduction

Film and image plate-based thermal neutron imaging for nondestructive inspection has found many useful engineering applications in the past three decades, as demonstrated by the establishment of an American Society for Testing and Materials (ASTM) standard (E1316) for its routine implementation at various research reactors (Chapter 5). However, in the past decade the use of thermal and cold neutrons for material science and engineering applications has rapidly evolved due to a dramatic improvement in the availability of large-area amorphous silicon flat panels, high-resolution (2048×2048) and large charge-coupled device (CCD) chip-based digital cameras, sophisticated and inexpensive optical lenses, and bright and thin (50–200 μm) neutron scintillation screens. This has

D. Penumadu (✉)
University of Tennessee-Knoxville, Joint Institute for Advanced Materials (JIAM)
Chair of Excellence, Department of Civil and Environmental Engineering, Knoxville, TN 37996-2010, USA
e-mail: dpenumad@utk.edu

I.S. Anderson et al. (eds.), *Neutron Imaging and Applications*,
Neutron Scattering Applications and Techniques, DOI 10.1007/978-0-387-78693-3_12,

also enabled imaging of materials in three dimensions using tomography, which requires obtaining a series of radiographic projections of the object as it is rotated in fixed angular steps through 180° (Chapter 6).

Imaging using X rays has had an impact on many fields of science and engineering and is now used routinely for studying materials. There are many user facilities in the United States and Europe that can provide a significant number of options for both source and detector configurations for various types of imaging possibilities. In general, the imaging includes the following modes of operation: transmission, diffraction, fluorescence, spectroscopy, and polarization. In transmission mode, the attenuation or phase shift of the X-ray beam by the sample is measured. Absorption contrast is then used to map the sample density or features of interest. Particular elemental constituents can be located using measurements on each side of an absorption edge to give an element-specific difference image with moderate sensitivity. Phase-contrast imaging can be sensitive to internal structure even when absorption is low and can be enhanced by tuning the X-ray energy. In X-ray tomography, one of these modes is combined with sample rotation to produce a series of two-dimensional projection images to be used for reconstructing the internal three-dimensional structure of the sample. This is particularly important for observing the morphology of microstructures. Many textbooks are currently published dealing with the vast amount of literature on this rapidly evolving topic. These are too extensive to cite in this chapter.

Imaging with neutrons is an evolving and powerful imaging technique for the internal evaluation of materials that complements the X-ray technique very well. The interactions of neutrons and X rays with matter are fundamentally different, and this difference forms the basis of many unique and complementary applications. As is discussed in Chapter 11, the sensitivity of neutrons to hydrogen and other unique properties make it an ideal probe for characterizing the structure, morphology, and hydrogen content in the development of next-generation fuel cells and hydrogen-storage materials. A comparison of the interaction of X rays and neutrons with matter has already been covered in earlier chapters; the following particular observations are important for material science and engineering applications:

- sensitivity to light elements and contrast for neighboring elements in the periodic table,
- variation of attenuation coefficients as a function of wavelength for thermal and cold energy spectrum,
- the ability to distinguish isotopes (for example hydrogen and deuterium),
- large penetration distance for many commonly used structural elements,
- the ability to detect small amounts of water, and
- interaction with magnetic moments.

The complementary aspect of imaging with both neutrons and X rays is demonstrated in Fig. 12.1 for a computer floppy disk. The option of using both X rays and neutrons now exists at recently established neutron imaging facilities

(a) (b)

Fig. 12.1 Radiograph of a computer floppy disk using neutrons (**a**) shows the polymeric components clearly while X-rays (**b**) are sensitive to the metallic components

such as that at the Paul Scherrer Institute in Switzerland (http://neutra.web.psi.ch/).

As is discussed in other chapters in this book, the use of neutron imaging is finding important applications for nondestructive evaluation (NDE) inspections, studying water flow management and coupled performance of fuel cells, studying structural materials such as steel and concrete, many applications in geology and paleontology, studying working engines using stroboscopic imaging, and more recently in biomedical and agricultural engineering. This chapter summarizes the use of neutron imaging for additional novel material science and engineering problems.

12.2 Phase Change and Transport During Metal Casting

Casting is one of the oldest methods of manufacturing, dating back to 3000 B.C., and is often used because of its ability to produce large and durable volumes of complex shaped objects with desirable features such as internal cavities and hollow sections. The lost foam casting (LFC) method, also known as expandable polystyrene (EPS) casting, began high-production application in the early 1980s. The LFC process has unique advantages for metal casting compared with competing technologies. It has no size limit and the near-net-shaped casting products can be obtained; complex shapes can be cast in one piece by using EPS foam patterns glued together. It is starting to be widely used by many casting companies as a replacement for conventional processes such as green sand and investment castings, especially in the automobile industry. For example, General Motors has nine different parts made using LFC, including aluminum cylinder head, aluminum cylinder block, aluminum-driven sprocket support for automatic transmission, iron crankshaft, iron-differential case, and iron clutch housing for automatic transmission. Castings from this process can have surface finishes from 5 to 20 μm; dimensional accuracy is better than traditional sand castings. Almost no further modification is required after the LFC process, the machining and assembly being eliminated or reduced [1]. Unlike other casting processes, no parting lines, cores, or riser systems are needed. The process is inexpensive, with the polystyrene, sand, and containing units being relatively cheap. Compared with conventional sand casting using resin-bonded sand, LFC reduces environmental waste because it can use unbonded and recycled sand.

Two major steps are involved in the LFC process. The first is to produce a polymeric foam pattern typically using EPS. The second step is the process of casting, i.e., replacing the polymer pattern by molten metal. After the coated polymer pattern is ready, it is placed in a mold around which silica sand is poured to fill all the cavities. Molten metal is then introduced into the mold through the gating, causing the polymer foam pattern to decompose into gaseous and liquid pyrolysis products which escape into the loose sand. The molten metal flow fills the space originally occupied by the foam pattern, thus yielding the target shape. The casting product is obtained by removing the sand and refractory coating. During the process of metal filling, a gas layer, also called the pyrolysis front, exists between the metal and the polymer fronts; it mainly consists of the thermal degradation products of styrene. The behavior of the pyrolysis front during the casting process is important and has a significant impact on the quality of the casting products [2]. However, the pyrolysis front is difficult to identify using X-ray-based radiography techniques and consequently has not been studied in the past.

Radiography has long been considered a valuable tool for detecting internal features and flaws. Shrinks, blowholes, gas or slag inclusions, and other flaws could be identified quickly using X-ray radiography, and gating and risering appropriately adjusted [3]. Traditional radiography using X rays has also been successfully used in practice to image aluminum flow to aid gating design and turbulence reduction and determine flow patterns [4, 5]. However, it is also crucial to characterize the behavior of the pyrolysis fronts of polymer foam patterns as they recede during the LFC process. X-ray radiography has difficulty in visualizing the polymer foam and pyrolysis products because they have low attenuation, but for neutron imaging they give high attenuation. Then other issues involved in LFC, such as the effect of the properties of the refractory coatings and unbonded granular materials on the filling behavior of molten metal flows, can also be investigated. In addition, an in-situ quality control approach for LFC can be developed based on neutron imaging.

The macroscopic cross section and the calculated thickness for 50% neutron attenuation of the typical materials involved in LFC are listed in Table 12.1. This shows that there is remarkable difference in the attenuation between aluminum and styrene, which creates the contrast mechanism in neutron radiography for studying the LFC process *in situ*.

Table 12.1 Material properties and the calculated thickness for 50% attenuation for thermal neutrons

Material	Chemical formula	Density (g cm^{-3})	Molecular weight (g mol^{-1})	Macroscopic cross-section (cm^{-1})	Thickness for 50% attenuation (cm)
Aluminum	Al	2.70	27	0.0984	7.05
Styrene	C_8H_8	0.9	106	3.12	0.22
Water	H_2O	1.0	18	2.62	0.26
Silica	SiO_2	2.65	60	0.259	2.68

Neutron radiography experiments were performed at the research facility located in the McClellan Nuclear Radiation Center (MNRC) in the United States [6]. As shown in Fig. 12.2, MNRC has a two megawatt TRIGA reactor with four neutron beam tubes all tangential to the reactor core. The neutron beam is approximately 9 inches (22.5 cm) in diameter and has an intensity of approximately 10^7 n cm^{-2} s^{-1}. The neutron spectrum is highly thermalized, which

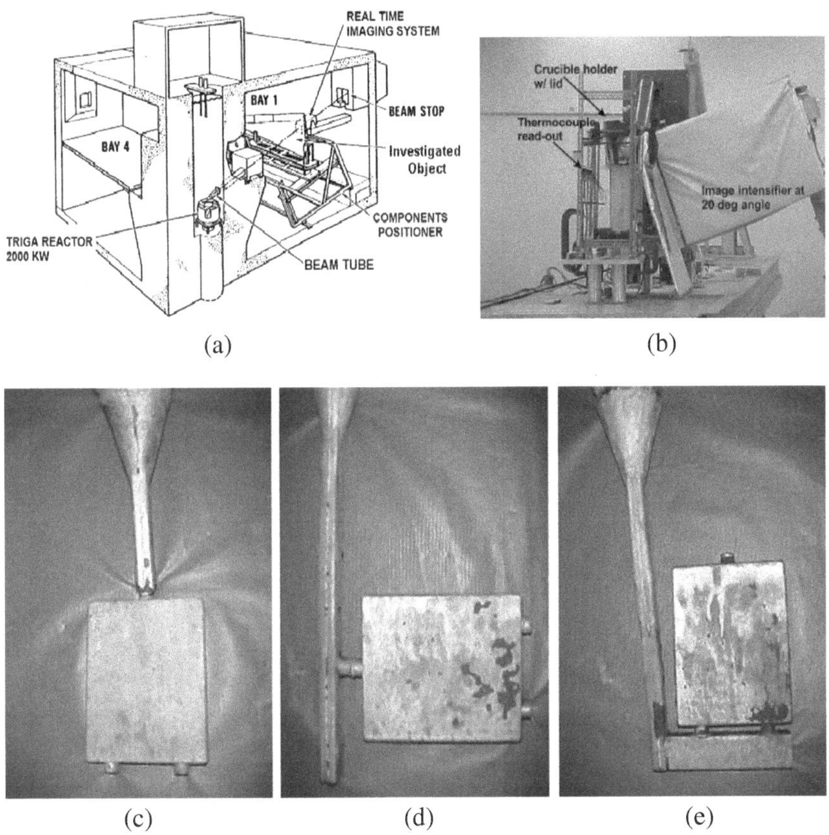

Fig. 12.2 (**a**) Neutron radiography facility at MNRC. (**b**) Casting setup and image intensifier. Three types of gating systems: (**c**) Top-gating, (**d**) side-gating, and (**e**) bottom-gating

allows high-quality imaging. The neutron beams are collimated with an L/D ratio of 50–400, resulting in good spatial resolution for digital still radiographs. The radioscopic systems use a CF Thompson tube as the neutron camera, which combined with a SIT (silicon intensified tube) camera allows image capture at a rate of 30 frames/s. The radiography setup for LFC is shown in Fig. 12.3. The polymer foams used in the study are EPS foams with a density of 22.4 kg/m^3. The casting metal used in the study was aluminum alloy. This was heated to 1400° F/760°C before pouring into the EPS patterns. Real-time neutron radiography was used to visualize the molten metal and pyrolysis foam fronts during the whole LFC process. The casting samples were plates of size 100 × 150 mm with varying thicknesses (4, 12, and 24 mm), and various gating orientations (bottom, side, and top), as shown in Fig. 12.2.

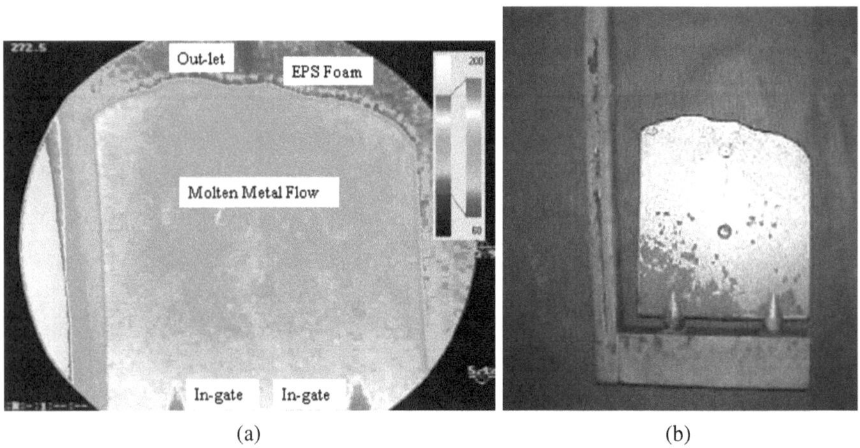

Fig. 12.3 Bottom-gating casting with pattern thickness of 4 mm. (**a**) Neutron radiograph (at 19 s). (**b**) Optical picture of the final casting

Real-time digital videos were recorded during the entire process of metal entry, foam decomposition, and casting solidification. Specific images of interest were then extracted from the digital videos, further processed, and analyzed using Image-Pro Plus V4.5 [7]. Figure 12.3 shows an example neutron radiograph of a bottom-gated casting that did not fill completely at 19 s since the hot aluminum metal was introduced to the down-sprue. There is a wealth of information in such radiographs as a function of time that is valuable for understanding the pyrolysis aspects of foam and the associated phase transitions of both polystyrene (from foam to liquid to gaseous monomer) and metal (liquid to solid state).

Only neutrons provide this outstanding ability to visualize and model complex multiphase and multiscale processes, penetrating metal and at the same time clearly showing light elements such as the byproducts of polymer degradation. The neutron images shown in Fig. 12.4 indicate the effect of foam thickness on the degradation process during LFC. It is of great interest to note that for thinner

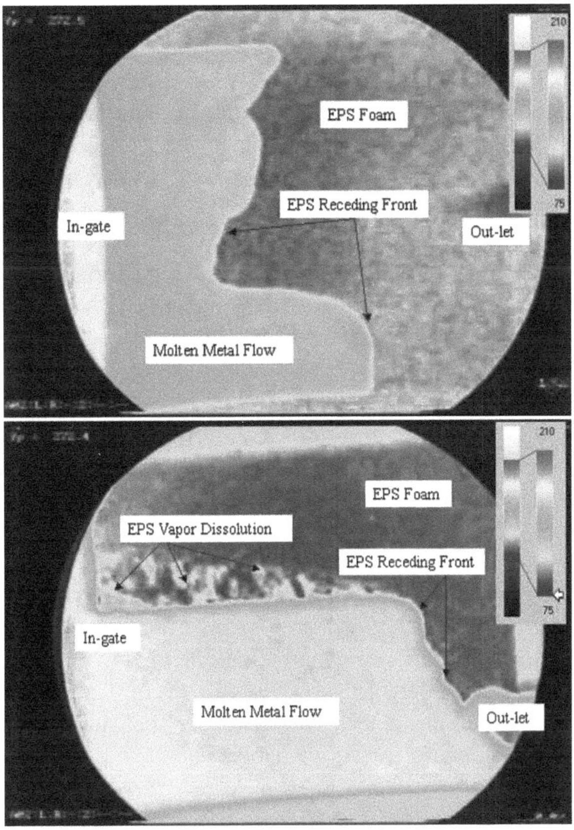

Fig. 12.4 The effect of EPS foam pattern thickness on degradation; 4-mm thick plate (*top*) and 24-mm thick (*bottom*)

foam patterns, one does not see diffusion pathways, as was noted for thicker foam, and often thicker castings have defects from incomplete pyrolysis of the foam in certain regions. Data of this nature are of profound significance for understanding the physics of the problem related to multiphase dynamic flow and transport. These data can be used to develop suitable engineered solutions to produce near-net-shaped castings with lower scrap rate. The spatial and temporal information on a given phase during casting solidification is valuable for modeling and verification of the dynamic process using computational fluid dynamic (CFD) codes such as Flow3D® and Magma®.

12.3 Particulate Materials and Porous Media

The stress–strain relationship of particulate materials is complex and depends on the initial state of packing, the past stress history, and the applied stress path (compression, shear, or extension). This section illustrates the development of an experimental approach using neutron imaging to provide a fundamental understanding of the macroscopic stress and strain (including the meaning of the terms "stress" and "strain" in particulate materials) and its relation to the microstructure (particle contacts and void size distribution). Images were acquired using the CONRAD instrument at the Berlin Neutron Scattering Centre in tomography mode with a white beam cold source [8].

The response of a granular assembly to a global applied force (usually applied via a loading platen) is an important topic of interest in the study of particulate materials. The commonly observed global stress–strain response of granular materials must be related to the local force and displacement variations at the scale of the particles. For continuum materials, the stress–strain relation can be well described macroscopically, based on continuum mechanics. However, for a granular assembly it is well known that the macroscopic stress–strain relation is quite complex and depends on the initial state of the assembly (local and global porosity and particle coordination numbers), the past stress history, and the loading path. Granular assemblies show anisotropy, inhomogeneity, and nonlinearity in the distribution of both the contact network and the contact forces among particles (see references [9–12]). Typical X rays (from table-top sources rather than synchrotrons) have difficulty in penetrating silica sand for more than a few millimeters and are not sensitive to the degree of saturation with water. Neutrons are ideal for such applications, where the intent is to characterize the initial geometric arrangement of particles and its evolution as a function of applied stress.

Figure 12.5 shows images of two slices of a silica sand core, obtained from a tomographic reconstruction using a high-resolution neutron imaging setup. It is

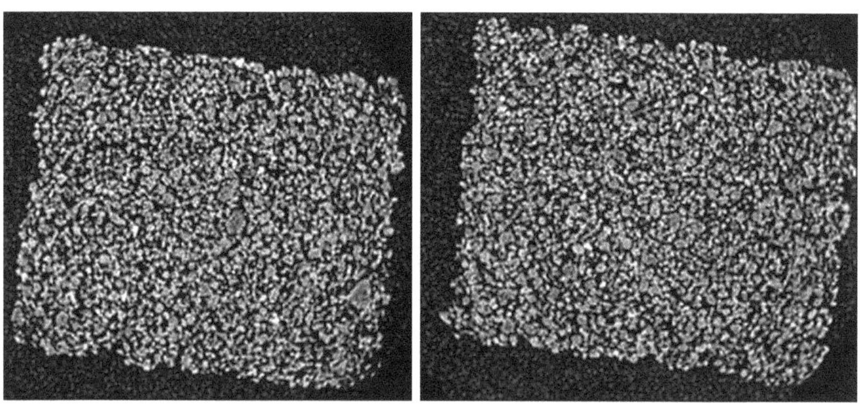

Fig. 12.5 Neutron images of a sand core specimen ($\sim 1 \times 1.24$ cm)

relatively easy to identify individual grains; the particles shown are mostly smaller than 800 μm. The sand core, obtained from a foundry, was produced by a specialized technique which utilizes a vaporized or gaseous chemical that operates as a catalyst or co-reactant in order to cure (polymerize) a resin binder film. The resin is spread uniformly over the surface of the sand and compacted into a mold (rectangular in the study illustrated). When the coating has fully cured, each sand grain is firmly joined to the others in contact with it. An understanding of the stability and durability of sand cores is important for the foundry industry, but a realistic fundamental model of core behavior from particle scale mechanics and numerical modeling is still a dream. Neutron imaging offers the unique possibility to provide actual information on particle contacts and voids with appropriate resolution. Using suitable thresholding techniques with advanced image processing algorithms, it will in the future be possible to describe the three-dimensional void and solid particle fabric and its evolution as a function of externally applied mechanical or thermal stress.

Many practical issues, such as the fate of contaminants in the sub-surface of the earth, or of nutrients for plant systems (Chapter 17), can benefit from information that is uniquely obtainable from neutron imaging. Using H_2O and D_2O alternately, or in clever miscible modes, neutron imaging experiments can be designed to extract information on porous media filled with fluid and gas phases, and the arrangement of solid phases, with relatively good precision and ease. They can also provide much needed information for the study of partially saturated porous media and suction characteristic curves that are presently unresolved.

The micromechanical analysis of granular assemblies using computer simulations, through numerical methods such as the distinct element method (DEM), can be evaluated by comparing the results for discrete particulate model systems with carefully controlled neutron imaging experiments that involve a mechanism to apply a known state of stress or strain using a portable loading system. For simulations, the material microproperties such as particle shear and tangential stiffness can be evaluated from nanoindentation test results on individual particles. The particle contact behavior, together with the nature of force chain structure and other related micromechanics associated with one-dimensional compression on granular materials having different shape and mechanical properties, is then analyzed in conjunction with experimental data generated using both neutron diffraction and imaging. The goal of the DEM is not to match the macroscopic response but to use the fundamental properties of the particles to predict the behavior of the assembly, and to evaluate the basic mechanisms contributing to the strength and deformation behavior of particulate systems in terms of particle force-chain structure (Fig. 12.6).

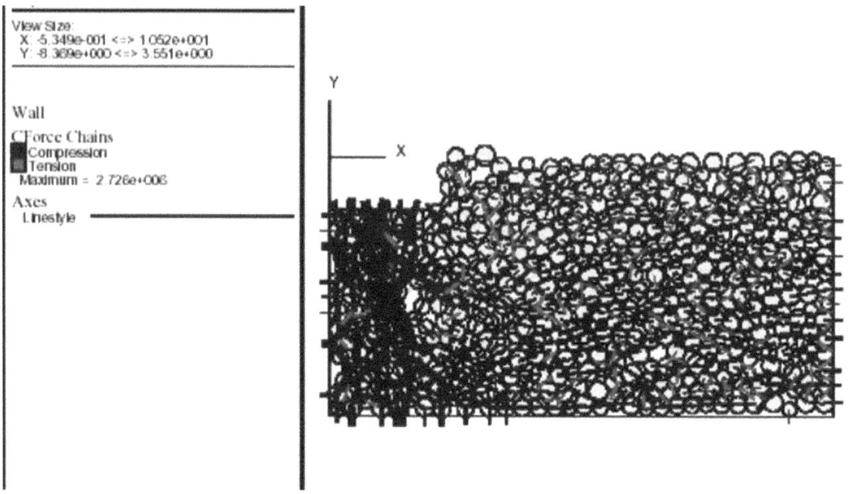

Fig. 12.6 Example force distribution of a granular assembly under a punch obtained using a particle flow code

12.4 Stroboscopic and Dynamic Neutron Imaging

There are some common engineering applications that require continuous fast imaging with neutrons to obtain detailed information related to a fast process. To solve such problems, major difficulties must be overcome related to the low number of neutrons detected in one measurement time frame and the number of pixels that can be read. To meet this challenge requires a high-flux neutron source, low decay time neutron scintillation materials, and the use of stroboscopic imaging for fast but periodic processes. Using the stroboscopic approach, the number of neutrons in one time window is still very low, but many exposures of the same time window of a periodic process may be accumulated on the detector before readout, thus increasing the available intensity (though not improving the signal-to-noise ratio).

The stroboscopic technique can be illustrated by a study of fuel injection and oil flux in a combustion engine. A commercial combustion engine was stripped of the intake and exhaust system and coupled to a 2 kW electric motor with a transmission belt [13]. The spark plugs were also removed, so that the engine would not generate compression in its cylinders. The electric motor was capable of turning the engine at low rotation speeds and, due to the reduced heat production, the water cooling could be omitted. A Hall sensor close to the transmission wheel on the camshaft generated a pulse which was used to trigger the image intensifier of a cooled CCD camera. The trigger signal could be delayed in order to shift the time window within the cycle of the engine. Measurements were made at two facilities: the Neutrograph facility at the Institut Laue-Langevin (ILL) reactor in Grenoble with a high thermal neutron flux of 3×10^9 n cm^{-2} s^{-1}and low L/D value of 170, and the ANTARES facility at the FRM-II reactor of the Technische Universitat Munchen with a lower thermal neutron flux of 1×10^8 n cm^{-2} s^{-1} but a high L/D value of 400. At ILL, the detection system was a MCP-intensified CCD camera with a Peltier-cooled chip (1300×1024 pixels) with 16 bit digitization.

The full cycle of the four-stroke engine running at 1000 rpm was split into 120 individual frames over two rotations; 150 individual images were recorded as an on-chip accumulation of 200 ms exposure each. In this way, the total exposure time for the full run was only of order 18 min. The field of view was 24×24 cm. At FRM-II the measurements were repeated with an Andor MCP intensified CCD camera with 1024×1024 pixels. Because of the lower available flux, the time window was extended to 1 ms and the rotation speed was reduced to 600 rpm. The images thus show more motional blurring but better definition of the stationary parts. Figure 12.7 shows typical results that demonstrate the differences in spatial and temporal resolution for two facilities related to stroboscopic imaging. As can be seen, the lubrication of a running engine can be visualized; this can have important applications related to the design and modeling of a future automotive powertrain.

Possible options for detection and triggering systems for short-time stroboscopic neutron imaging include LCD shutters, interline CCDs, CCDs with a gated image intensifier, and electronically shuttered CCDs [14]. Stroboscopic neutron radiography with a time resolution down to a few microseconds, as well as dynamic tomography on a sub-minute time scale, can allow the visualization of thick and strongly absorbing materials [15], thus providing the ability to also investigate large samples with extremely low contrast as required in some fields of engineering.

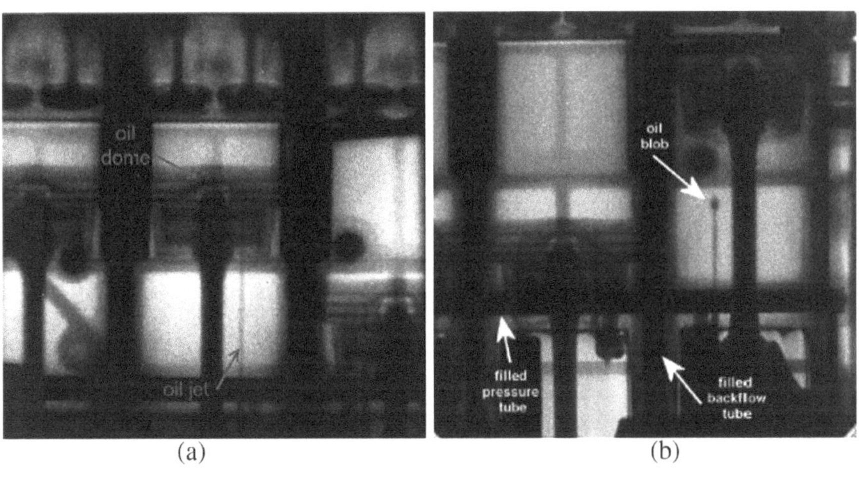

(a) (b)

Fig. 12.7 Dynamic radiography of a running engine, (**a**) acquired at the Neutrograph facility and (**b**) acquired at the ANTARES facility [13]

The requirements for dynamic imaging are clearly more severe since the number of neutrons recorded in a single time frame is very low, even for the highest flux sources, so a low background is extremely important as well as a fast frame rate. A recently developed prototype detector for this type of application is based on electron multiplying charge-coupled device (EMCCD) technology [16]. This CCD technology evolved from the original development of the high-resolution γ-ray imaging detector by RMD [17], which uses a back-illuminated, thermoelectrically cooled, 512 $\times$ 512 pixel EMCCD as a readout sensor coupled to a LiF/ZnS:Ag scintillator. The scintillator screen is 225 μm in thickness and provides bright emission for short exposure times (2 ms) with an imaging area of $\sim$ 25 $\times$ 25 mm^2 and an effective pixel size of $\sim$ 50 μm. The camera permits a minimum of 2 ms data acquisition and can provide 30 frames per second (fps) imaging at full pixel resolution. In binned mode, the frame rate can be increased to 225 fps. While the internal gain effectively offsets read noise, reducing it to $< 1\ e^-\ pixel^{-1}\ s^{-1}$, the dark noise is also minimized to $< 1\ e^-\ pixel^{-1}\ s^{-1}$ by cooling the EMCCD to –35°C.

Using this detector, dynamic imaging in integrated mode was conducted at the National Institute of Standards and Technology reactor facility using the thermal neutron port BT2. Figure 12.8 shows neutron images of water droplets from a plastic bottle with a small hole, imaged at 30 fps to compensate for the low flux and relatively low detection efficiency of the LiF/ZnS:Ag screen. It is important to consider the scattering effects for such images using tools such as

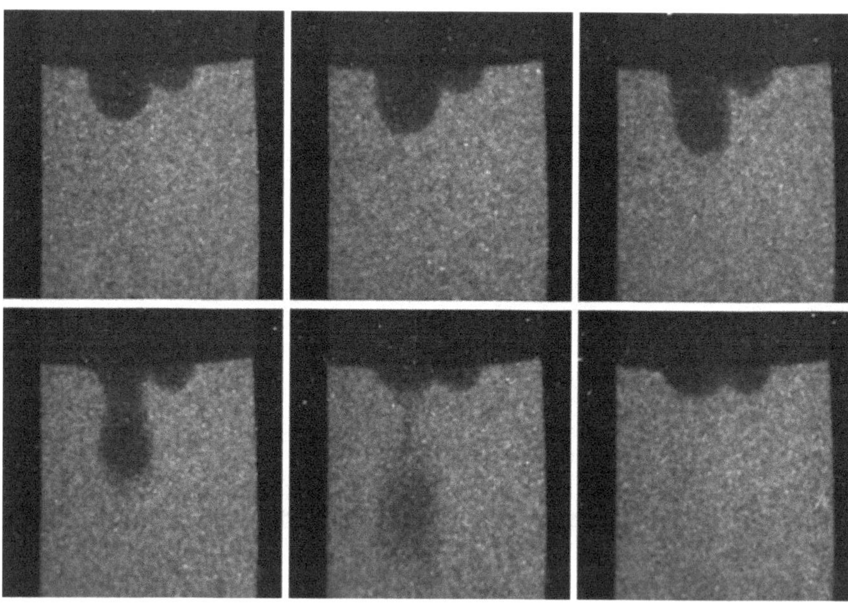

Fig. 12.8 Dynamic imaging (single frames) of a water droplet using the EMCCD detector

the Monte Carlo method to extract contrast from linear attenuation associated with absorption versus scattering.

The ability to use EM CCD technology now opens up many possibilities for pseudo real-time neutron imaging applications that require the study of flow or transport of fluids through porous media or channels. The following are examples:

- optimization of the channel geometry for microfluidic-based devices for thermal management of high-performance electronic chips and related applications,
- optimization of catalytic converters,
- understanding of physico-thermal issues related to the conversion of low-yield hydrocarbon-rich geo-materials for next-generation oil extraction,
- study of the environmental degradation of polymeric composites due to moisture and humidity exposure, and
- development of novel drainage materials for hydraulic management and filtration applications.

12.5 Residual Stress Imaging

Residual stress is one of the more common phenomena occurring in materials. It arises during processes such as machining, grinding, rolling, heat treatment, and welding. Residual stress can be defined as self-equilibrating internal stress existing in a free body without constraints or external force added on the boundary, or as an elastic response of the material to an inhomogeneous distribution of inelastic strains such as plastic strains, precipitation, phase transformation, misfit, or thermal expansion strain. It can originate from shape misfit between the unstressed shapes of different parts, regions, or phases of components. Residual stress has a significant impact on the mechanical behavior and durability of materials. It is not immediately apparent in many cases, as it is difficult to measure or predict, and can lead to premature failure of materials if not considered in the design.

The application of neutron diffraction to measure/image stresses and strains in metallic and other materials under *in situ* and *ex situ* loading conditions is now well established and has found important applications in many fields of engineering (see, e.g., [18]). The approach is to use a fixed-wavelength neutron source (typically at a reactor) or a multi-wavelength pulsed source (typically at a pulsed spallation source) to obtain a diffraction pattern for a gauge volume ($\Delta x \Delta y \Delta z$) which is typically in the range of 0.5–1000 mm^3. Analysis of the diffraction pattern, using single Bragg peak fit or full pattern Rietveld refinement, gives an accurate value for the inter planar spacing, d, from which the lattice strain ε can be determined:

$$\varepsilon_{\psi\phi}(x,y,z) = \frac{d_{\psi\phi}(x,y,z) - d_0}{d_0}\,, \tag{12.1}$$

where d_0 is the inter planar spacing under stress free conditions. ε is then determined as a function of position (x,y,z) and orientation (ψ,φ) in the specimen by translation/rotation through the neutron beam, and the strain tensor

$$\begin{bmatrix} \varepsilon_{xx} & \varepsilon_{xy} & \varepsilon_{xz} \\ \varepsilon_{xy} & \varepsilon_{yy} & \varepsilon_{yz} \\ \varepsilon_{xz} & \varepsilon_{yz} & \varepsilon_{zz} \end{bmatrix} \tag{12.2}$$

at each location can be determined from measurements in several orientations (ψ,φ) (a minimum of six in the general case) by applying the relation

$$\begin{aligned} \varepsilon_{\phi\psi} &= \varepsilon_{xx}\cos^2\phi\sin^2\psi + \varepsilon_{xy}\sin 2\phi\sin^2\psi \\ &+ \varepsilon_{yy}\sin^2\phi\sin^2\psi + \varepsilon_{zz}\cos^2\psi \\ &+ \varepsilon_{xz}\cos\phi\sin 2\psi + \varepsilon_{yz}\sin\phi\sin 2\psi \quad . \end{aligned} \tag{12.3}$$

The stress tensor is then derived from the strain tensor using Hooke's law.

A particular advantage of neutrons for residual stress imaging is the ability to penetrate deep into large components, nondestructively, as illustrated in Figs. 12.9 and 12.10. Hard (high energy) X rays can now offer similar penetration (depending on the material) but the relevant diffraction angles are then very small, so the gauge volumes are "needle-like," which is problematic for

Fig. 12.9 Wingspar from an Airbus A340 aircraft mounted for stress imaging on the Engine-X diffractometer at the ISIS spallation neutron source

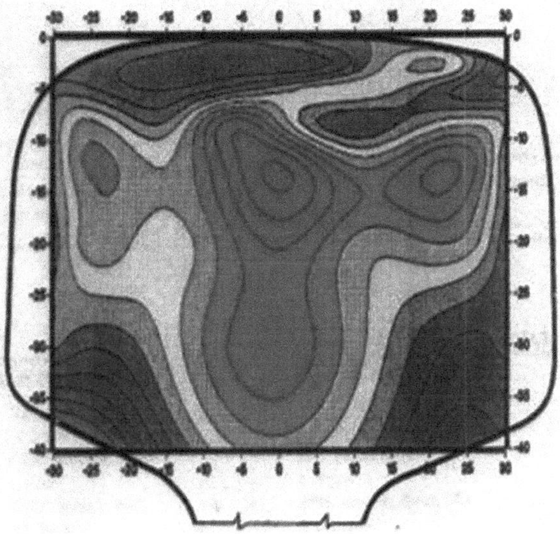

Fig. 12.10 Image of the stress distribution in the head of a railway rail (see e.g., [19])

measuring one component of the stress. The two techniques are therefore highly complementary and are increasingly used in combination.

12.6 Bragg Edge-Based Energy Selective Neutron Imaging

The principles of energy-dispersive and Bragg edge neutron imaging have been described in Section 4.1 in Chapter 6 of this book. When these are fully implemented at pulsed neutron sources, they will enable new and exciting applications of energy-selective neutron radiography and tomography for various aspects of materials science and engineering. However, there are significant challenges for detector development since both high spatial resolution and high time resolution (to measure the neutron wavelength via its time of flight) are required; to refine the position of a Bragg edge to an accuracy of 5×10^{-5}, equivalent to 50 microstrain, needs a time resolution of the order of 10 μs. Current detectors with an acceptable spatial resolution of ~50 μm and reasonable duty cycle can only achieve ~100–1000 μs. Also, no beamlines have yet been built that are optimized specifically for this type of measurement, but some experiments have been carried out that fully demonstrate the potential of the technique.

The transmission cross section as a function of wavelength shown in Fig. 1.4 of Chapter 1 is for an "ideal" polycrystalline iron powder. In a "real" solid metal component, the sharp Bragg edges can either be shifted due to residual stress, or change in height due to texture (nonisotropic distribution of crystallite orientation). If the edge shifts can be measured with sufficient spatial and wavelength resolution, then they can be used to image the strain or stress, as is shown in Fig. 12.11.

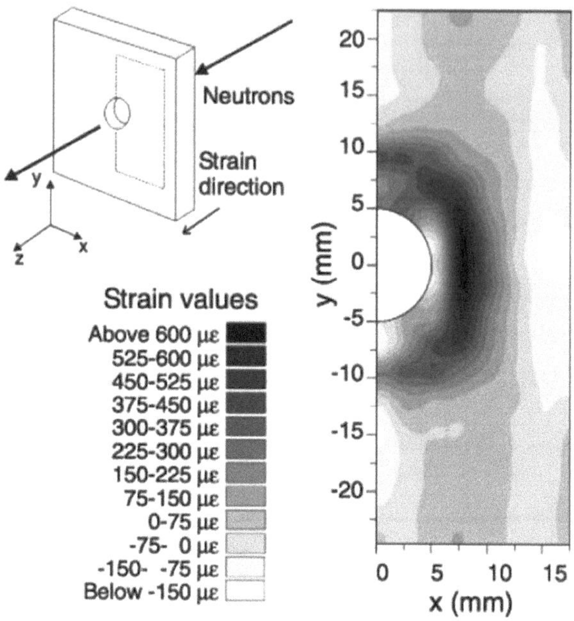

Fig. 12.11. Image of the residual elastic strain distribution around a cold expanded hole in a 12-mm thick steel plate. The measured strains correspond to the through-thickness average of the out-of-plane strains. The in-plane spatial resolution $\sim 2\times2$ mm^2. [20]

In practice there can often be a combination of edge shift and height change, averaged over the crystallites within the gauge volume, leading to significant broadening and distortion of the sharp Bragg edges. Two samples of the same material can therefore have quite different transmission spectra, which can be used as a signature of the method used to create the samples, as illustrated in Fig. 12.12.

The Bragg edges also obviously differ between the different materials that might make up a macroscopic system (Fig. 12.13). The ratio of images measured below and above a Bragg edge for one of the component materials gives a good contrast for that material, while the other material effectively disappears (provided that it does not also have a Bragg edge in the same wavelength region).

In principle, full tomographic imaging can be carried out as a continuous function of wavelength, with the possibility to, then, reconstruct images with highlighted contrast to illustrate stress, texture, composition, or even complex combinations. Currently, the available neutron fluxes and detector technologies mean that the resolution has to be relaxed in some respect to achieve reasonable images; for example, either positional resolution is relaxed to achieve high time resolution for high Bragg edge contrast, as in Fig. 12.13, or wavelength resolution is relaxed to achieve higher spatial resolution, as in Fig. 12.14. However, with the advent of higher-flux neutron sources, such as

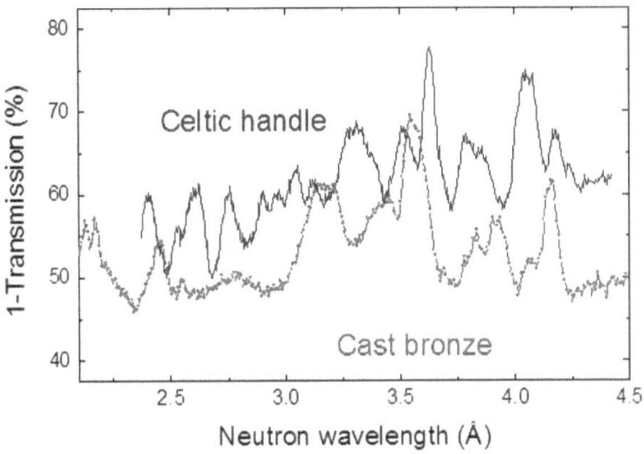

Fig. 12.12 Neutron transmission spectra of the following two bronze samples: one an archaeological specimen and the other a cast piece. The significant differences indicate that the "Celtic handle" has been worked after casting [21]

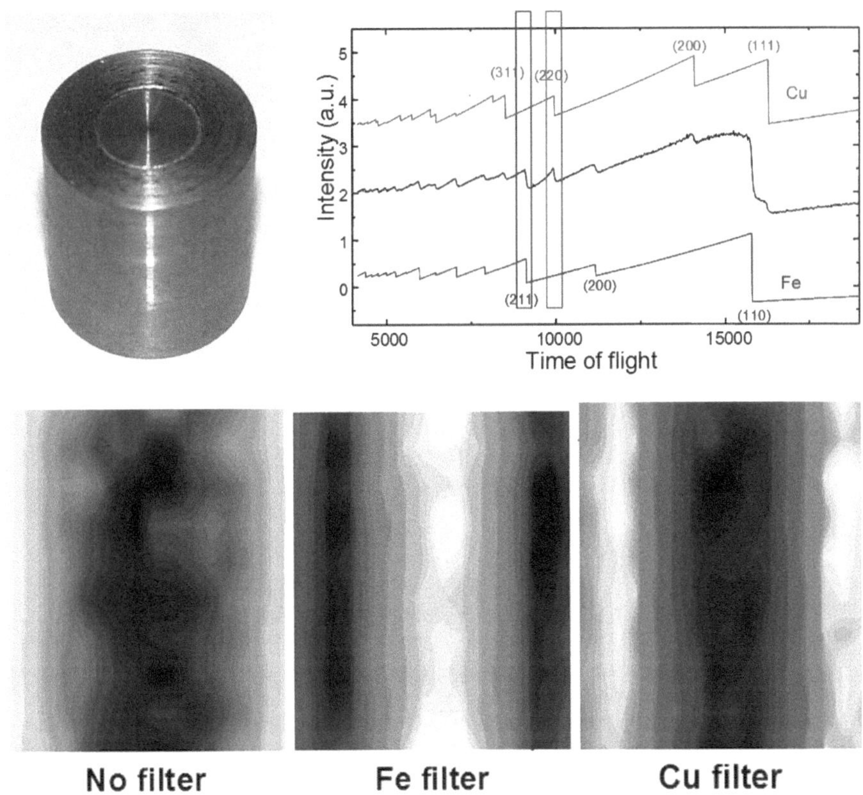

Fig. 12.13 Neutron transmission images for a copper cored iron slug, with Bragg edge filtering used to enhance the contrast for either copper or iron [22]

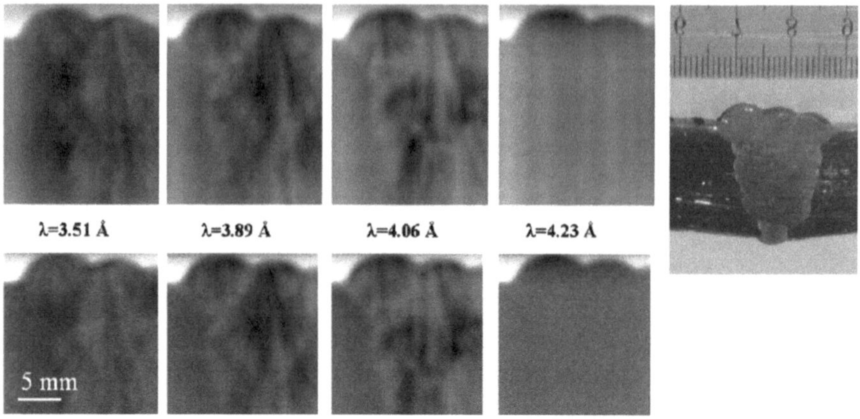

Fig. 12.14 Neutron transmission radiographs of a steel welding as a function of neutron energy, which was measured using two different detectors systems, PI-MAX (*upper row*) and ANDOR (*lower row*) [22]

the Spallation Neutron Source in the United States, and/or the design of specialized instruments using advanced neutron optics and detectors, such as the IMAT instrument proposed for the ISIS source in the United Kingdom, these restrictions are likely to be overcome within the next 5 years. Then, this will open the way for a much broader range of applications of neutron imaging in materials science and engineering.

References

1. J. Campbell, The concept of net shape for castings. *Mater. Design* **21**(4), 373–380 (2000).
2. X. Yao and S. Shivkumar Mould filling characteristics in lost foam casting process. Mater. Sci. Technol. **13**, 841–846 (1997).
3. R. E. Turner, Radiography: Major contribution to the metalcasting industry. *Modern Casting* **61**(1), 44–45 (1972).
4. B. Sirrell, M. Holliday, et al., *The Benchmark Test 1995. 7th Conference on Modelling of Casting*. Welding and Advanced Solidification Processes, London, UK.
5. M. Cox, R. A. Harding, et al., Optimised running system design for bottom filled aluminium alloy 2L99 investment castings. *Mater. Sci. Technol.* **19**(5), 613–625 (2003).
6. X. Luo. *Study on Infrastructure Materials by Using Neutron Radiography and Diffraction*. PhD dissertation, University of Tennessee, Knoxville, TN, USA (2007).
7. X. Luo and D. Penumadu, Visualization of the multiphase flow in the lost foam casting process by neutron radiography and image processing. *Automotive Light Metal Castings: Technology and Applications, Materials Science and Technology (MS&T)*, Detroit, MI, USA (2007).
8. D. Penumadu and N. Kardjilov, High resolution neutron tomography of sandcores used in metal casting (2008). To be submitted.

9. T.S. Majmudar and R.P. Behringer, Contact force measurements and stress-induced anisotropy in granular materials. *Nature* **435**(7045), 079–1082 (2005).
10. N.P. Kruyt, Contact forces in anisotropic frictional granular materials. *Int. J. Solids Struct.* **40**(13–14), 3537–3556 (2003).
11. M. Otto et al., Anisotropy in granular media: Classical elasticity and directed-force chain network. *Phys. Rev. E* **67**(3), 031302-1 to 24 (2003).
12. P. Lehmann, P. Wyss, et al., Tomographical imaging and mathematical description of porous media used for the prediction of fluid distribution. *Vadose Zone J.* **5**(1), 80–97 (2006).
13. B. Schillinger, J. Brunner, and E. Calzada, A study of oil lubrication in a rotating engine using stroboscopic neutron imaging. *Physica B* **385–386**, 921–923 (2006).
14. B. Schillinger, H. Abele J. Brunner, G. Frei, R. Gahler, A. Gildemeister, A. Hillenbach, E. Lehmann, and P. Vontobel, Detection systems for short-time stroboscopic neutron imaging and measurements on a rotating engine. *Nucl. Instrum. Meth. Phys. Res. A* **542**, 142–147 (2005).
15. A. Hillenbach, M. Engelhardt, H. Abele, and R. Gahler, High flux neutron imaging for high-speed radiography, dynamic tomography and strongly absorbing materials. *Nucl. Instrum. Meth. Phys. Res. A* **542**, 116–122 (2005).
16. V.V. Nagarkar, I. Shestakova, S. C. Thacker, S. R. Miller, D. Penumadu, J. F. Ankner, H. Z. Bilheux, and C. E. Halbert, Time-resolved high resolution neutron imaging studies at the ornl spallation neutron source. Session N14-4, Nuclear Science Symposium, Honolulu, Hawaii, USA, Oct. 27-Nov. 3 (2007).
17. V.V. Nagarkar, I. Shestakova, V. Gaysinskiy, S.V. Tipnis, B. Singh, W. Barber, B. Hasegawa, and G. Entine, CCD-Based Detector for SPECT. *IEEE Trans. Nucl. Sci.* **53**(1), 54–58 (2006).
18. M.T. Hutchings, P. J. Withers, et al., *Introduction to the Characterization of Residual Stress by Neutron Diffraction*. Boca Raton, CRC Press (2005).
19. P. J. Webster, X. Wang, G. Mills, G. A. Webster, Residual stress changes in railway rails. *Physica B* **180–181**, 1029 (1992).
20. J.R. Santisteban, L. Edwards, M.E. Fizpatrick, A. Steuwer, and P.J. Withers, *Appl. Phys. A* **74**, S1433–S1436 (2002).
21. J.R. Santisteban, S. Siano, and Daymond, M.R, Neutron strain scanning of archaeological bronzes, *Mater. Sci. Forum* **524–525**, 975–980 (2006).
22. W. Kockelmann, G. Frei, E.H. Lehmann, P. Vontobel, and J.R. Santisteban, Energy-selective neutron transmission imaging at a pulsed source. *Nucl. Instrum. Meth. Phys. Res. A* **578**, 421–434 (2007).

Chapter 13
Novel Neutron Imaging Techniques for Cultural Heritage Objects

C. Andreani, G. Gorini, and T. Materna

Abstract The use of neutrons for cultural heritage (CH) research is illustrated with special reference to neutron tomography (NT) methods, providing three-dimensional (3D) images of neutron attenuation, and the analysis techniques known as prompt gamma-ray activation analysis (PGAA) and neutron resonance capture analysis (NRCA), providing the elemental composition of an object. PGAA and NRCA are well-established nondestructive methods for bulk analysis of CH objects, with sensitivities that can reach the parts-per-million range. By improving the spatial resolution of PGAA and NRCA it will be possible to measure the composition of small parts inside a large object or even to provide a full 3D map of the elemental composition of an artifact. The imaging techniques under development are called prompt gamma-ray activation imaging (PGAI), neutron resonance capture imaging (NRCI) and neutron resonance transmission (NRT) tomography. The NRCA experience at the GELINA neutron source is the starting point for the development of NRCI/NRT now taking place at the 100 times more powerful ISIS pulsed neutron source.

PGAI, NRCI, and NRT are unlikely to achieve the spatial resolution obtained by conventional NT. It is the combined use of these new imaging techniques with neutron and X-ray tomographies that is proposed here as the way forward in neutron imaging. There are parallel developments in the use of Bragg edges to improve the elemental sensitivity of NT (dichromatic tomography). Another foreseen development is 3D phase imaging (neutron diffraction imaging, NDI), which can essentially be undertaken using existing diffraction instruments. The ongoing developments are promising and suggest that neutron-based techniques will provide the basis for integrated analysis protocols leading to important advances in the scientific characterization of the materials that constitute a CH object.

C. Andreani (✉)
University of Milano-Bicocca, Department of Materials Science, Via Roberto Cozzi 53, 20125 Milano, Italy

I.S. Anderson et al. (eds.), *Neutron Imaging and Applications*,
Neutron Scattering Applications and Techniques, DOI 10.1007/978-0-387-78693-3_13,

Keywords X-ray and neutron scattering · Neutron diffraction · Neutron imaging · Ancient Roman marbles · Textures · Experimental methods of materials testing and analysis

13.1 Imaging, Neutrons, and Cultural Heritage

The use of X-ray and neutron facilities for cultural heritage (CH) research is rapidly growing [1, 2]. Detailed three-dimensional (3D) maps of neutron and X-ray attenuation in CH objects are produced by the tomographic methods described earlier in this book. X-ray and neutron facilities also provide a number of nondestructive analytical techniques that are used in the CH field. X-ray fluorescence (XRF) is a well-known example, in particular because of the widespread use of portable XRF devices, a clear advantage for analysis of valuable objects at (or close to) museums. More generally, the application of nuclear-based techniques to CH research has expanded in recent years as laboratories with small ion accelerators have been installed close to museums, facilitating the study of CH objects by particle-induced X-ray and gamma-ray emission (PIXE and PIGE) and Rutherford backscattering [3].

With XRF and PIXE, imaging is possible down to micrometer resolution; but because both techniques are based on the detection of low-energy X-rays, only information on the surface of an object is retrieved. Bulk analysis techniques are alternatively available (see below) but do not offer a spatial resolution adequate for measuring the composition of small parts inside a large object well. Ideally, one would like to combine the spatial resolution of tomographic methods with the power of nondestructive analytical techniques to provide a 3D map of the elemental composition of a large object. Can neutrons be used for such a challenging task? The question is addressed in this chapter by first reviewing some general features of neutron (and X-ray) techniques as relevant for CH studies.

X-rays and neutrons conveniently complement each other in CH research. This is because CH objects can be made from diverse materials sampling the Mendeleev table at both low and high atomic numbers, making the attenuation of X-rays and neutrons very different. X-rays have a lower penetration depth at high atomic numbers. Thus X-ray radiography (XR) is suitable for probing the interior of objects made from wood or bone, but it has problems with bulky metals such as gold, silver, and lead, which are nearly opaque to X-rays. On the other hand, neutron radiography (NR) can probe metal artifacts (e.g., copper, tin, iron, bronze, lead) but does not penetrate thick layers of organic materials because of strong neutron beam attenuation by hydrogen. An introduction to NR and its applications can be found elsewhere in this book.

Neutron tomography (NT) is performed by taking a large set of radiographs of an object at different orientations and reconstructing all data offline in a virtual 3D image. For further details, see Chapters 6 and 7 in this book. Only a few NT setups exist in the world (see [4] for a review of facilities in Europe) because of the high neutron fluxes required to perform a full scan of an object in a reasonable time. This has not prevented the successful application of NT to CH objects. Generally CH objects benefit from NT developments driven, example, by engineering applications. Figure 13.1 shows an example where the image contrast of an ancient Roman brooch is improved by using a cold neutron beam: the proportionality of neutron capture cross section to neutron wavelength is exploited to select the neutron wavelength that best matches the thickness and composition of the object. Imaging CH objects with neutrons offers unique opportunities: it permits the search for hydrogenous materials inside sealed metal objects, which is impossible with X-rays; it allows differentiation between elements close in the Mendeleev table, which would show the same contrast to X-rays; and finally, it enables differentiation between isotopes. The elemental information is inferred from the neutron attenuation. A typical example is a metal artifact, such as a statuette or a vase, in which different materials can be found in inserts or welds and provide useful information on the fabrication methods.

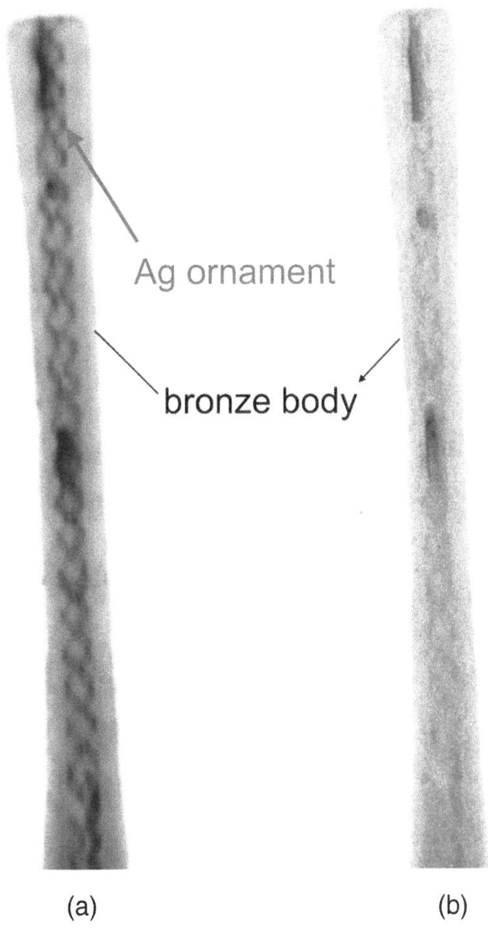

Fig. 13.1 Radiographs of an ancient Roman brooch taken with (**a**) cold neutron (5.5 Å) using a velocity selector at the Prompt Gamma-Ray Activation (PGA) station of the Paul Scherrer Institut (PSI), Switzerland [5] and (**b**) standard thermal (1.8 Å) neutron radiography technique at the Neutron Transmission Radiography (NEUTRA) station at PSI [6]

In the cold/thermal neutron energy range, the elemental sensitivity of neutron imaging can be enhanced using Bragg cutoffs (or Bragg edges). At a so-called Bragg cutoff wavelength, the neutron attenuation coefficient drops drastically. If the drop in attenuation is important and one can monochromatize the beam and select its wavelength, this behavior can be exploited to change the material contrast. Indeed, if two radiographs are taken, one with a wavelength just below the Bragg cutoff of a selected element and one with a wavelength just above it, the ratio will be sensitive to this element only. On the other hand, if two wavelengths are selected where the attenuation coefficient is equal, the element disappears in the ratio of the two radiographs as well as in the tomographic result (Fig. 13.2) [5]. This dichromatic tomography method has been demonstrated both at reactor sources and at pulsed neutron sources. It is similar to what can be done with X-rays using the rise in attenuation at the K-edge energy [7] (Fig. 13.3). Both methods can in principle provide a 3D map of a selected element. However, not all elements have Bragg edges, and only elements with high atomic number, which have the K-edge at high X-ray energy, can be imaged in large samples. Thus the use of Bragg edges and K-edges represents a step forward, but with limited scope.

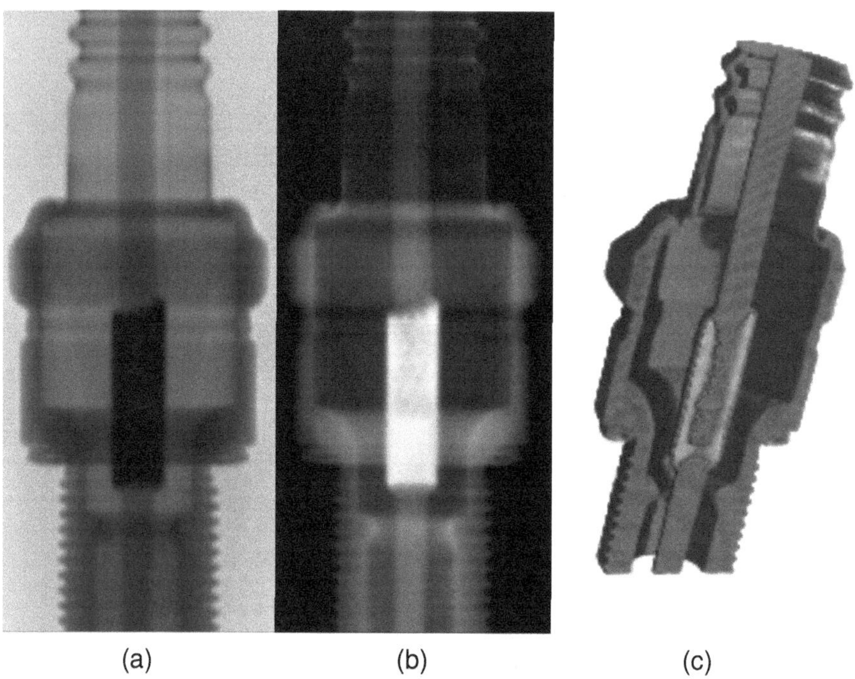

Fig. 13.2 (**a**) Neutron radiograph of a spark plug at 6.9 Å; (**b**) result of the division of two radiographs, one at 6.9 Å and the other at 3.2 Å, showing that the steel cladding (central part) became transparent; (**c**) 3D result of dichromatic tomography. Detailed explanations can be found in [5]

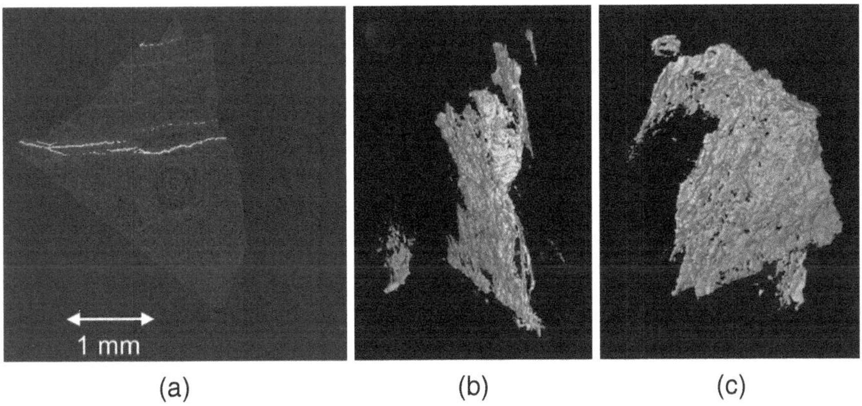

Fig. 13.3 A gold-sensitive tomography of an ore of Ashanti (Ghana) origin performed at the high-energy beam line (ID15) of the European Synchrotron Radiation Facility (Grenoble, France) shows that thin (about 40 μm) but extended layers (gold veins) appear in the quartz beads [7]: (**a**) shows a horizontal cut in the tomography made at 80 keV, where the gold veins and the quartz matrix are visible on top of ring-shaped reconstruction noise; (**b**) and (**c**) are two different 3D views of the result of the dichromatic tomography, sensitive to gold only.

Bragg edges may be more useful for CH applications because of the texture information embedded in their detailed shapes. Texture in the phases distorts the Bragg edges, so it can be used as a fingerprinting technique with the potential for texture mapping [8]. Appropriate transmission detectors need to be developed before this technique can be used routinely at pulsed neutron sources. On the other hand, phase-sensitive neutron diffraction (ND) techniques are in regular use for phase and microstructure characterization of ceramic and bronze artifacts, including texture analysis (see below). Texture provides important clues to the deformation history and thus to historic production steps, a prominent example being the neutron texture analysis of the early Copper Age Iceman axe [9].

Three features of neutron interactions with nuclei are exploited to provide analytical tools to determine the elemental compositions of CH objects:

1. radioactive nuclei produced by thermal/cold neutron capture,
2. prompt gamma rays from thermal/cold neutron capture, and
3. resonant neutron absorption at element-specific neutron energies in the epithermal energy range.

The first feature has long been exploited by a method known as neutron activation analysis (NAA). It started with the detection of radioactive isotopes through their half-lives. Now, these isotopes are observed and quantified with high-resolution gamma-ray detectors on the basis of gamma-ray energies as well as gamma-ray half-lives. This technique, known as instrumental neutron activation analysis (INAA), is now a valuable analytical tool for CH studies in use at many facilities [10, 11]. INAA may be invasive because it may require the preparation of a small sample (e.g., powder) of the object before irradiation in a reactor. A related noninvasive imaging method is neutron activation autoradiography (NAAR) [12], in which two-dimensional (2D) objects are exposed to cold neutrons and the map of the neutron-induced gamma activation is recorded at different times by radiographic methods. NAAR is limited to 2D objects but is a unique tool for the investigation of paintings.

For the noninvasive elemental imaging of 3D objects, features (2) and (3) in the list above, which provide the basis for recently developed analytical methods, should be considered. Prompt gamma-ray activation analysis (PGAA) is based on the detection of characteristic gamma-ray energy spectra during thermal/cold neutron capture. It provides a nondestructive determination of the elemental (isotopic) composition of CH objects with good sensitivity to both major and trace elements. Mainly developed by G. Molnár and coworkers at the Budapest research reactor [13], PGAA has found many applications and has reached a high level of sophistication. Neutron resonance capture analysis (NRCA) was developed as a new analytical tool by H. Postma and coworkers at the GELINA pulsed neutron source [14]. NRCA makes use of the unique resonance absorption properties of epithermal neutrons. Many elements found in common CH materials have neutron absorption resonances in the energy range below 1 keV, including copper, cobalt, silver, tin, antimony, and gold. The gamma rays from resonant neutron capture are detected as a function of time, that is of neutron velocity. The time spectrum has emission peaks at element-specific resonance energies. Calibration of the gamma-ray detector array using reference samples is used to convert the resonant gamma-ray yield into absolute elemental composition. An alternative approach is also possible in which the neutron transmission through an object is measured instead of the gamma-ray emission. The transmitted neutron spectrum of a pulsed neutron beam has characteristic resonance dips in the time spectrum that can be used to determine the elemental composition of the object in the same way as the resonance peaks of NRCA.

PGAA and NRCA are well-established nondestructive methods for bulk analysis of CH objects, with sensitivities that can reach the parts-per-million range. There is no important reason why the spatial resolution of PGAA, NRCA, and neutron transmission methods should not be improved, eventually leading to nondestructive measurement with good spatial resolution of the composition (major and trace elements) of small parts inside a large object. These 3D extensions of PGAA and NRCA are presented in this chapter as a new development that can fill the gaps between 3D imaging, surface analyses, and bulk analyses. In the following discussion, the 3D extensions will be called prompt gamma-ray activation imaging (PGAI), neutron resonance capture imaging (NRCI), and neutron resonance transmission (NRT) tomography. PGAI, NRCI, and NRT are unlikely to achieve the spatial resolution obtained by conventional NT. It is the combined use of these new imaging techniques with neutron and X-ray tomographies that we foresee as the way forward in neutron imaging. Another development foreseen is 3D imaging of crystalline phases (neutron diffraction imaging, NDI), which can essentially be undertaken using existing diffraction instruments. These methods together are expected to provide a new, comprehensive neutron-based approach that could be used for 3D imaging of the elemental and phase compositions of CH objects.

Interestingly, this is a rare case of a new development in neutron instrumentation that is motivated by CH applications. The new imaging techniques are being developed within the Ancient Charm (AC)[1] project, a collaboration of ten universities, central laboratories, and cultural heritage institutions across Europe [15]. A technical challenge of the project is represented by the need to perform high-efficiency gamma-ray measurements in a neutron environment, with the associated background, and to extend spatially resolved neutron transmission measurements into the epithermal range. The underlying techniques are presented in more detail in this chapter together with highlights of recent developments and CH applications.

13.2 Two Neutron Beam Analytical Techniques: Neutron Resonance Capture Analysis and Prompt Gamma-Ray Activation Analysis

13.2.1 Neutron Resonance Capture Analysis

Neutron resonance capture analysis makes use of the unique resonance absorption properties of epithermal neutrons for analyzing the bulk composition of materials and objects. The absorption cross sections of most elements are well known; they vary from one element to another and actually vary between isotopes of the same element [16]. Neutron absorption is followed by the prompt emission of a gamma-ray cascade with total cascade energies of up to about 8 MeV. The detailed energy spectrum of the gamma emission is also well known in the case of thermal neutron capture, but it can vary among resonances. NRCA is used at the GELINA pulsed neutron source for analysis of CH artifacts to determine the concentration of neutron-sensitive elements [17–25].

So far NRCA has been applied to a large number of artifacts, mainly bronze objects. A typical NRCA spectrum looks like the one in Fig. 13.4 [25] obtained by exposing an Etruscan votive object to the GELINA neutron beam. A number of resonance peaks of different elements (copper, lead) are clearly visible, corresponding to neutron energies in the range from 3,250 to 3,650 eV. The energies and intensities of the resonance peaks characterize the elemental composition of the sample. Figure 13.4 shows the 3,357 eV resonance of lead-206 and three copper resonances at 3,310, 3,503, and 3,588 eV. Because

[1] Ancient Charm (Analysis by Neutron Resonant Capture Imaging and other Emerging Neutron Techniques: Cultural Heritage and Archaeological Research Methods), http://ancient-charm.neutron-eu.net/ach.

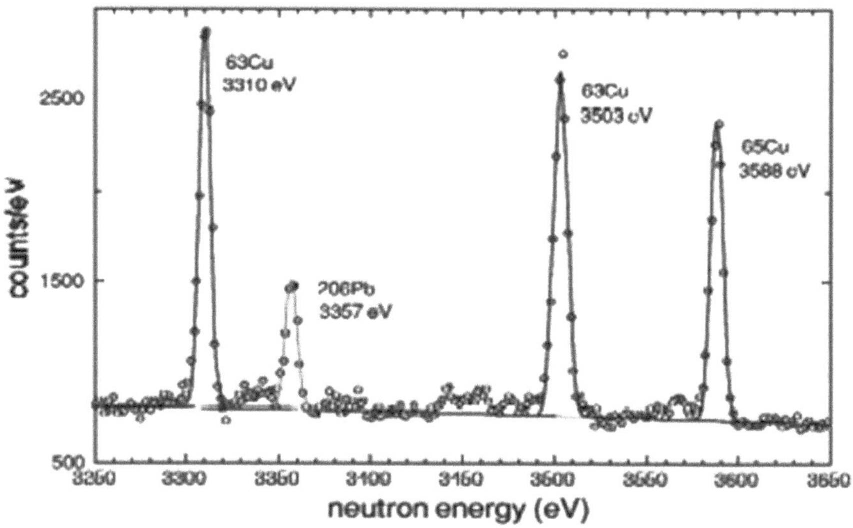

Fig. 13.4 Example of neutron resonance capture analysis NRCA spectrum featuring three copper resonances and one lead resonance in the region of 3,250–3,650 eV used to determine the Pb/Cu weight ratio of an Etruscan votive object [25]

these four resonances are weak, the Pb/Cu weight ratio can be obtained after small self-shielding corrections. A set of Etruscan statuettes from a collection originally owned by Earl Corazzi of Cortona (Tuscany), now at the National Museum of Antiquities in Leiden (Netherlands), are among the objects investigated by NRCA. In this case, NRCA was used to distinguish suspected fakes from genuine statuettes, solely on the basis of the elemental compositions of the artifacts [20]. The conclusion was based on the observation that minor quantities (up to several percent) of zinc occurred in some of these statuettes. Given the melting techniques available to the Etruscan blacksmiths, no more than a fraction of a percent of zinc would occur in a genuine artifact (Fig. 13.5).

13.2.2 Prompt Gamma-Ray Activation Analysis

Prompt gamma-ray activation analysis is a nondestructive nuclear method for performing both qualitative and quantitative multielement analysis of major, minor, and trace elements in samples. The technique is useful for the analysis of light elements such as H, B, C, N, Si, P, S, and Cl, as well as for heavy elements such as Cd, Sm, Gd, and Hg [26]. When a thermal or cold neutron is absorbed by a target nucleus, the compound nucleus is in an excited state with energy equal to the binding energy of the added neutron. Then the compound nucleus will almost instantaneously (in less than 10^{-14} s) de-excite into a more stable configuration through the emission of characteristic prompt gamma rays. In many cases, this new configuration yields a radioactive nucleus that also de-excites (or decays) by

Fig. 13.5 Photograph of two Etruscan statuettes, one genuine and one false [20]

emission of characteristic delayed gamma rays. PGAA is based on the detection and quantification of the prompt gamma rays emitted by the sample during neutron irradiation, while NAA uses the delayed gamma rays from the radioactive daughter nucleus (Fig. 13.6). Consequently, PGAA complements NAA by allowing determination of elements that do not form radioactive products after irradiation (e.g., hydrogen and boron) and elements for which the half-life is too long to be conveniently measured by NAA (e.g., carbon). Boron is the only exception to the

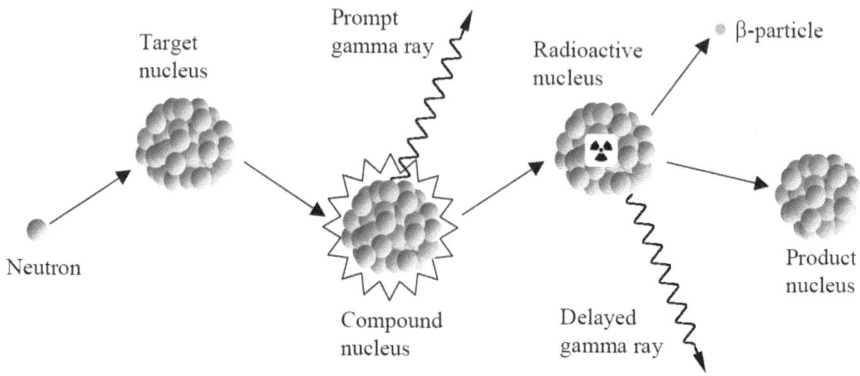

Fig. 13.6 Nuclear processes involved in prompt gamma-ray activation analysis (PGAA) and neutron activation analysis (NAA): neutron capture is immediately followed by emission of prompt gamma rays. The compound nucleus decays by beta emission followed by the emission of delayed gamma rays

usual prompt gamma-ray measurements in that the gamma measured is not due to the (n, γ) reaction. Indeed, boron (like lithium-6) reacts with neutrons by emission of α-particles [i.e., via the reaction $^{10}B(n, \alpha)^{7}Li$]. Most of the lithium-7 is formed in an excited state and de-excited instantaneously (within 10^{-14} s) by the emission of 477 keV gamma rays.

The PGAA technique requires mainly a source of thermal or cold neutrons and a high-resolution spectrometer for measurement of gamma rays with energies over a range from about 100 keV to 11 MeV. The energies of the prompt gamma rays identify the neutron-capturing elements, while the intensities of the peaks at these energies are used to determine their concentrations (Fig. 13.7). An advantage of the PGAA method is that nothing special is required in the way of sample preparation: the sample can be in a solid, liquid, or gaseous state. Depending on the amount of the investigated material,

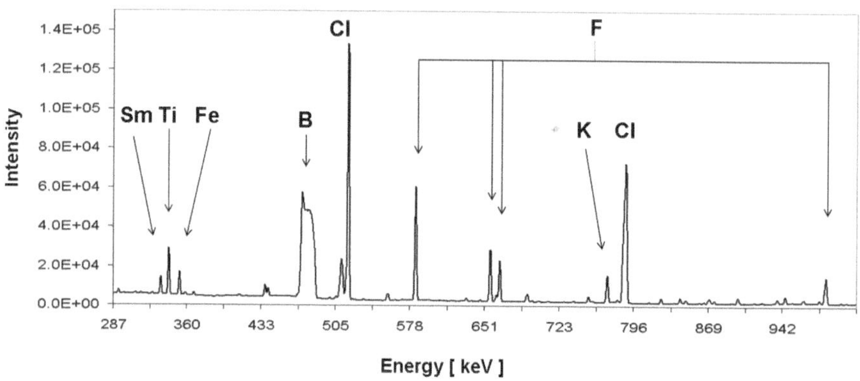

Fig. 13.7 Example of prompt gamma-ray activation analysis (PGAA) spectrum from an estuarine sediment

element, and target matrix, the detection limit for a PGAA installation can be as low as 10 ppb (for B, Sm, Gd). The activity of the object after irradiation is low or decreases rapidly in most cases; therefore, the object can be returned within a few days of the measurement. Compared with NAA, where the sample is placed close to the core of the reactor, the neutron beam intensity at the sample is six orders of magnitude lower.

A typical setup for PGAA consists of a sample box, a shielding, and a detection system [13, 27–30]. The object is exposed to an intense thermal or, preferably, cold neutron beam on the order of 10^6–10^8 n/cm^{2}s, originating from a neutron guide. The neutron guide is preferably curved to avoid a direct view to the moderator or cold neutron source, thus reducing the fast-neutron and gamma-ray background. The huge number of peaks in a spectrum (up to about a thousand) requires a high-purity germanium (HPGe) detector with high-energy resolution (<2 keV at 1,332 keV). Moreover, many high-energy gamma rays will not deposit their full energy inside the HPGe detector and thus build up large Compton tails that will hide peaks at lower energies or generate (unwanted) additional escape peaks. To reduce these effects, the HPGe detector is surrounded by a fast gamma-ray detector, usually made of an annular bismuth germanate oxide (BGO) or NaI(Tl) scintillator that serves to reject events simultaneously detected in the HPGe detector and the scintillator. This anti-Compton shield acts as an active shield against unwanted, external high-energy gamma rays that deposit part of their energy in the scintillator before scattering in the HPGe detector. The whole detection system is then covered with lead bricks and additional neutron shielding to further improve the signal/background ratio. The sample box is usually covered for the same reasons with lithium-6, except at the entry of the neutron beam, to prevent scattered neutrons from reaching the detector. The detection system is normally placed at 90° to the neutron beam.

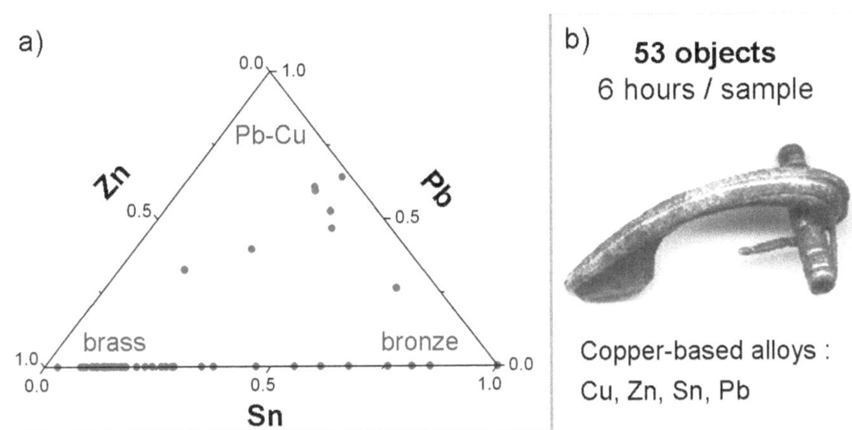

Fig. 13.8 Main composition of 53 brooches studied by S. Baechler et al. [31] (left) and a photograph of one of them (right)

The count rate of the acquisition system is limited to about 10 kHz by adjusting the neutron beam size with lithium-6 collimators.

PGAAs have been performed on various samples, in a wide range of fields, including material science, medicine, nuclear science, geology, and archeology. In one of the first CH applications, S. Baechler and coworkers [31] studied 53 La Tene and Roman brooches from western Switzerland and identified four different groups of copper-based alloys (Fig. 13.8).

13.3 Combined Prompt Gamma-Ray Activation Analysis Scanning and Neutron Tomography

The scope for upgrading a PGAA station to PGAI can be assessed by an estimate of the experimental time needed for a full scan of an object. The time needed to measure a $5 \times 5 \times 5\ cm^3$ object with a 1-mm^3 spatial resolution would be equal to 125,000 times the acquisition time to measure a 1-mm^3 sample. Limiting the acquisition time to 4 s—which is already a huge compromise because it takes from minutes to a few hours to analyze most samples using PGAA (to detect trace elements)—and estimating the time to move the object at 1 s, the total beam time would be more than a week. Unfortunately, objects with cultural importance are typically manufactured objects, made of numerous small parts, each of them being to some extent homogeneous. A complete scan at high spatial resolution of such objects would be not only impossible but also of little use. A better approach to the problem is to use another technique to distinguish the different parts of the object and then to measure them by PGAI, with an adequate spatial resolution smaller than an ideal level of detail but large enough to measure the main and minor elements in a reasonable length of time. The full process can be optimized further in the following way:

1. the object is imaged by X-ray tomography with a high spatial resolution, typically 10 μm;
2. an NT of the object is performed with a resolution down to 100 μm;
3. a bulk PGAA of the whole object is performed to identify all the elements present; and
4. the regions of interest are measured by PGAI.

The acquisition time is regulated by the requirements for quantifying the elements seen in the bulk analysis. Another limitation may come from the need to keep the induced long-lived activation of the object to an acceptable level; for CH objects this means returning the object to a curator after a few weeks. This is not a problem for PGAA but should be reconsidered in the case of trace elements exposed to large neutron fluences in PGAI.

Alternative geometrical arrangements can be considered in the implementation of PGAI. The first and preferred one consists in collimating the neutron beam using a 1-mm-thick lithium-6 foil with a hole that lets a neutron beam of the order of 1 mm^2 to pass. A thick gamma-ray collimator made of lead, placed in front of the spectrometer, defines a pencil beam of gamma rays allowed to enter the spectrometer. The intersection of the neutron beam and the gamma-ray beam delimits a voxel, a region of the object measured by PGAI. With this solution, a 3D map of the composition of the object is obtained by scanning the object across the neutron beam and in front of the HPGe detector in three dimensions (XYZ). For each voxel, a gamma-ray spectrum is acquired and its analysis yields the composition of the small portion of the object, defined by the intersection of neutron beam collimation and HPGe collimation. An advantage of this solution is that one can in a simple way select the voxel and its size. Provided an X-ray and/or a neutron tomography of the object is performed before the PGAI measurements, the experiment can be as simple as the following: the user selects with the 3D visualization software the 3D coordinates of a region of interest (ROI) and a motorized 3D positioning stage places this ROI in front of the collimators for a PGAI measurement. In principle, the best orientation of the sample in the set of collimators can be found by minimizing the attenuation of neutrons in the path to the ROI and of the gamma rays in the path from the ROI to the detector.

This scanning solution presents the following drawbacks.

1. Neutrons scatter inside the object, especially if the object contains hydrogen, so the irradiated region is no longer a beam of 1 mm^2.
2. Collimation of high-energy gamma rays (>1 MeV) requires at least 20-cm-thick lead brick. This increases the distance between the object and the detection system, reducing its efficiency and thereafter increasing the acquisition time. A compromise again must be chosen between best collimation and measuring time. Fortunately, most emitted gamma rays have energies below 1 MeV.
3. Reducing the viewing cone of the spectrometer reduces not only the size of the ROI but also the efficiency of the detector. It does not, however, reduce the background.

An alternative solution that would improve the signal/background ratio of PGAI consists of removing the gamma-ray collimation. The object would be scanned in two dimensions with the neutron beam and rotated from 0 to 180°. This solution corresponds to prompt gamma-ray activation tomography in which the HPGe acquires spectra coming from lines though the object. Measuring the set of lines at different orientations of the object yields a 3D map of the composition of the object. The number of positions is larger than with the scanning solution. Indeed, keeping in mind our $5 \times 5 \times 5$ cm^3 object and a final spatial resolution of 1 mm^3, the required number of projections is slightly larger than for the scanning solution: $50 \times 50 \times \pi/2 \times 50 = 196{,}350$. The acquisition time is, on the other hand, more than 100,000 times shorter in a first approximation because the spectrometer views a $50 \times 1 \times 1$ mm^3 portion of the object, and the collimation in front of the detector can be released. The total measurement time would then be less than 3 days, dominated by the time to move the object.

The real disadvantage of this method is that it is not possible to determine the composition of a part of the object without measuring at least a full slice of it, as required by the reconstruction algorithm. The a priori knowledge of the object structure from neutron or X-ray tomography is not in this case directly useable — the whole object must be measured regardless. A third option for implementation of PGAI would consist of keeping the gamma-ray collimation and not the neutron collimation. This solution reproduces single photon emission computed tomography, with spectra as projections; the reconstruction of the composition maps consists of reconstructing each element separately. This solution, however, does not make use of the main advantage of using cold neutrons (i.e., that one can easily collimate them and irradiate only a selected part of the object without reducing the total measurement time).

S. Baechler and coworkers performed the first PGAI experiment at the PGA station of Paul Scherrer Institut (PSI), Switzerland, to study a piece of the 2-billion-year-old Oklo natural reactor [32]. The cold neutron beam was focused using a neutron lens and reduced to 1 mm^2 by a lithium-6 collimator at the focal point. No collimator was placed in front of the HPGe detector, and the sample was scanned only in two dimensions in front of the neutron beam without rotations. Although they gave the composition integrated over the neutron path only, the results agreed with the uranium-sensitive tomography (Fig. 13.9) taken at the European Synchrotron Radiation Facility (Grenoble, France) using high-energy X-rays [7]. Besides the combinative use of two techniques to interpret PGAI results, this experiment showed that one could correct PGAI data for gamma-ray attenuation using the tomography results.

It is obvious from the simple calculation on the required measuring time that an intense cold neutron beam should be available to study large objects. A new PGAA facility [33] is now available at FRM-II[2] in Germany [34]. One of its major features is a super-mirror elliptic tapered neutron guide that should focus cold

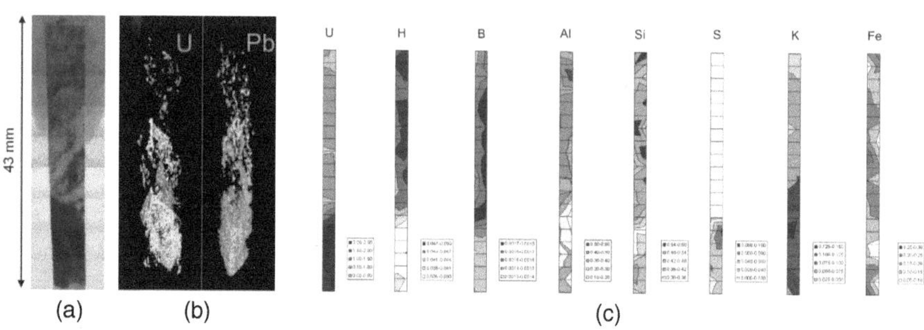

Fig. 13.9 (**a**) X-ray radiograph at 87 keV of a piece from the Oklo natural reactor (Gabon); (**b**) 3D distributions of lead and uranium resulting from tomographies taken at the high-energy beam line (ID15) of the European Synchrotron Radiation Facility (Grenoble, France) [7]; (**c**) spatial distributions of selected elements in the same piece, obtained by prompt gamma-ray activation imaging at the Paul Scherrer Institut (Villigen, Switzerland) [32]

[2] Forschungsreaktor München II (translation: Munich Research Reactor II), Munich Technical University.

neutrons to a spot size of $4 \times 7\,\text{mm}^2$ at 10 cm from the end of the guide. The cold neutron flux is expected to exceed 10^{10} n/cm^{2}s at the focal point [35]. One of the aims of the facility is to test different PGAI geometry options on CH objects.

13.4 Imaging with Neutron Resonances

The NRCA experience at the GELINA neutron source [17–25] is the starting point for the development of an imaging method based on resonant neutron capture. Imaging applications require high neutron fluxes, and the 100 times more powerful ISIS pulsed neutron source (Chilton, UK) is now used for the development of NRCI and NRT [15]. Figure 13.10 shows a sketch of the NRCI/NRT setup that is under construction at ISIS. In the case of NRCI, the object is scanned in front of a collimated neutron beam while a set of scintillator detectors surrounding the object record the capture gamma rays. The composition of the object must be retrieved by a tomographic reconstruction corrected for neutron self-absorption. In practice the spatial resolution is limited by the need to collimate the beam of epithermal neutrons while keeping the signal/background at an acceptable level. The best compromise at the moment seems

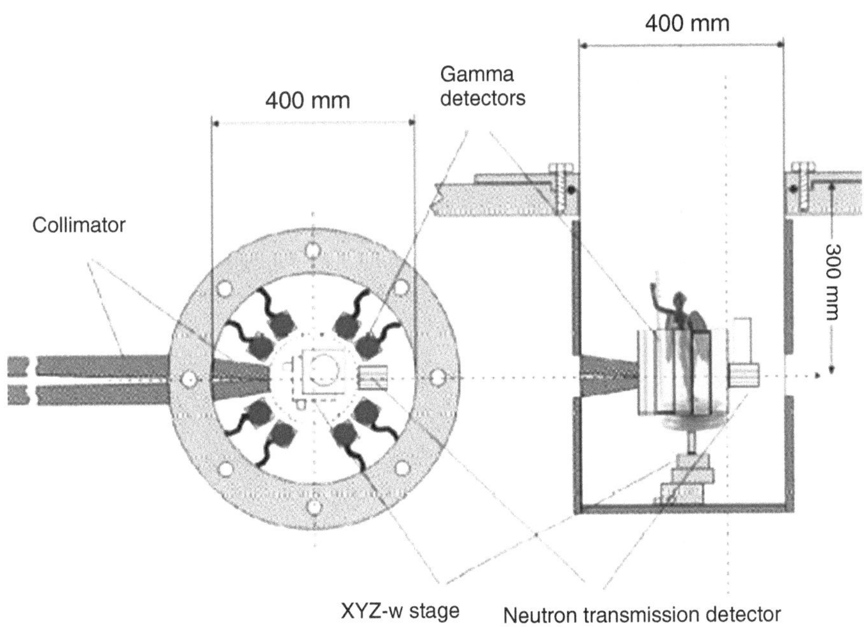

Fig 13.10 Sketch of a setup for performing neutron resonance capture imaging and neutron resonance tomography. The sample is attached to a frame allowing translation and rotation motion relative to the beam and detectors, which are fixed. The gamma-ray detectors for NRCI are placed around the sample. A position-sensitive detector for NRT is placed in the neutron beam behind the sample

to be a collimated beam diameter of 5 mm, which is also the anticipated spatial resolution to be achieved by NRCI in the ISIS experiment.

For NRT measurements, the neutron collimator is removed and the object is scanned and rotated in front of the NRT detector. Indeed the main challenge in NRT is the realization of a position-sensitive detector for epithermal neutrons suitable for time-of-flight (TOF) measurements. It is obvious that the spatial resolution of such a detector will not reach, in the near future, the resolution of the ones used in thermal or cold NT. The detector prototype tested at ISIS [36] features 2×2 mm^2 pixels made from lithium glass. A length of 8 mm for each pixel provides adequate efficiency for epithermal neutron detection, and an array of 10×10 pixels can be managed. Thus NRT is expected to provide better spatial resolution than NRCI in a much shorter time. It may, however, have lower sensitivity to trace elements, as suggested by the GELINA experience, where transmission measurements are in regular use for total neutron cross section measurements. Compared with NT, NRCI and NRT have a coarse spatial resolution but provide direct information on the elemental composition; NT provides composition information only indirectly via energy analysis of neutrons. An interesting observation is that the information acquired by the NRT detector is much richer than a radiograph: a full TOF spectrum per pixel with several resonances per element is measured, yielding several element-sensitive images at once. The NRT potential is better understood in the case of large samples. With thermal or cold NT, when the sample is "black" (i.e., too thick or too absorbing), there is no way to image it. With NRT, there are often many resonances for one or more elements in the object. An object may be black for one resonance but could still be imaged with another resonance in the spectrum. The NRT detector can also be used in the thermal energy range to provide, for example, strain and texture information using Bragg edge analysis to complement neutron diffraction imaging (NDI) studies. The ISIS experience will tell us to what extent the potential of the new NRCI/NRT device under construction can be exploited to provide 3D maps of the elemental and phase composition of CH objects.

From the descriptions of the different techniques PGAA, NRCA, PGAI, NRCI, NT, and NRT, one may conclude that an appropriate combination of all of them, along with NDI, can be the basis for integrated analysis protocols for the study of CH objects. The integrated use of imaging and other neutron-based techniques for CH studies is indeed an area of ongoing research, as illustrated by the applications discussed below.

13.5 Applications to Cultural Heritage Studies

In the past few years, neutron techniques have increasingly been used for the study of CH materials, in particular to address their nature and authenticity, their provenance and diffusion, the manufacturing techniques used in their fabrication, and their state and conservation [37]. In the following discussion, we show first results from the AC project [38–41] and examples of applications of thermal and cold neutron tomographic techniques to the study of phase and microstructure characterization of ceramic artifacts, ancient bronzes, and marble artifacts. A common feature in the selected examples is that the use of neutrons was motivated either by the general aim of developing new imaging techniques or by the need to perform a nondestructive analysis. Cast and restored metal objects were inspected to elucidate the production technology and the corrosion damage. Marble and ceramic artifacts were inspected to probe the presence of adhesives, glued parts, epoxy filler materials, or epoxy protective coatings.

For testing purposes, the AC archaeologists prepared "black boxes." These closed metal cubes contain objects of various shapes, forms, and elemental composition representative of the composition of real archaeological objects. The task was to recognize the elemental composition and the rest–strain of the inner objects by PGAI, NRCI, and NDI [38–41]. Neutron and X-ray radiography and tomography were used to visualize the inner content of the black boxes and to help with targeting the coordinates of the interesting spots for the new 3D neutron techniques. NT was performed at the ANTARES[3] facility [42] at FRM-II, and the X-ray tomographies were performed at the Centre for X-ray Tomography at the Ghent University [43]. Figure 13.11 shows a 3D visualization (tomograph) of the reconstructed X-ray data (a) and the corresponding neutron data (b) for one of the black boxes. Using solely the image content, it can be seen that neither X-ray tomography nor NT alone allows the derivation of both types of materials and the elemental composition of the scanned object. However, the

Fig. 13.11 X-ray tomograph (**a**) and neutron tomograph (**b**) of black box H-IX from the Ancient Charm project [39]

[3] Advanced Neutron Tomography and Radiography Experimental System.

combined use of the two techniques allows areas of differing materials and their shapes to be distinguished and separated. At a later stage, to determine precisely the elemental composition of these objects, PGAI analysis was performed directly at the coordinates provided by the NT image. Figure 13.12 shows an X-ray tomograph and the corresponding neutron tomographs for another black box. Here three separate layers of materials can be visually distinguished. The upper row shows corresponding horizontal slices of the object (looking "from above"); the bottom row shows vertical slices (looking "from the side"). A suitable isocontour of the X-ray slice helps in visualizing the initial misalignment of the NT image and its correct alignment after the registration. This example demonstrates in various ways the value added by the registration and the subsequent point-to-point comparison.

The neutron tomograph in Fig. 13.13 refers to laminated bronze artifacts from rich sepulchral complexes in the necropolis of *Osteria-Poggio Mengarelli* and *Cavalupo* (eighth century B.C.) [44]. Quantitative phase analysis was performed by neutron diffraction at the INES[4] diffractometer installed at the ISIS neutron source. This result was complemented by the analysis of the spatial distribution of the components, obtained with an NT experiment (Fig. 13.13),

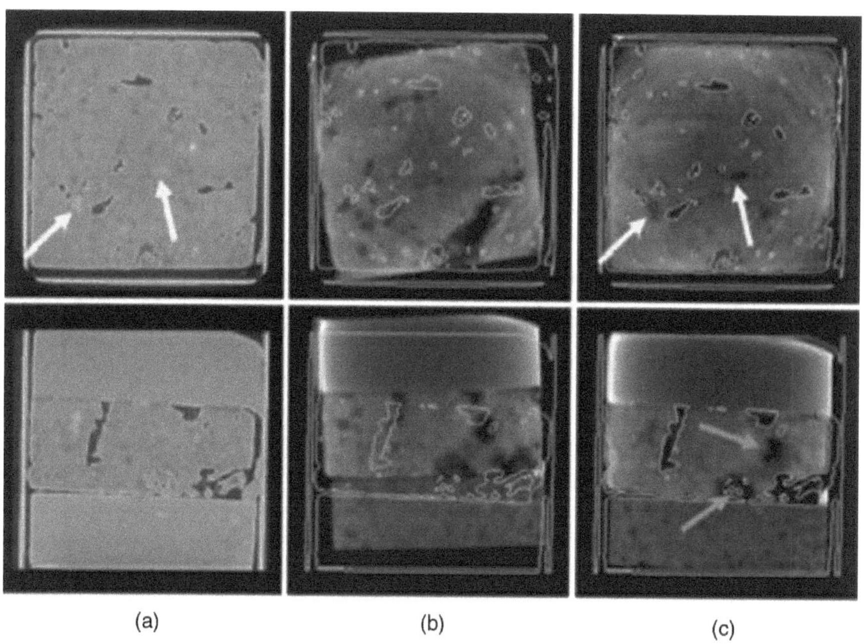

Fig. 13.12 **(a)** X-ray tomograph of black box D-VIII from the Ancient Charm project [39]; corresponding neutron tomograph in the initial position (**b**) and after automatic registration (**c**) [38]

[4] Italian Neutron Experimental Station, a collaborative project between ISIS and the Italian National Research Council [Consiglio Nazionale delle Ricerche].

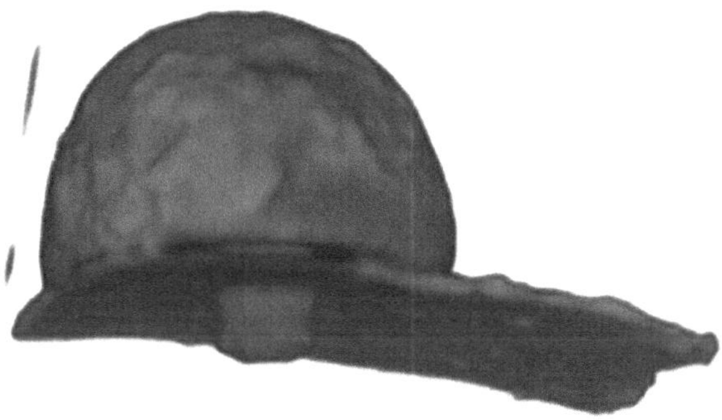

Fig. 13.13 Neutron tomography from bronzes [44] provided information on the spatial distribution of the components forming the samples. Measurements were performed at the COld Neutron RADiography (CONRAD) instrument at the Hahn-Meitner Institut (Berlin, Germany)

and additional information on elemental composition through XRF. A surface photomicrograph and infrared scan were used to complete the characterization of the bronzes.

The results in Fig. 13.14 refer to marbles from Villa Adriana [45, 46] for which the geological setting is a priori unknown. Marble is typically composed of carbonate minerals (such as calcite and dolomite), silicates (such as quartz, plagioclase and micas), oxides (such as rutile and magnetite), and phosphates (such as apatite) can be present as trace or minor phases. Natural calcite rocks and marbles commonly exhibit a preferred orientation (texture), which forms during growth or deformation and which is modified during recrystallization and plastic tectonic deformations. These geological textures, formed over millions of years, are not affected by quarrying or stone sculpting. Thus, in the archaeological context, the texture analysis of marble helps in attribution and provenance determination. It may further be useful in the field of conservation: for instance, the weathering of marble by anisotropic thermal expansion is also influenced by the crystallographic textures [47]. Based on references [45, 46], it was possible to distinguish among the different marble types purely on the basis of the mineral phase compositions and the crystallographic textures (geological stress). In particular, the phase analysis performed on the Villa Adriana marble fragments allows for a clear separation into four different groups. The same marbles were also investigated by thermal NT (Fig. 13.15). From this study, it was possible to distinguish distinct components and their relative spatial distributions in the marbles. These are indicated with different colors in the tomographic picture in Fig. 13.15. (a) darkest shade (most abundant component), (b) medium shade (intermediate component), and (c) lightest shade (smaller component) (courtesy of R. Triolo et al. [48]).

Finally, ceramic artifacts dated from the second to the third century (A.D.) from a large collection found in recent excavations (1988) in Puglia (Italy) were investigated. Their provenance is different areas of the central Mediterranean. Neutron diffraction and NT measurements [48] performed respectively at the ISIS pulsed neutron source and at PSI [49] provided valuable information on composition, provenance, and fabrication methods. Samples were found to be made of the same type of clay, with similar percentages of calcite, quartz, and

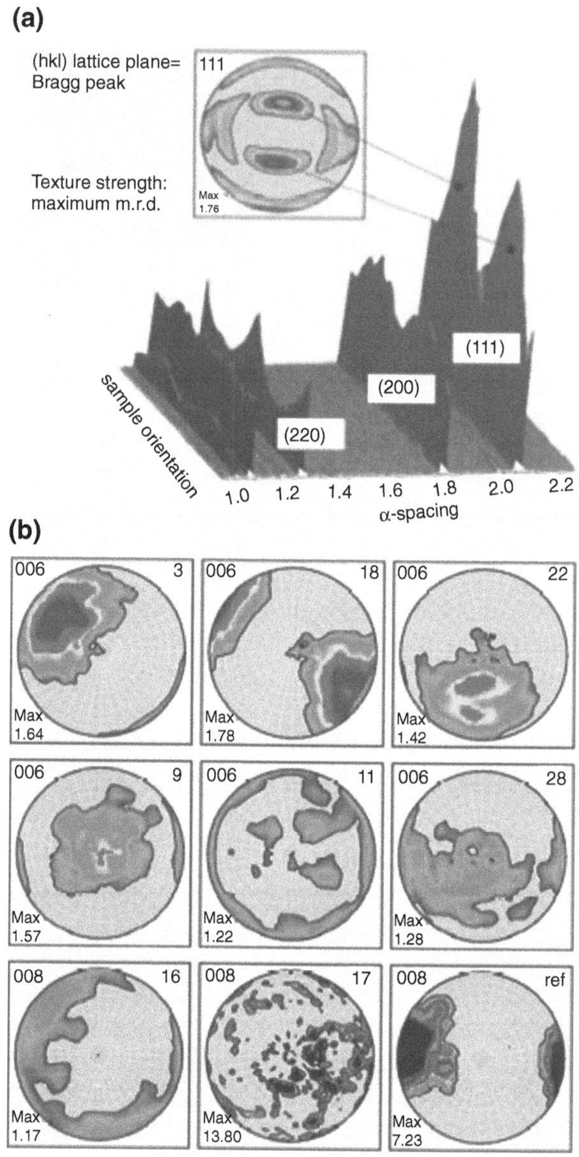

Fig. 13.14 (**a**) Example of texture analysis of a copper reference sample derived from diffraction data on the ROTAX instrument installed at ISIS (courtesy W. Kockelmann); (**b**) Maps of the c-axis distributions, represented by the (006) pole figures, of eight distinct marble samples from Villa Adriana [45] compared with a geological calcite reference sample from Syros, Greece. For each pole figure, the maximum multiples of random distribution are given

plagioclase. The examples in Fig. 13.16 demonstrate that a 3D inside view of objects a few centimeters thick can be achieved within a relatively short time (from a few minutes to a few hours) with spatial resolution of as low as 100 μm. The characterization of a ceramic artifact and the analysis of the composition of alloys in a metallic object provided an additional piece of information to aid in establishing whether the object belonged to the history of the local people or was imported from another culture [50].

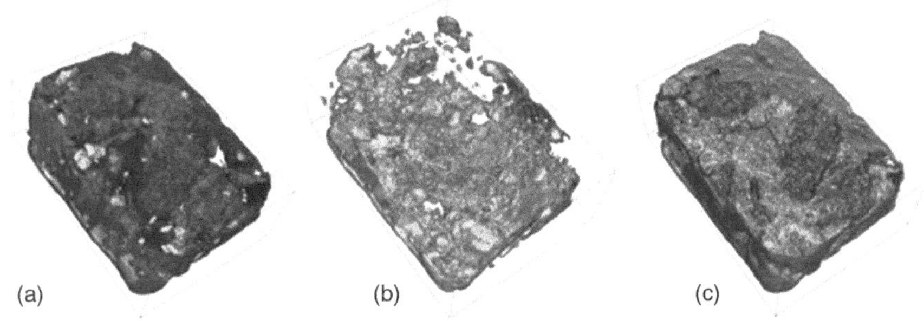

Fig. 13.15 Example of neutron tomography of marble fragments from Villa Adriana [48]

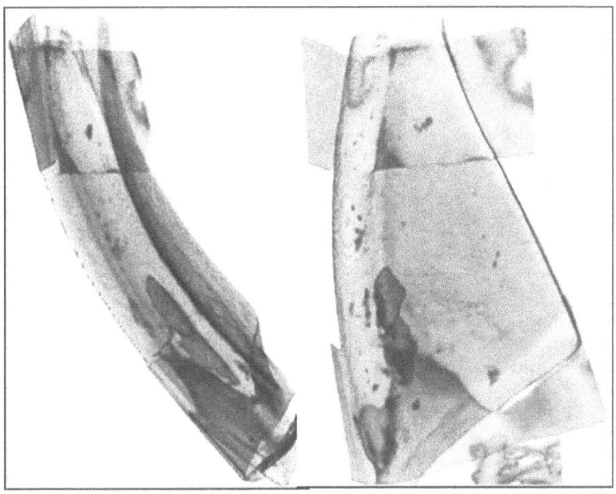

Fig. 13.16 Example of neutron tomography from ceramic artifacts [48]

The characterization by neutron diffraction of ancient bronze, marble, and ceramic artifacts illustrates how the penetrating power of neutrons, combined with imaging analysis, leads to advances in the scientific characterization of the materials that constitute a CH object. The techniques used are truly nondestructive, an important advantage if a unique and precious artifact is to be investigated without risk of damaging it in the analysis. The ongoing developments in applications of neutrons in CH research are promising and suggest that neutron-based techniques will have a lasting impact on the way CH analyses are performed. This will be especially the case if comprehensive analysis protocols are developed and adopted as best practices by the CH community.

Acknowledgment This work was performed with financial support from the European Community under contract HPRI-2001-50043 and within the CNR-Council for the Central Laboratory of the Research Councils (UK) agreement. Carla Andreani and Giuseppe Gorini acknowledge Consiglio Nazionale delle Ricerche (CNR), Italy, for financial support for the experiments performed at ISIS. Thomas Materna acknowledges financial support from the

Swiss National fund, the University of Fribourg (Switzerland), and the University of Cologne (Germany) for the experiments performed at PSI and the European Synchrotron Radiation Facility. We also wish to thank Giulia Festa for critical revision of the text.

References

1. Portal (News and developments from the CCLRC). *Looking forward to the past: science and heritage*. Special issue Autumn/winter (2006). http://www.scitech.ac.uk/Publications/nl/Portal/.
2. ESRF Newsletter. *Science and art*. **44** (December 2006).
3. M. Menu et al., The dedicated accelerator-based IBA facility AGLAE at the Louvre, *Nucl. Instrum. Meth. Phys. Res. B* **45**, 610–614 (1990).
4. E.H. Lehmann, P. Vontobel, B. Schillinger, T. Bücherl, S. Baechler, J. Jolie, W. Treimer, R. Rosa, G. Bayon, S. Legoupil, S. Körner, V. Micherov, M. Balasko, in *Proceeding of the 15th World Conf. on Non-Destructive Testing*, Rome, Italy, October 15–21, (2000). Available from http://www.ndt.net/article/wcndt00/papers/idn804/idn804.htm.
5. N. Kardjilov, S. Baechler, M. Bastürk, M. Dierick, J. Jolie, E. Lehmann, T. Materna, B. Schillinger, P. Vontobel, *Nucl. Instrum. Meth. Phys. Res.* A **501**, 536 (2003).
6. NEUTRA: Thermal neutron facility for imaging at the Paul Scherrer Institute, Villigen, Switzerland http://neutra.web.psi.ch/facility/index.html.
7. T. Materna et al., Uranium-sensitive tomography with synchrotron radiation, *J. Synchrotron Rad.* **6**, 1059–1064 (1999).
8. W. Kockelmann, G. Frei, E.H. Lehmann, P. Vontobel and J.R. Santisteban, *Nucl. Instrum. Meth. Phys. Res. A* **578**, 42 (2007).
9. G. Artioli, M. Dugnani, I. Angelini, L. Lutterotti, A. Pedrotti, A. Fleckinger, *Proc. Archaeometallurgy Europe* **2**, 19–27, (2003); G. Artioli, M. Dugnani, T. Hansen, L. Lutterotti, A. Pedrotti, G. Sperl, in: A. Fleckinger (ed.), *La mummia dell'età del rame. 2. Nuove ricerche sull'uomo venuto dal ghiaccio*, Vol. 3, Folio Verlag, Bolzano, 9–22, (2003).
10. E. Ciliberto and G. Spoto (eds.) *Modern Analytical Methods in Art and Archaeometry*, John Wiley & Sons, Inc. New York (2000).
11. D.R. Brothwell and A.M. Pollard (eds.) *Handbook of Archaeological Sciences*, John Wiley & Sons, Inc., New York (2001).
12. C.O.Fischer et al, Neues zur Neutronen-Aktivierungs-Autoradiographie, *Restauro* **105**, 426 (1999) (in German).
13. Zs. Révay and T. Belgya, Principles of the PGAA method, in G.L. Molnár ed., *Handbook of Prompt Gamma Activation Analysis with Neutron Beams*, Kluwer Academic Publishers, pp. 1–305 (2004).
14. H. Postma, M. Blaauw, P. Bode, P. Mutti, F. Corvi and P. Siegler, Neutron-resonance capture analysis of materials, *J. Radioanal. Nucl. Chem.* **248**(1), 115–120 (2001).
15. G. Gorini for the Ancient Charm collaboration, Ancient Charm: A research project for neutron-based investigation of cultural-heritage objects, *Il Nuovo. Cimento. C*, **30**, 47–58 (2007).
16. S.F. Mughabghab et al., *Neutron Cross-sections*, vol.1, Academic Press, New York (1981). Available online at http://atom.kaeri.re.kr/cgi-bin/endfplot.pl.
17. H. Postma, M. Blaauw, P. Schillebeeckx, G. Lobo, R. Halbertsma and A.J. Nijboer, Non-destructive elemental analysis of copper-alloy artefacts with epithermal neutron-resonance capture, *Czech. J. Phys.* **63**(Suppl. A), A233–A240 (2003).
18. M. Blaauw, H. Postma, P. Mutti, An attempt to date an antique Benin bronze using neutron resonance capture analysis, *Appl. Radiat. Isot.*, **62,** 429–433 (2005).
19. M. Blaauw, H. Postma and P. Mutti, Quantitative neutron capture resonance analysis verified with instrumental neutron activation analysis, *Nucl. Instrum. Meth. Phys. Res. A* **505**, 508–511 (2003).

20. H. Postma, P. Schillebeeckx, and R.B. Halbertsma, Neutron resonance capture analysis of some genuine and fake etruscan copper-alloy statuettes, *Archaeometry* **46** 635–646 (2004).
21. H. Postma and P. Schillebeeckx, Non-destructive analysis of objects using neutron resonance capture, memorial issue Gabor Molnár, *J Radioanal. Nucl. Chem.* **265,** 297–302 (2005).
22. R. C. Perego, H. Postma, M. Blaauw, P. Schillebeeckx, A. Borella, Neutron Resonance Capture Analysis: improvements of the Technique for resonances above 3 keV and new applications, *J. Radioanal. Nucl. Chem.* **271**(1) 89–94 (2007).
23. H. Postma, R.C. Perego, P. Schillebeeckx, P. Siegler, and A. Borella, Neutron resonance capture analysis and applications, *J. Radioanal. Nucl. Chem.* **271**(1) 95–99 (2007).
24. H. Postma, D. Fontijn, P. Schillebeeckx, R.C. Perego, and J.J. Butler, Neutron Resonance Capture Analysis of Geistingen axes, LUNULA, *Archaeologia protohistorica* **XIII**, 41–46, (2005).
25. H. Postma and P. Schillebeeckx, Neutron-resonance capture as a tool to analyse the internal compositions of objects non-destructively, *Notiziario Neutroni e Luce di Sincrotrone* **11**(2), 14–18 (2006).
26. Z.B. Alfassi, C. Chung (Ed.), *Prompt Gamma Neutron Activation Analysis*, CRC Press, Boca Raton, Florida, 1995.
27. M. Crittin et al. The new prompt gamma-ray activation facility at the Paul Scherrer Institute, Switzerland, *Nucl. Instr. Meth. Phys. Res. A* **449**, 221 (2000).
28. Zs. Revay et al., Cold neutron PGAA facility at Budapest, *Nucl. Instrum. Meth. Phys. Res. B* **213**, 385–388 (2004).
29. H-J Cho et al., Study on prompt gamma-ray spectrometer using Compton suppression system, *Nucl. Instrum. Meth. Phys. Res. B* **229**, 499–507 (2005).
30. Zs. Revay et al., Construction and characterization of the redesigned PGAA facility at the University of Texas at Austin, *Nucl. Instrum. Meth. Phys. Res. A* **577**, 611–618 (2007).
31. S. Baechler et al. The k0-method in cold-neutron prompt gamma-ray activation analysis, *J. Radioanal. Nucl. Chem.* **256**, 239–245 (2003).
32. S. Baechler et al., Non-destructive analysis of a bulky sample from a natural fossil reactor, *J. Radioanal. Nucl. Chem.* **250**, 39–45 (2001).
33. P. Kudejova et al., On the construction of a new instrument for cold-neutron prompt gamma-ray activation analysis at the FRM II, *J. Radioanal. Nucl. Chem.* **265**, 221–227 (2005).
34. Research reactor II (FRM-II), Technische Universität München, 2007 http://www.frm2.tum.de/en/index.html.
35. P. Kudejova, *Two new installation for non-destructive sample analysis: PIXE and PGAA*, Ph.D. Thesis, University of Cologne (2005).
36. E. Perelli Cippo et al, *Meas. Sci. Technol.* **19**, 034027 (2008).
37. W. Kockelmann, L.C. Chapon, R. Engels, j. Schelten, C. Neelmejer, H.M. Walcha, G. Artioli, S. Schalev, E. Perelli Cippo, M. Tardocchi, G. Gorini, and P. Radaelli, Neutrons in cultural heritage research, *J. Neutron Res.* **14**(1), 37–42 (2006).
38. P. Kudejova, J. Cizek, R. Schulze, J. Jolie, B. Schillinger, K. Lorenz, M. Mühlbauer, B. Masschaele, M. Dierick, J. Vlassenbroeck, A marker-free 3D image registration for the Ancient Charm project. Case study with neutron and X-ray tomography datasets, *Notiziario Neutroni e Luce di Sincrotrone* **12**(2), 6–13 (2007).
39. P. Kudejova for the Ancient Charm collaboration, Neutron and X-ray imaging of the 'black boxes' for the Ancient Charm project, *Archeometriai Mũhely*, 35–40 (2008/1).
40. G. Festa, W. Kockelmann, A. Kirfel and the Ancient Charm collaboration, Neutron diffraction analysis of 'black boxes', *Archeometriai Mũhely* 2008/1 61–72.
41. Z. Kis et al., Prompt gamma activation imaging on 'black boxes' in the 'Ancient Charm' project, *Archeometriai Mũhely*. 41–60 (2008/1).
42. ANTARES *Facility at the research reactor II (FRM-II)*, Technische Universität München (2007). http://www.physik.tu-muenchen.de/antares/.

43. *X-ray Radiography and Tomography Facility at the Ghent University (UGCT)* (2007) http://www.ugct.ugent.be.
44. G. Festa, P.A. Caroppi, A. Filabozzi, C. Andreani, M.L. Arancio, R. Triolo, F. Lo Celso, V. Benfante, S. Imberti, Composition and corrosion phases of Etruscan Bronzes from Villanovan Age, *Measur. Sci. Technol.* **19**(3) 034004 (7 pp) (2008).
45. A. Filabozzi, A. Pietropaolo, C. Andreani, M. P. De Pascale, G. Gorini,.W. Kockelmann, L.C. Chapon, Non invasive neutron diffraction analyses of marbles from the "Edificio con Tre Esedre" in Villa Adriana, *Il Nuovo Cimento C* **29**, 237–252 (2006).
46. A. Filabozzi, C. Andreani, M.P. De Pascale, G. Gorini, A. Pietropaolo, E. Perelli Cippo, R. Senesi, M. Tardocchi and W. Kockelmann, Texture and structure studies on marbles from Villa Adriana via neutron diffraction technique, *J. Neutron Res.* **14**, 55–58 (2006).
47. S. Siegesmund and K. Ullemeyer, T. Weiss, and E.K. Tschegg, *Int. J. Earth Sci.* **89**, 170 (2000).
48. R. Triolo, V. Bonfante, and F. Lo Celso, private communication (2007); see also N. Kardijlov et al, Neutron tomography in modern archaeology, *Notiziario Neutroni e Luce di Sincrotrone* **13**(2), 6–9 (2008).
49. G. Kühne et al., CNR—the new beamline for cold neutron imaging at the Swiss spallation neutron source SINQ, *Nucl. Instrum. Meth. Phys. Res. A* **542**, 264–270 (2005); ICON: Cold neutron facility for imaging at the Paul Scherrer Institute, Villigen, Switzerland, described in http://neutra.web.psi.ch/facility/index.html.
50. G. Festa for the Ancient Charm collaboration, ANCIENT CHARM: A new project for neutron-based 3D imaging with applications to CH research. Poster Section (Ist prize winner): *Looking Forward to the Past: Science and Heritage*, Tate Modern London (28 November 2006). Available online at http://ancient-charm.neutron-eu.net/ach/conferences.

Chapter 14
Probing the Potential of Neutron Imaging for Biomedical and Biological Applications

K.L. Watkin, H.Z. Bilheux, and J.F. Ankner

Abstract Neutron imaging of biological specimens began soon after the discovery of the neutron by Chadwick in 1932. The first samples included tumors in tissues, internal organs in rats, and bones. These studies mainly employed thermal neutrons and were often compared with X-ray images of the same or equivalent samples. Although neutron scattering is widely used in biological studies, neutron imaging has yet to be exploited to its full capability in this area. This chapter summarizes past and current research efforts to apply neutron radiography to the study of biological specimens, in the expectation that clinical and medical research, as well as forensic science, may benefit from it.

Keywords Biology · Neutron imaging · X-ray imaging · Biomedical · Contrast agent · Thermal neutrons · Epithermal neutrons · Small gadolinium particulate oxide (SPGO)

14.1 Introduction

Although carbon dominates the chemistry of biological tissues, hydrogen atoms are the most abundant. Protons (^1H) scatter thermal neutrons more strongly than any other nuclei. These two facts ensure that hydrogen is the primary contributor to neutron contrast in biological materials. In addition, since neutrons interact with the nucleus, deuterium (^2H) exhibits quite different scattering properties from ^1H. These contrast differences are already exploited in neutron scattering and imaging and can readily be applied in a biomedical context.

The principle of neutron imaging is based on the attenuation, via both scattering and absorption, of a directional neutron beam by the matter through which it passes. Neutron imaging is complementary to, rather than competitive with, X-ray techniques. X-rays are scattered and absorbed by electrons, so absorption and

K.L. Watkin (✉)
University of Illinois, College of Applied Health Science, 901 So. Sixth, Champaign, IL 61821, USA
e-mail: watkin@uiuc.edu

I.S. Anderson et al. (eds.), *Neutron Imaging and Applications*,
Neutron Scattering Applications and Techniques, DOI 10.1007/978-0-387-78693-3_14,

scattering increase monotonically with atomic number. Neutrons, on the other hand, interact with nuclei and their scattering power does not scale in any regular way with atomic number, as explained in Chapter 1, in this volume.

Neutron radiography and tomography have flourished for engineering and materials science applications, as illustrated in Chapters 11 and 12. Recent studies have been carried out in more biological contexts, such as water transport in plants, as referenced in Chapter 17, and resin uptake in wood [1].

Current probes for medical imaging such as X-rays, ultrasound, magnetic resonance imaging (MRI), and positron emission tomography (PET) principally determine differences in structural anatomy. Ultrasound and functional magnetic resonance imaging (fMRI) also enable the study of functional activity such as changing blood flow, whereas PET highlights specific metabolic functions by measuring the concentration and distribution of a radiopharmaceutical tracer similar to glucose. Neutron imaging is essentially similar in spatial resolution to these techniques, but it exhibits different elemental and isotopic contrasts, which may be advantageous in many cases.

14.2 First Experimental Measurements on Biological Tissues

Initial neutron imaging experiments were conducted in Germany in the mid-1930s [2], soon after the discovery of the neutron by Chadwick in 1932. Thermal neutron imaging of biological tissues was first conducted in the mid 1950s, and work continued through the 1960s. Due to a lack of efficient detectors for epithermal and fast neutrons, experiments during this period were limited to thermal neutron imaging and so used thin biological samples. These were often imaged by passing a beam of thermal neutrons through the sample and exposing a dysprosium- or indium-loaded plate that created beta-emitting radioactive isotopes following neutron capture. The activated plate was then developed using conventional autoradiographic techniques, by placing the plate in close contact with an X-ray-sensitive photographic film.

As a proof-of-principle, plant tissues were successfully imaged with thermal neutrons [3], as illustrated in Fig. 14.1. This work demonstrated the ability of neutrons to detect ^{1}H nuclei in biological tissues. Neutron radiographs of a boral (boron–aluminum alloy) sheet were also produced and compared with their X-ray counterparts to illustrate the complementarity of the two techniques.

Several-millimeter-thick biological samples were imaged with thermal neutrons, as referenced in [4–6]. However, when a much thicker sample such as a human chest was studied, the image (Fig. 14.2) was seriously distorted by both multiple scattering, arising mainly from ^{1}H nuclei, and poor detector spatial resolution. Furthermore, the resolution of the image was mediocre compared with that of an equivalent X-ray radiograph [7].

Although the strong attenuation of thermal neutrons drastically limited both the choice of biological tissues and specimen thickness, various animal tissues were exposed to thermal and cold neutrons and compared with X-ray

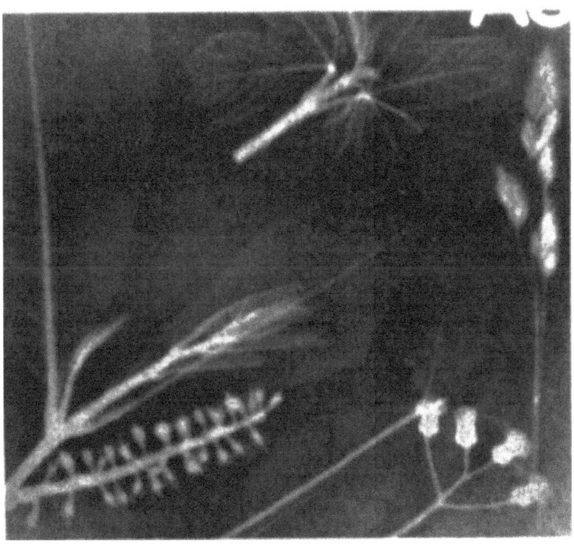

Fig. 14.1 Neutron radiography of plant tissues demonstrates the sensitivity of neutrons to hydrogenous materials [3]

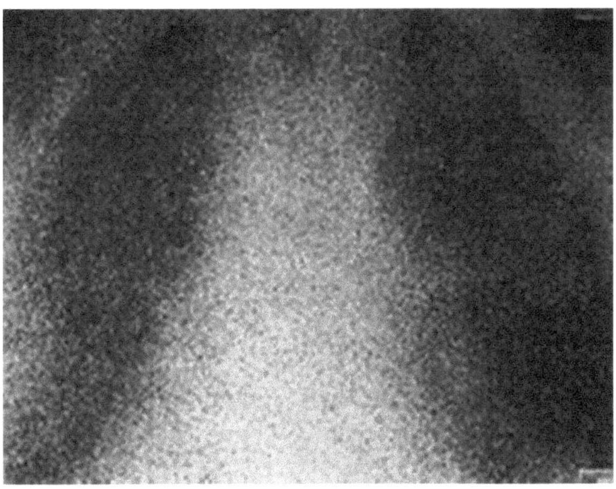

Fig. 14.2 An early thermal neutron radiograph of a human chest was limited by multiple scattering and poor detector spatial resolution [7]

radiographs [6]. Thermal neutron dose was estimated to be equivalent to X-ray dose, whereas incident epithermal dose was predicted to be lower for the same neutron exposure time – epithermal neutrons generally have lower elastic scattering cross sections than thermal neutrons. Gadolinium and indium were commonly used as contrast agents for thermal and epithermal neutron imaging, respectively. A brief discussion of radiation damage due to neutrons can be found in a later section of this chapter.

Figure 14.3 shows images of thin tumor specimens: (A) adenocarcinoma, (B) liposarcoma, and (C) leiomyoma [6]. Thermal neutron images are displayed in the upper row. The control images for each tumor, 20-kV roentgenograms (X-ray images), are shown in the lower row. The greater contrast seen in the neutron images probably arises from differences in hydrogen density, since many tumors appear to have greater hydrogen content than healthy tissues [6].

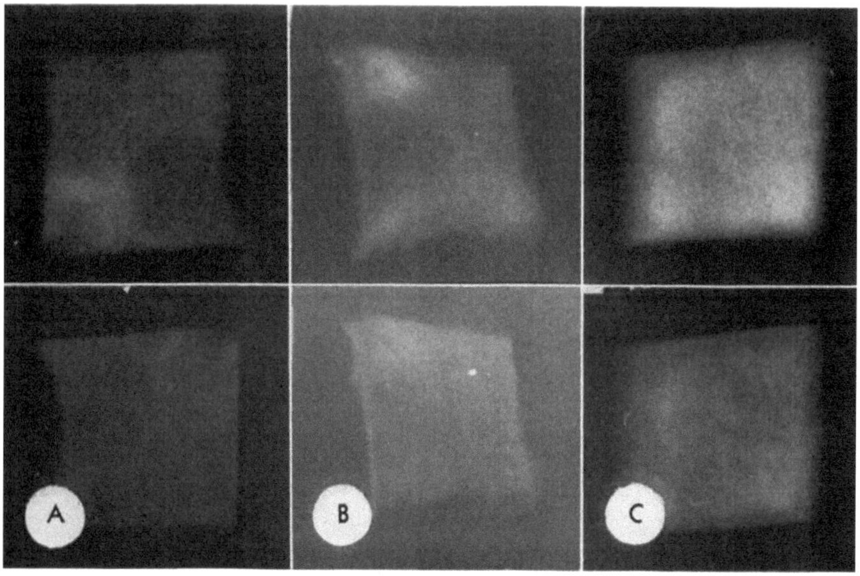

Fig. 14.3 Comparison of tumors imaged with thermal neutrons (*top*) and 20-kV X-rays (*bottom*). The contrast in the neutron radiographs is more pronounced because of higher hydrogen concentrations in tumorous tissue [6]

The complementarity between neutrons and X-rays is better illustrated by images of a rat paw and tail, shown in Fig. 14.4. Although bones are clearly visible in the X-ray radiograph, blood vessels are more prominent in the neutron image. The nasopharynx, which is obscured by bone in the X-ray radiograph, is clearly visible in the epithermal neutron image, as illustrated in Fig. 14.5. Depending on the biological sample and the region(s) of interest, there are significant differences in visible features between X-ray and neutron radiography.

Much of the early interest in neutron imaging of biological tissues was generated by the need to determine the distribution of boron delivery agents being developed for treatment of cancers using boron neutron capture therapy (BNCT). In this method, boron atoms are loaded into cancerous cells, which are then irradiated by neutrons; boron absorbs a neutron and produces an energetic alpha particle and ^{7}Li nucleus, which kill the cell. The thermal neutrons used were either strongly absorbed by the body before reaching the cancerous cells, resulting in high radiological dose, or required highly invasive surgery to expose the cancer more directly to the neutron beam [8]. In the

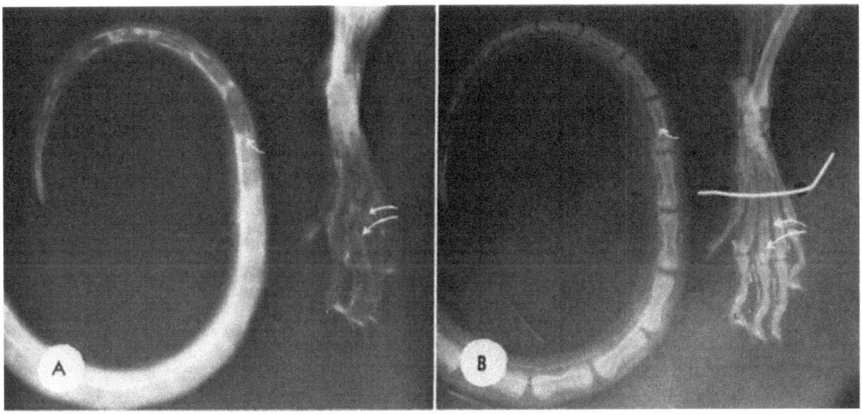

Fig. 14.4 Radiographs of a rat paw and tail using (**A**) thermal neutrons and (**B**) 42-kV X-rays. Gd_2O_3 in oil was used as the contrast agent [6]

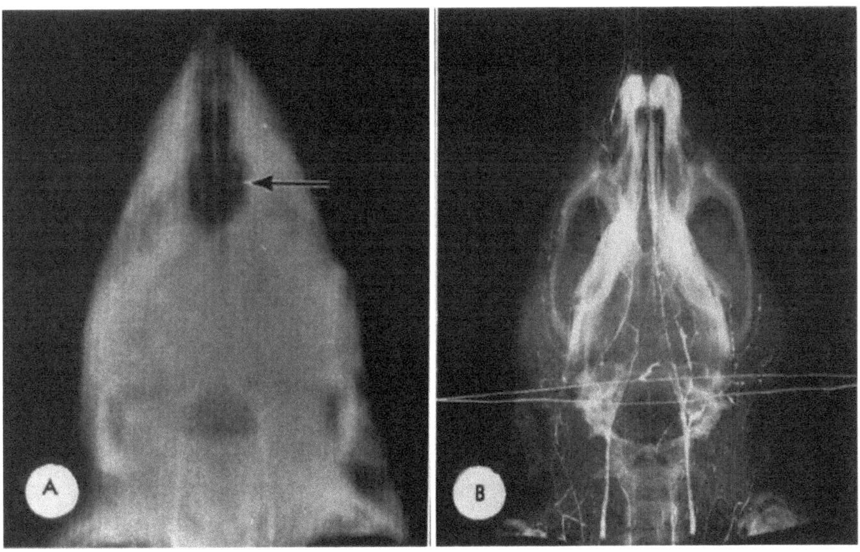

Fig. 14.5 Radiographs of a rat head using (**A**) epithermal (i.e., 0.5 to 2 eV) neutrons and (**B**) 42-kV X-rays. The rat nasopharynx is visible in (A), as indicated by the arrow. Although In_2O_3 was used as a contrast agent, the epithermal neutron flux was not sufficient to demonstrate its contrast capabilities [6]

1990s it was realized that epithermal beams should be more favorable for BNCT, since body tissues act as a natural neutron moderator. Interest in clinical applications reawakened, as did interest in thermal and epithermal imaging of biological tissues.

14.3 Neutron Dosimetry: A Short Overview

As with any imaging method, there is a need to understand and evaluate the benefits and risks of exposing a living subject to neutrons. In order to estimate the potential radiological hazard that neutron imaging poses to soft biological tissues, Monte Carlo simulations of neutron exposure of a simulated human arm were conducted [9]. The conclusions were quite pessimistic. The simulations predicted no effective contrast at given neutron energies of 1 keV, 120 keV, fission spectrum, and 14 MeV. However, the study provided useful information on dose deposition between tissue, bone cortex, and marrow. For example, the calculated neutron dose at 120 keV energy to bone was twice the dose to contiguous soft tissue.

While epithermal (1 eV–0.1 MeV) and fast (>0.1 MeV) neutrons are required to penetrate thick tissues, they do not pose a direct radiological hazard. Rather, hydrogenous tissue moderates these neutrons down to the thermal energy range (<1 eV) at which they can be captured and thereby cause biological damage. Fast neutrons also knock off protons from hydrogenous tissues, creating an additional radiological hazard.

Recently, Monte Carlo simulations for neutron stimulated emission computed tomography (NSECT), an in vivo tomographic spectroscopy using fast neutrons (Chapter 15), demonstrated that the radiation dose deposited in a human abdomen or breast using NSECT would be comparable to the dose from an abdominal CT (computed tomography) exam or a mammogram, respectively [10–12].

Clinical, i.e., small-animal, and pharmaceutical research may benefit the most from epithermal neutron imaging. The exposure of the specimen is a lesser concern in those fields, and imaging may aid medical diagnosis without the risk of radiation exposure.

Recent data on the biological effects of low radiation exposures can be found in Ref. [13]. Information on the biological effects of fast neutrons on biological tissues can be found in Refs. [14, 15].

14.4 Recent Trends

Neutron beams generated by high-energy medical linacs have recently been shown to be sufficiently intense to produce images exhibiting soft-tissue contrast [16]. It was reported that direct thermal neutron imaging was not as contrast-sensitive as indirect neutron imaging (Chapter 5) or fast neutron imaging. Later, neutron radiography of rabbit VX-2 liver cancer using thermal and cold neutron radiography permitted clear observation of hepatic vessels and the VX-2 tumor, with cold neutron imaging being especially effective. The image contrast of this modality enhanced tumor identification compared with absorption-contrast X-ray radiography [17].

Very recently (Figs. 14.6 and 14.7), a study of rat femurs demonstrated the relative strengths of X-ray and neutron imaging. Heavy elements (predominantly calcium and phosphorus in bone) determine X-ray contrast, so the X-ray images highlight the hard bone framework, with the hydrogenous matter being essentially invisible. Neutrons, on the other hand, highlight a relatively uniform distribution of hydrogen throughout the interior and exterior regions of the bone and do not resolve the filaments of trabecular hard bone running through the marrow. Interestingly, the neutron images (Fig. 14.7) show an obvious bone reduction in the decalcified samples – after inorganic compounds are dissolved by soaking in a 15% ethylene diamine tetra acetic acid solution, as expected

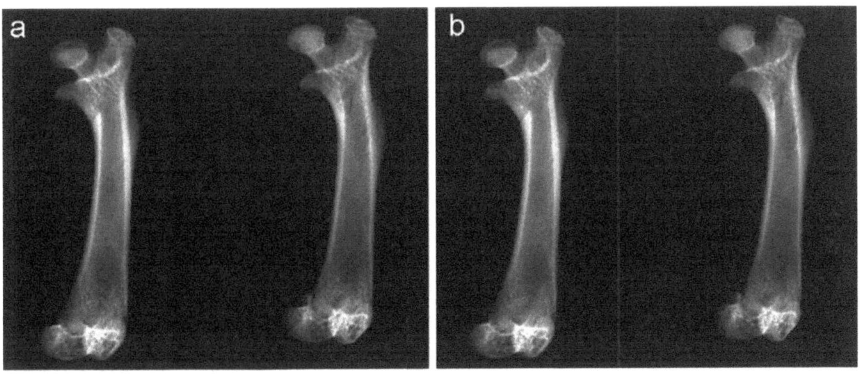

Fig. 14.6 Images made by 40-keV X-ray radiography of (**a**) calcified bones (i.e., heated at 600°C for 5 h) and (**b**) natural bones. The distinction between the cortical (surface) and the trabecular (inner) bone regions is evident [18]

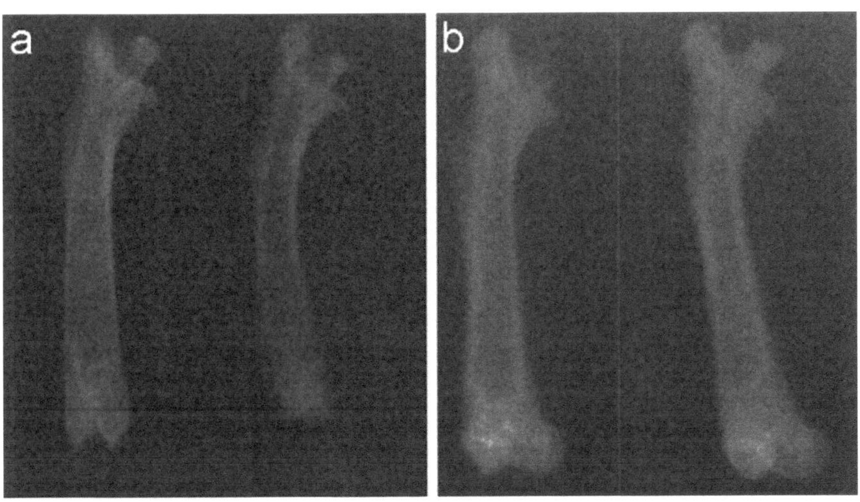

Fig. 14.7 Neutron radiographs of rat bones from (**a**) decalcified and (**b**) natural bones [18]

from the egression of mineral material from inside the bone. Post-mortem bone desiccation was also evaluated as a possible forensic tool [18].

Several recent neutron imaging experiments on biological samples have been published, e.g., the remains of an ichthyosaur embryo imbedded in a shell [19], arachnids [20], and insects [21]. Examples of radiographs of other biological specimens can be found at http://neutra.web.psi.ch/gallery/biological.html.

Adult male porcine lung tissue (1.5 mm thick, 1 cm by 1 cm sample size) immersed in formalin was successfully imaged using thermal neutrons. These slices were first extracted from frozen samples from the anatomical region illustrated in Fig. 14.8.

Standard histological techniques were used to allow comparison of neutron and optical images (Fig. 14.9). After neutron exposure, the samples were stained using hematoxylin and eosin stain, which is known to provide good resolution of lung tissue components such as muscle and cartilage.

Although the images in Fig. 14.9 look very similar, the distribution of their pixel intensities can supply a true quantitative comparison. Recently developed software [22] enables automatic integration of multiscale volumes from different imaging modalities through a cyber-based portal imaging system. It was found that 92% of the pixel intensities of these two images match, meaning the pixel-by-pixel intensity between the histological and the neutron images is very similar. These initial results for fixed porcine lung tissue provide significantly more detail than any previously reported images taken with a thermal neutron beam [7]. To compare thermal neutron imaging with X-ray imaging, the specimen was also imaged using microCT, as shown in Fig. 14.10. The microCT measurement was conducted before histological staining and subsequent optical imaging. The gross morphology in all the images is very similar. Cartilage

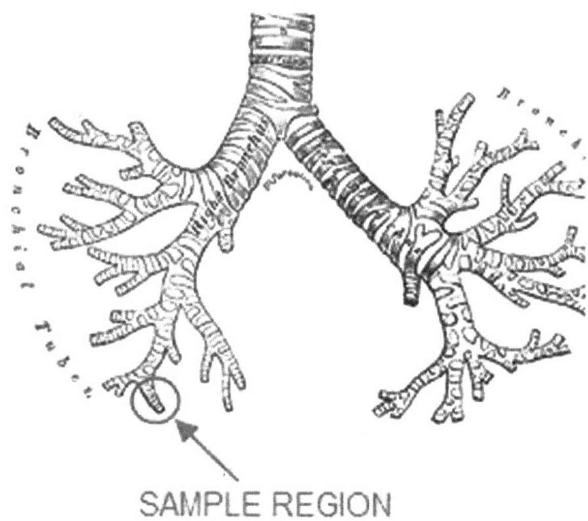

Fig. 14.8 Approximate location of the thin porcine lung specimen imaged using thermal neutrons. Only the lung airways are illustrated

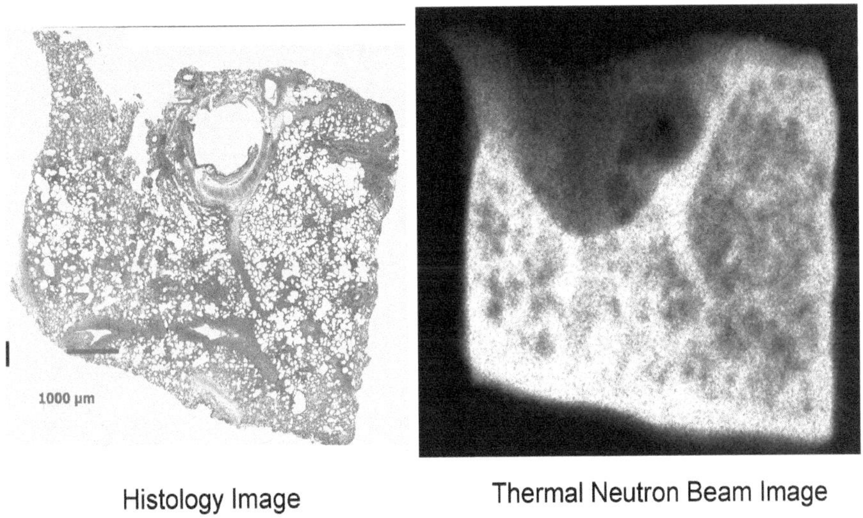

Fig. 14.9 Histological and thermal neutron images of the same section of porcine lung, with a 92% pixel agreement (see text)

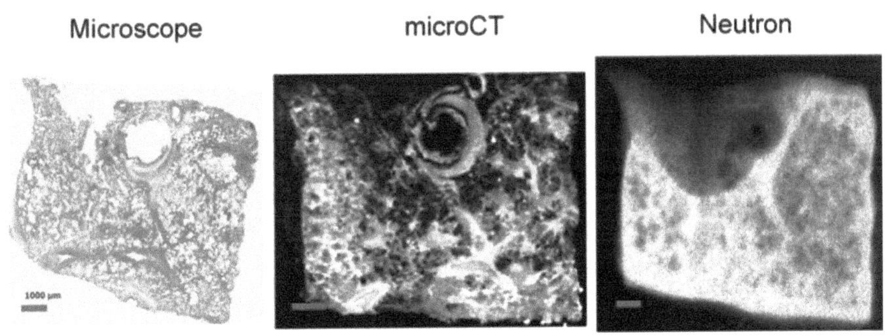

Fig. 14.10 Multimodal images of the same specimen of porcine lung tissue

supporting the airway is clearly visible in each modality, along with differentiation of the alveolar regions, although the greatest definition is seen in the histology images.

In a recent study, the anatomical changes in corn kernels affected by *Aspergillus flavus* fungus were analyzed quantitatively using neutron tomography [23], demonstrating the effectiveness of the technique in identifying the degradation of the biological tissues compared with tissues in uninfected control kernels. Resistant kernels showed no sign of anatomical change. This interesting research shows the sensitivity of neutron radiography to the effects of diseases on biological tissues. The most recent and promising studies are illustrated by tomography of a rat lung [24].

14.5 Small Particulate Gadolinium Oxide (SPGO) Nanoparticles for Targeted and Nontargeted Contrast Enhancement for Future Biomedical Neutron Imaging Applications

There is a great and largely untapped potential for the use of neutrons in biomedical imaging using isotopically labeled marker compounds. Medical compounds incorporating neutron-absorbing isotopes can be targeted to specific organs, such as the heart, to enable neutron imaging of anatomical structure and organ function as well as to deliver radiation doses specifically to tumor sites. Such a system could provide higher-resolution images than PET and more precision in targeting disease. Neutrons may also be used with water-stable SPGO (small particulate gadolinium oxide) nanoparticles [25–27] for targeted or non-targeted contrast enhancement. Neutron imaging through metal – e.g., stents and other metallic objects used for biological purposes – is possible. Small-animal imaging studies, isotopic imaging, and targeted contrast-enhanced imaging are all potential applications.

Recent experiments highlight the biomedical potential of the neutron-sensitive SPGO contrast agent. Dextranized SPGO (10 nm) was injected into the portal vessel of a porcine liver specimen. A thermal neutron radiograph of the thin specimen (~1.5 mm) with the dextranized SPGO highlighting the lumen of the vessel is shown in Fig. 14.11. SPGO particles appear dark because of the high thermal neutron capture cross section of gadolinium.

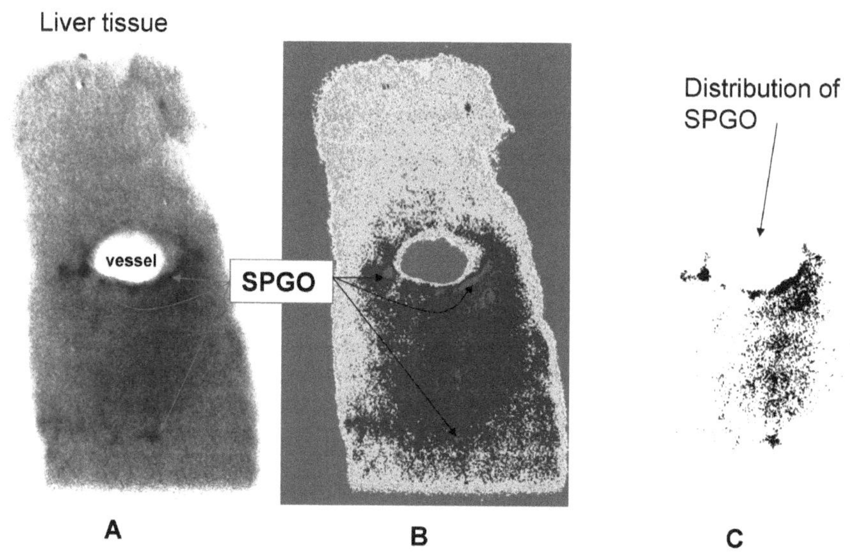

Fig. 14.11 Neutron images of SPGO-doped liver tissue. **A**: original liver specimen; **B**: color-enhanced image of A; **C**: distribution of the SPGO nanoparticles created using simple thresholding techniques

14.6 Summary

Many recent advances in medical imaging are based upon technological advances in computer visualization, refined image projection techniques, and advanced charge-coupled device systems, among others. Although there have been several recent attempts to create small neutron generating systems that may be useful at the laboratory level, the use of larger reactor- and accelerator-based neutron sources with dedicated imaging systems remains the most effective and efficient way to use thermal neutrons for biological tissue imaging.

Neutron radiographs of biological tissues accompanied the first neutron imaging experiments. Neutron imaging of biological samples flourished during the 1960s and 1970s and then largely disappeared until the 1990s. The recent emergence of new techniques and improved detection (e.g., for epithermal neutrons – Chapter 4) are leading to a rejuvenation of the field of neutron imaging.

Neutron imaging of living human subjects still provokes obvious radiological concerns. However, its potential for applications such as clinical research, small-animal imaging, and forensic science needs to be investigated further. The high sensitivity of neutrons to hydrogen atoms should be of benefit for forensic applications, where quantification of water changes in dead tissues is highly important.

Acknowledgments This work is supported by the U.S. Department of Energy. The Oak Ridge National Laboratory Spallation Neutron Source is managed by UT-Battelle, LLC, for the Department of Energy under contract DE-AC05-00OR22725. We would like to thank Dr. John Nemeth of Oak Ridge Associated Universities for financial support of part of this research, under contract DE-AC05-06OR23100. We would also like to thank our many colleagues, especially Daniel Hussey at the National Institute of Standards and Technology Center for Neutron Research and Candice Halbert from Oak Ridge National Laboratory.

References

1. E. Lehmann, S. Hartmann, P. Wyer, *Nucl. Instrum. Meth. A* **542**, 87 (2005).
2. H. Kallman, *Research* **1**, 254 (1947).
3. J. Thewlis et al., *Br. J. Appl. Phys.* 7, 345 (1956).
4. H.L. Atkins, Biological application of neutron radiography, *Mat. Eval.* **23**, 453–458 (1965).
5. Barton, J. P., Some possibilities of neutron radiography, *Phys. Med. Biol.* **9**, 33–42 (1964).
6. M. Brown and P.B. Parks, *Am. J. Roentg.* **106**, 472 (1969).
7. J. Anderson et al., *Br. J. Radiol.* **37**, 957 (1964).
8. R.F. Barth, Coderre, M.G.H.Vicente and T.E.Blue, *Clin. Can. Res.* **11**, 3987–4002 (2005).
9. T.F. Budinger et al., *Phys. Med. Biol.* **16**(3), 439–450 (1971).
10. C.E. Floyd et al., *Phys. Med. Biol.* **51**, 3375–3390 (2006).
11. J.E. Bender, A.J. Kapadia, A.C. Sharma, G.D. Tourassi, B.P. Harrawood, and C.E. Floyd, *Med. Phys.* **34**, 3866–3871 (2007).
12. A.J. Kapadia, A.C. Sharma, J.E. Bender, G.D. Tourassi, C.R. Howell, A.S. Crowell, M.R. Kiser, B.P. Harrawood, R.S. Pedroni, and C.E. Floyd, *IEEE Trans. Nucl. Sci.* **55**(1), 501–509 (2008).

13. F. Zolzer, C. Streffer, *Radiat. Res.* **169**, 207–213 (2008).
14. Health effects of exposure to low levels of ionizing radiation: BEIR V, National Academic Press, ISBN 0-309-03997-5.
15. Introduction to Radiological Physics and Radiation Dosimetry, Wiley-VCH, ISBN 0-471-01146-0.
16. N. Adnani and B.G. Fallone, Neutron imaging using Medical Linacs, *Proceedings of the 22nd Annual EMBS International Conference*, July 23–28, Chicago IL. (2000).
17. Y. Tsuchiya, M. Matsubayashi, T. Takeda, T. Lwin, J. Wu, A. Yoneyama, A. Matsumura, T. Hori and Y. Itai, Imaging of rabbit VX-2 hepatic cancer by cold and thermal neutron radiography *Jpn. J. Appl. Phys.* **42**, 7151–7153 (2003).
18. K.K. Moghaddam, T.T. Taheri, and M. Ayubian, Bone structure investigation using X-ray and neutron radiography techniques, *Appl. Radiat. Isot.* **66**, 39–43 (2008).
19. D. Schwarz, P. Vontobel, E.H. Lehmann, C.A. Meyer, and G. Bongartz, Neutron tomography of internal structures of vertebrate remains: A comparison with X-ray computed tomography, *Palaeontol. Electron.* **8** (2): Art. No. 30A (2005).
20. J.T. Cremer, M.A. Piestrup, H. Park, C.K. Gary, *Appl. Phys. Lett.* **87**, 161913 (2005).
21. B. Allman et al., *Nature* **408**, 158–159 (2000).
22. http://www.ncsa.uiuc.edu/Conferences/2007Meeting/agenda.html, see Bajcy and Watkin presentation.
23. T.E. Cleveland et al., *J. Cereal Sci.* **1** (2008).
24. B. Schillinger et al., private communication.
25. M.A. McDonald and K.L. Watkin, *Acad. Radiol.* **13**(4), 407–536 (2006).
26. M.A. McDonald, K.L. Watkin, *Invest. Radiol.* **38**(6), 305–310 2003.
27. K.L. Watkin, M.A. McDonald, *Acad. Radiol.* **9**(S2), 285–289 (2002).

Chapter 15
Neutron Stimulated Emission Computed Tomography: A New Technique for Spectroscopic Medical Imaging

A. J. Kapadia

Abstract Neutron stimulated emission computed tomography (NSECT) is being developed as a new medical-imaging technique to quantify spatial distributions of elements in a sample through inelastic scattering of fast neutrons and detection of the resulting gamma rays. It has the potential to diagnose several disorders in the human body that are characterized by changes in element concentration in the diseased tissue. NSECT is sensitive to several naturally occurring elements in the human body that demonstrate concentration changes in the presence of diseases. NSECT, therefore, has the potential to noninvasively diagnose such disorders with radiation dose that is comparable to other ionizing imaging modalities. This chapter discusses the development and progress of NSECT and presents an overview of the current status of the imaging technique.

Keywords Inelastic scatter · Medical imaging · NSECT · Neutron imaging · Spectroscopy · Tomography

15.1 Introduction and Background

Neutron spectroscopic techniques have been widely used in medical and biomedical research to detect the presence of elements in the body. Techniques, such as instrumental neutron activation analysis (INAA), prompt-gamma neutron activation analysis (PGNAA), and delayed neutron activation analysis (DNAA), have been used since the 1960s to identify elements in healthy and diseased tissue, and techniques based on inelastic scattering of fast neutrons have been used frequently in body composition studies to measure whole body carbon and oxygen content.

The use of NAA techniques for medical applications was first reported in 1964 by Anderson et al. [1] for measurement of sodium in the body. Between

A.J. Kapadia (✉)
Duke Advanced Imaging Laboratories, Department of Radiology, Duke University Medical Center, Durham, NC 27705, USA
e-mail: anuj.kapadia@duke.edu

I.S. Anderson et al. (eds.), *Neutron Imaging and Applications*,
Neutron Scattering Applications and Techniques, DOI 10.1007/978-0-387-78693-3_15,

1968 and 1972, Chamberlain reported the measurement of body calcium and sodium in the body [2–4] and described techniques for whole-body NAA [5–7] and pulsed NAA [8]. In 1971, Cohn and Dombrowski reported the measurement of calcium, sodium chlorine, nitrogen, and phosphorus in the human body through in vivo NAA [9]. Since then, NAA and PGNAA have been used for a variety of applications, such as the measurement of nitrogen [10], carbon and oxygen [11], cadmium [12], and manganese [13] in the body and in trace element research to identify cancerous tissue [14–17]. The development of NAA has been documented in several review articles [18–21], which the reader may refer to.

Inelastic neutron scatter analysis (INSA) using fast neutrons was initially reported by Kyere et al. in1982 [22], who used 14 MeV neutrons from a (d,T) sealed-tube neutron generator to determine whole body carbon content as a measure of energy expenditure in the body. Following this experiment, Kehayias et al. reported methods to determine fat content in the body through measurement of whole body carbon and oxygen through inelastic neutron scatter [23–25], which they later combined with hydrogen and nitrogen measurement through in vivo NAA [26]. Until recently, most experiments utilizing fast-neutron inelastic scatter focused primarily on in vivo measurement of body fat.

The use of nuclear resonance scattering (NRS) was reported by Wielopolski et al. for detection of iron in the liver [27] and in the heart [28] using an indirect method of nuclear excitation by gamma rays generated through neutron capture. Recently, these authors reported the use of 14 MeV neutrons for in vivo measurement of liver iron through INSA and NRS [29].

In this chapter, we would discuss the development of a tomographic technique that uses inelastic scattering of fast neutrons and detection of the resulting gamma rays to quantify spatial distributions of elements in a sample. The technique, called neutron stimulated emission computed tomography (NSECT) [30, 31], was pioneered at Duke University in 2003 by the late Dr. Carey E. Floyd Jr. for the purpose of diagnostic medical imaging. NSECT uses a beam of fast neutrons to excite stable isotopes of elements in a sample to determine their concentration and spatial distribution within the sample. Since it is sensitive to a wide variety of elements that naturally occur in the human body, it has the potential to noninvasively diagnose several disorders that are characterized by changes in element concentrations in the diseased tissue [32–34].

Here we present an overview of the NSECT technique and discuss the current status of the research. The reader may note that while medical imaging of the human body is the ultimate goal of NSECT, the technique is still in development stages, with most experiments being performed on phantoms and excised tissue specimens. The methods and results presented in the chapter are mainly for the purpose of demonstrating feasibility and proof-of-concept for future medical applications.

15.1.1 Principle

The principle of NSECT can be described briefly as follows. A neutron incident on a sample travels freely along its projected path until it collides with an atomic nucleus of an element present in the sample. If the collision with the atomic nucleus results in inelastic scatter, the nucleus can get excited to one of its quantized higher-energy states. The excited nucleus is often unstable and will rapidly decay to a lower energy state, emitting a gamma-ray photon with energy equal to the difference of the two states. These energy states are well established for most elements and isotopes and are mostly unique for the elements commonly found in the body. Therefore, the energy of the emitted gamma photon can be treated as a unique signature of the emitting element. Tomographic detection and analysis of gamma lines in the emitted spectrum provide quantitative information about the spatial distribution of the element in the sample.

While the principle of NSECT is similar to that of gamma-ray emission computed tomography (G-ECT) [35], NSECT provides the advantage of being able to detect stable isotopes of elements that naturally occur in the body. Unlike G-ECT, NSECT does not require to inject radioactive tracers to the patient to produce emission decay. Stable isotopes of naturally occurring elements can be excited through inelastic scatter and detected to obtain spatial and quantitative information about their distribution in the body.

While NSECT can be used over a wide range of elements and isotopes, the elements of interest for medical applications show characteristic gamma lines with energies ranging from 350 to 6500 keV [34]. To induce gamma emission from these transitions, the energy of the incident neutrons is required to be greater than the largest energy state to be excited in the target. For most medical applications, neutron energies of approximately 7–8 MeV are sufficient to image elements found commonly in the body. NSECT applications at Duke University typically use neutron energies approximately 1 MeV greater than the highest excited state desired to compensate for energy loss in propagating through the body toward the targeted element in the tissue. For example, an investigation of ^{12}C at 4.4 MeV can be performed with a 5–6 MeV beam, whereas detection of ^{56}Fe at 847 keV can be performed with a 1.5–2 MeV beam.

As the name suggests, NSECT is a computed tomography application of neutron inelastic scatter spectroscopy. Tomography is performed using a single-beam, fixed-detector geometry similar to the acquisition mode of first-generation computed tomography systems where a single beam is translated and rotated around the object in discrete steps until the entire volume of the object is sampled [36]. In NSECT, a thin neutron beam is used to illuminate a single line through the object being scanned. Each line, also called a projection, represents the location of characteristic gamma intensities within the object. Tomography is performed by translating the object through the beam, rotating it through a fixed angle, and repeating the translation and rotation until the entire volume of the object has been scanned. The set of projections obtained from the tomographic scan can then be reconstructed to obtain a two-dimensional image of the spatial distribution and concentration of each element in the object. It can be noted that while first-generation CT translated the beam and detector, the use of Van-de-Graaff sources in NSECT requires that the sample be moved. In both cases, the final outcome is the same (i.e., the entire volume of the sample object is scanned with the beam).

Owing to the high gamma energies observed in NSECT, it is impractical to use standard gamma cameras that are normally used in G-ECT studies. The G-ECT gamma cameras are unable to provide sufficient spatial or energy resolution for NSECT imaging. NSECT therefore uses high purity germanium (HPGe) detectors, which exhibit excellent energy resolution within the range of energies observed in body elements [37]. Unfortunately, HPGe detectors with adequate efficiency are usually large single-crystal detectors that do not have the ability to provide spatial information. Spatial information is obtained by tracking the position of the neutron beam as it illuminates a single projection path at a time through the sample. Consequently, the current implementation of NSECT requires a collimated neutron beam to facilitate tomographic acquisition.

15.1.2 Applications in Medical Imaging

NSECT has the potential for use in a variety of applications in both medical and biological imaging research. It is sensitive to a wide variety of elements that naturally occur in the human body and has the ability to provide a three-dimensional map of the elemental distribution within the target object from a single scan. These characteristics of NSECT have the potential to diagnose several disorders in the human body that are characterized by changes in element concentrations in the diseased tissue. Disorders such as iron and copper overload in the liver [38–41] and several types of cancers including breast [14, 16, 17, 42–48], prostate [49–51], and brain [52, 53] have been associated with changes in element concentrations. NSECT has the potential to diagnose these disorders through a noninvasive in vivo scan without the use of radioactive isotope tracers. Three potential medical and biological applications of NSECT are discussed below.

15.1.2.1 Cancer Diagnosis

Several studies have indicated that changes in trace element concentrations in human tissue may be a precursor to malignancy in several organs, such as the brain, prostate, and breast. In vivo measurement of such cancer-marker trace elements has the potential to differentiate between malignant and benign tissue at very early stages and enable in vivo cancer diagnosis through element quantification. Concentration differences measured through NAA have been reported in the breast [16, 17, 42, 45, 46, 48], colon [54–61], prostate [49, 51, 62–66], kidney [54], liver [50, 54, 67, 68], stomach [54], urinary bladder [54], testis, and female reproductive organs [54]. NSECT cancer diagnosis can be performed by measuring the trace element composition of the tissue and analyzing this information to identify combinations of elements that are markers of cancer. Owing to NSECT's tomographic ability, given adequate sensitivity, it can identify both the location of the tumor as well as its cancer state from a single scan.

15.1.2.2 Noninvasive Measurement of Liver Iron and Copper

A second application of NSECT is in the diagnosis of liver disorders such as hemochromatosis [38–40] and Wilson's disease [41], which exhibit changes in iron and copper concentrations in the liver, respectively. Iron overload is a serious concern for patients with thalassemia, sickle cell anemia, and hemochromatosis mutations and for other patients who receive chronic blood transfusions. Similarly, copper overload is observed in patients with Wilson's disease. The current definitive diagnostic technique for these disorders is invasive liver biopsy followed by chemical analysis of iron to determine the severity of the disorder [38, 40, 69]. NSECT provides a noninvasive alternative with the potential to obtain the same information without the need for a biopsy. Several different techniques (reported in [33]) have attempted to detect and quantify iron concentration in the liver in vivo, for example, X-ray computed tomography (X-ray CT), MRI, magnetic resonance spectroscopy (MRS), super-conducting quantum interference susceptometry (SQUID), and nuclear resonance scattering (NRS) [27, 70–88]. While each has had varying degrees of success, few have found widespread acceptance in the clinical environment. Both X-ray CT and MRI lack the sensitivity required for detecting mild to moderate degrees of iron overload, and MRI also shows reduced sensitivity in cases of severe iron overload [75, 76, 80, 89]. In MRS, the low signal from bound iron makes quantification studies difficult. MRS also has limited resolution (larger than 1 cm) [90]. While SQUID has found some success in detecting liver iron, there are only about four SQUID facilities in the world that offer iron overload testing, making it difficult to translate into a widely used clinical alternative. NRS has achieved some success in patients with iron overload but with high levels of dose [27]. NSECT has the potential to overcome many of these limitations and develop into a clinical technique for noninvasive measurement of liver iron and copper [33].

15.1.2.3 Small Animal Imaging

A third potential area of application for NSECT is in the spectroscopic imaging of small animals [91, 92] to understand the structural and functional mechanisms of metabolism and transport of elements in normal and disease models. For example, NSECT can be used to follow the metabolic migration of nutrients in normal and in genetically modified mice, measurement of iron and calcium levels in the cardiac wall, and molecular transport and drug delivery studies where molecules of interest are tagged with stable isotopes rather than radioactive tracers.

NSECT represents a technique that combines both anatomic and functional imaging. Through CT acquisition, it is possible to obtain a three-dimensional map of element distribution within a specific organ, thereby providing anatomic information about the location and extent of the diseased tissue. Simultaneously, information about elements within the tissue can provide information about the functional (disease) state of the tissue.

15.2 Imaging Facility and Apparatus

All NSECT experiments performed so far have been conducted at the Triangle Universities Nuclear Laboratory (TUNL) accelerator facility located at Duke University. TUNL is a low energy nuclear physics research laboratory with a 10 MeV tandem Van-de-Graaff accelerator capable of producing 20 MeV proton or deuteron beams and a host of other light ion beams. The shielded neutron source provides researchers with an intense pulsed collimated mono-energetic neutron beam with energies between 5 and 23.2 MeV. A detailed description of the facilities and apparatus used in NSECT has been provided in [32, 34].

Three key components are required in an NSECT acquisition system: (a) neutron source, (b) gamma detector, and (c) tomographic acquisition and reconstruction technique. Each of these components is readily available at TUNL. Figure 15.1 shows a schematic of an NSECT system with the major components. A description of the key components is given below.

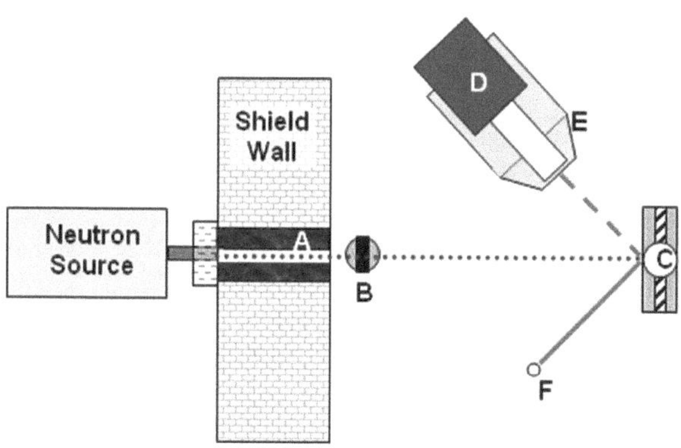

Fig. 15.1 Schematic of an NSECT system used in tomographic measurements. Major components of the system are labeled as A – Neutron Collimator; B – Instantaneous Neutron Flux Monitor; C – Sample Manipulation Tomographic Gantry; D – HPGe detector; E – Anticoincidence Compton Shield; and F – Inelastically Scattered Neutron. The neutron beam (shown by the line of dots) travels from the source to the sample (C) and scatters inelastically. Gammas emitted in the inelastic scatter interaction (shown as line of dashes) are detected by the HPGe detector, while the scattered neutron (F) continues along another trajectory (shown as a solid line).
(Figure modified from [94] pp. 2313–2326. © Institute of Physics and IOP Publishing Limited 2008)

15.2.1 Neutron Source

NSECT can be performed with a variety of neutron sources, some of which are available commercially, while others are specialized sources available only at state-of-the-art neutron facilities. An ideal source for an NSECT application should produce a narrow collimated beam of fast neutrons with high neutron flux and allow beam pulsing and energy tuning with a nearly monochromatic energy profile. High neutron flux is required to achieve minimal scan durations, while a collimated beam is required to attain spatial resolution in the tomographic acquisition. Beam pulsing or chopping is necessary to identify the time-resolved inelastic scatter signal and separate it from other neutron and gamma effects that occur later in time. Without beam pulsing (i.e., with a continuous-beam source), the inelastic scatter signal can be obscured by radiation detected from room and sample-related background effects that are observed at different times, such as radioactive decay following neutron capture in the sample. These effects can be reduced through time-of-flight (TOF) correction, which needs beam pulsing. The TOF correction technique for NSECT has been described in detail in a separate publication [93].

Energy tuning, while highly desirable, is not an essential requirement for a neutron source. A requirement of inelastic scatter spectroscopy is that the energy of the neutron beam must be higher than the maximum energy state to be excited. However, unnecessarily high neutron energies will simply excite higher unwanted states in both desired and undesired elements, leading to a busy spectrum cluttered with noise. Further, as NSECT detectors are susceptible to Compton noise from high gamma energies, exciting unwanted high energy states in the sample will also produce higher Compton background noise signals in the detectors, which can potentially obscure elements with low-lying characteristic gamma lines. A monochromatic energy source producing neutrons with energy slightly higher than the highest desired excited state is the most suitable. With a tunable neutron source, a beam can be produced with energy that is sufficient to excite the elements of interest and avoid other unwanted elements in the sample. For example, to image carbon in a biological sample (at 4.439 MeV), a neutron energy of 5.0 – 5.5 MeV is ideal as it excites the 4.439 MeV energy state in ^{12}C but avoids excitation of the 6.13 MeV excited state in ^{16}O. A monochromatic source also facilitates lower patient dose by restricting neutron illumination to the required range of energy. Higher neutron energies that contribute to dose by depositing more energy per interaction and lower thermal energies that contribute to dose through neutron capture and subsequent radioactive emission are eliminated in a true monochromatic source.

Table 15.1 lists some neutron sources that have been identified for use in NSECT: Van-de-Graaff accelerators, deuterium-deuterium (DD) neutron tubes, and deuterium-tritium (DT) neutron tubes. Van-de-Graaff accelerators provide tunable collimated neutron beams that are suitable for tomography. NSECT experiments performed thus far have used a Van-de-Graaff accelerator

Table 15.1 A comparison of three neutron sources suitable for NSECT applications

Characteristic	Van-de-Graaff	DD Neutron Tube	DT Neutron Tube
Energy	5 MeV to 23.2 MeV	3.2 MeV	14 MeV
Flux	$10^4 - 10^6$ n/s.cm^2	$10^8 - 10^{11}$ n/s	$10^{10} - 10^{13}$ n/s
Pulsing	2 ns @ 2.5 MHz	100 μs	100 μs
Collimation	Available	Available	Available
Monochromatic Beam	Yes	Yes	Yes
Energy Tuning	Yes	No	No

at Duke University as the source of neutrons. The accelerator produces a tunable, monochromatic, pulsed neutron beam, which can be tuned to energies between 5 and 23.2 MeV and produces collimated fluxes of 10^4–10^6 n/s.cm^2. The beam can be collimated to any desired shape and size using collimator inserts and is pulsed to provide 2 ns wide bunches at the target to allow measurement of neutron and gamma TOF. The repetition rate of the beam pulses is adjustable from 2.5 MHz down to 39 kHz [32]. Beams measuring 0.8 cm at full width at half maximum have been used to obtain image resolution of approximately 1 cm [94].

DD and DT neutron tubes produce high-flux monochromatic beams and possess the advantage of being commercially available. DD tubes produce neutrons at 3.2 MeV, while DT tubes produce neutrons at 14.2 MeV. Owing to the required neutron energy constraint, DD tubes are not suitable for applications that require the imaging of carbon or oxygen. Both tubes can be pulsed down to 100 μs or operated in continuous mode, and collimation is available at the cost of neutron flux.

A thin neutron plastic scintillator attached to a photomultiplier tube acts as an instantaneous neutron flux monitor to measure fluctuations in the neutron beam current. The attenuation of neutrons by the flux monitor in the energy range of current NSECT studies is typically less than 0.1%.

15.2.2 Gamma Detector

As mentioned earlier, most elements of interest for medical NSECT applications show characteristic gamma lines with energies ranging from 350 to 6500 keV. In some elements, gamma lines differ by as little as 4 keV (e.g., ^{27}Al at 843.7 keV and ^{56}Fe at 846.7 keV). NSECT gamma detectors must therefore possess excellent energy resolution as well as high efficiency at high gamma energies. Standard gamma camera detectors used in nuclear medicine studies are unable to provide the energy resolution and efficiency required and, therefore, cannot be used for NSECT applications.

Several gamma-ray detectors have been explored and evaluated for use in NSECT, including single-crystal and clover HPGe detectors and the comparatively inexpensive options NaI-Tl and BGO. For multielement applications, such as cancer diagnosis, only HPGe detectors are found to be suitable, providing the most desirable combination of energy resolution and efficiency. HPGe detectors typically possess energy resolution of 0.1%, providing approximately 1 keV resolution at 1 MeV. BGO and NaI-Tl detectors provide typical energy resolutions of approximately 7 and 12%, respectively, which is insufficient to separate close-lying gamma lines in a multielement analysis. However, these detectors are suitable for single-element evaluations where the element of interest lies in a region, which is free from close-lying gamma lines from other elements.

NSECT experiments performed at Duke University have used two HPGe segmented detectors in the two fold segmented clover detector configuration. In this configuration, each detector consists of four coaxial n-type germanium crystals mounted together in the shape of a 4-leaf clover. Each detector exhibits minimum relative efficiency of 22% (relative efficiency compares the efficiency of the detector at 1332 keV to that of a 3 inch cubic NaI detector) and full width at half maximum less than or equal to 2.25 keV for 1.332 MeV gamma rays of ^{60}Co. Detectors are typically calibrated against a ^{22}Na source and positioned at $\pm$135 degrees from the incident neutron beam. A majority of the elements of interest to NSECT decay through electric quadrupole transitions, whose distribution has maximum intensity at 45 and 135 degrees. Placing the detectors at the 135 degree orientation helps to maximize signal intensity and simultaneously prevent detector activation and damage from forward scattering neutrons. In Figure 15.1, the HPGe detector (labeled D) is visible surrounded by an anticoincidence Compton shield (labeled E), which is used to reduce the effects of Compton scattering in the detectors. These shields have a minimum peak-to-Compton ratio of 41 for ^{60}Co gamma rays at 1.332 MeV [32].

15.2.3 Tomographic Acquisition Gantry

As with all other tomographic imaging systems, NSECT performs tomography by acquiring a complete set of projections of line integrals through the sample. For NSECT, these projections are defined by the angular and spatial sampling intervals between successive positions of the beam on the sample. Acquisition is performed in a manner similar to first generation CT with a stationary beam and single element detector located at fixed positions with respect to each other. The sample is translated through the beam, rotated through a finite angle, and translated through the beam again. The process is repeated until the volume of interest in the sample has been scanned. While ideal tomography systems would acquire projections between 0 and 360 degrees, due to limitations in acquisition time and concerns about patient dose, the current NSECT tomography system scans the sample from 0 to 180 degrees and employs the maximum likelihood expectation maximization (MLEM) technique [95] for reconstruction. The

MLEM technique is able to take into account parameters such as nonuniform and partial sampling when reconstructing an image.

Tomography is performed using the single-beam, fixed-detector geometry of first-generation computed tomography systems where a thin neutron beam is used to illuminate a single line (projection) through the object being scanned. Tomography is performed by translating the object through the beam, rotating it through a fixed angle, and repeating the translation and rotation until the entire volume of the object has been scanned. The set of projections obtained from the tomographic scan are then reconstructed to obtain a two-dimensional image of the spatial distribution and concentration of each element in the object.

Figure 15.1 shows the tomographic computer-controlled gantry (labeled C), which enables tomographic acquisition by translating and rotating the sample with respect to the neutron beam.

Reconstruction can be performed through any available reconstruction algorithm. While a variety of algorithms are available, the MLEM algorithm is selected due to its superiority of performance in undersampled acquisition systems and its ability to incorporate the Poisson nature of photon counting [95], both characteristics important to NSECT. An analysis of the MLEM algorithm and its advantages over other iterative and Fourier techniques for image reconstruction in NSECT has been described in [96].

Data acquisition can be performed using any nuclear spectroscopy acquisition system that allows acquisition and storage of raw data and offline retrieval for data visualization and analysis. Examples of such software that has been used at Duke include CODA [97] and SpecTcl [98].

15.3 Current Applications

There are many diverse and exciting applications where NSECT can potentially be used to make a clinical diagnosis through noninvasive tomography. Disorders, such as cancer, hemochromatosis (liver iron overload), and Wilson's disease (liver copper overload), can be diagnosed by identifying the concentrations of the respective disease-marking elements in the tissue or organ. Several experiments have been performed to investigate the feasibility of diagnosing these disorders through NSECT, using both tomographic acquisition as well as nontomographic (single projection) spectroscopy. The results of some of these experiments are described below.

15.3.1 Experiment 1: Tomographic Acquisition of Multielement Phantom

Figure 15.2 shows an image of a solid iron-copper phantom containing bars of natural iron and natural copper arranged in the shape of the letter 'N.' Each bar

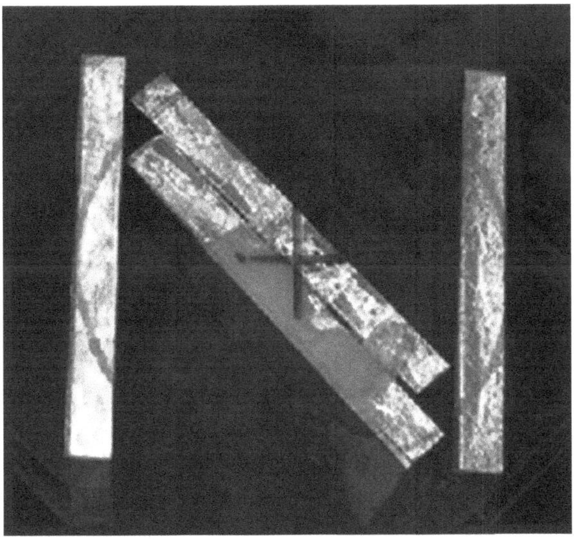

Fig. 15.2 Geometry of the phantom imaged in the tomography experiment. The vertical outer bars represent copper while the diagonal inner (gray) bars represent iron. Each bar measures 0.6 cm by 6 cm by 2.5 cm.
(Figure from Floyd et al, "Neutron Stimulated Emission Computed Tomography of a Multi-Element Phantom," *Phys Med Biol*, vol. 53, pp. 2313–2326. © Institute of Physics and IOP Publishing Limited 2008)

measures 6 mm in width, 60 mm in height, and 25 mm in depth. The phantom was scanned with an 8 mm wide 7.5 MeV neutron beam with 11 spatial projection at 8 angles. As can be seen in Fig. 15.3, the corresponding NSECT reconstructed image shows excellent agreement with the spatial locations of iron and copper within the image. Images were reconstructed individually for each of the two elements and then combined to generate the figure shown. Reconstructed resolution was observed to be approximately 1 cm, which is reasonable for a beam that measured 8 mm at full width at half maximum. The experiment demonstrates the tomographic ability of NSECT in identifying spatially individual elements in a multielement sample.

Figure 15.4 shows the corresponding gamma spectrum from the NSECT scan of the iron-copper phantom. Six spectral lines can be identified for energy transitions from iron and copper in the phantom.

Detailed analysis and a complete description of the experiment are available in [94].

15.3.2 Experiment 2: Diagnosis of Iron Overload in Human Liver Phantom

Figure 15.5 shows a human torso phantom with two chambers – an outer chamber corresponding to an adult torso and an inner chamber corresponding

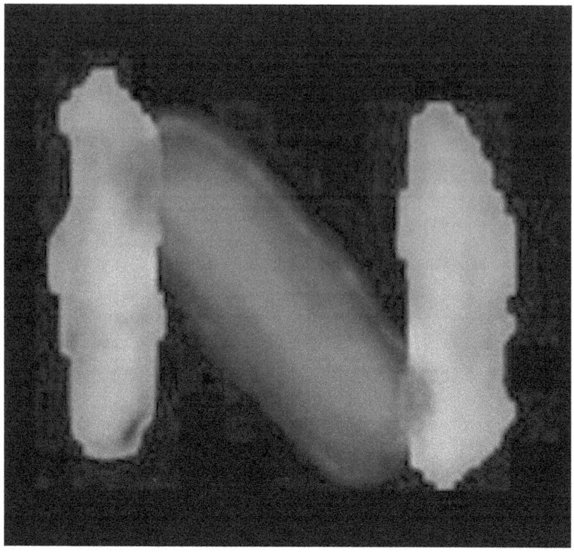

Fig. 15.3 Reconstructed image from the NSECT acquisition of the sample. The vertical outer regions represent copper while the diagonal inner region represents iron. Each element was reconstructed separately and then combined.
(Figure from Floyd et al, "Neutron Stimulated Emission Computed Tomography of a Multi-Element Phantom," *Phys Med Biol,* vol. 53, pp. 2313–2326. © Institute of Physics and IOP Publishing Limited 2008)

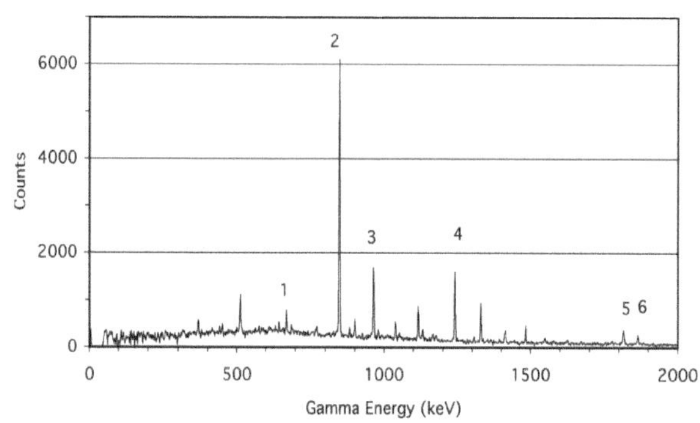

Fig. 15.4 Gamma energy spectrum from the iron-copper phantom showing spectral lines from six transitions in ^{56}Fe and ^{63}Cu:
1. ^{63}Cu from 1st excited state to ground state; energy 660 keV
2. ^{56}Fe from 1st excited state to ground state; energy 847 keV
3. ^{63}Cu from 2nd excited state to ground state; energy 962 keV
4. ^{56}Fe from 3rd to 2nd excited state; energy 1239 keV
5. ^{56}Fe from 4th to 2nd excited state; energy 1811 keV
6. ^{63}Cu from 6th to 1st excited state; energy 1864 keV
(Figure from [94] pp. 2313–2326. © Institute of Physics and IOP Publishing Limited 2008)

Fig. 15.5 A phantom of the human torso with two chambers – outer chamber corresponding to an adult torso, and inner chamber corresponding to an adult human liver. Both chambers can be filled separately with any desired material.
(Figure from Kapadia et al. [32], pp. 2633–2649. © Institute of Physics and IOP Publishing Limited 2008)

to an adult human liver. The phantom was filled with bovine liver tissue with artificially induced iron overload and was scanned using a 5 MeV neutron beam. The resultant spectrum shown in Fig. 15.6 was analyzed to quantify the concentration of iron in the liver. The gamma line at 847 keV was detected for 4.18 g of ^{56}Fe. Based on this spectrum, a clinically relevant projected sensitivity of 6 mg/g was obtained for iron overload diagnosis. In addition, gamma lines were identified for several other elements in the liver, including Cl, Cu, K, Na, and Zn, which were confirmed through NAA, and for ^{12}C from the phantom. Gamma lines were also identified for ^{74}Ge and ^{76}Ge from the gamma-ray detectors, ^{42}K from the room background, and ^{1}H (neutron capture) from the phantom. The dose from the scan was calculated using a Monte-Carlo simulation as 0.375 mSv [33], which is significantly lower than an abdominal X-ray exam that typically delivers 2 mSv [99]. The experiment demonstrates that NSECT has the potential to detect clinically relevant concentrations of iron in the human body through a noninvasive scan with reasonable dose. Although the scan time for this experiment was unreasonable at over 24 h, it

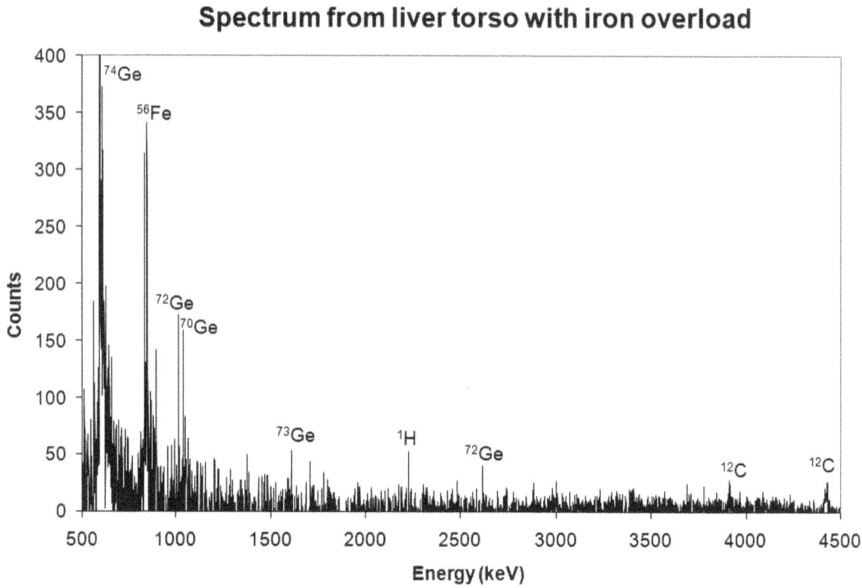

Fig. 15.6 Gamma energy spectrum from the uniform iron overload torso phantom, showing peaks corresponding to elements detected in the liver. The peak at 847 keV corresponds to ^{56}Fe. Peaks are also seen for Ge from the detector and ^{12}C and ^{1}H from the tissue. (Figure from Kapadia et al. [32], pp. 2633–2649. © Institute of Physics and IOP Publishing Limited 2008)

can be brought down to a few seconds using high-flux neutron sources (such as the aforementioned DD and DT tubes) and increasing the number of detectors.

Detailed analysis and a complete description of the experiment are available in [33].

15.3.3 Experiment 3: Detection of Breast Cancer

To investigate the feasibility of breast cancer detection through NSECT (Figs. 15.7 and 15.8), samples of benign and malignant breast tissue obtained through breast biopsy were scanned with a 6 MeV neutron beam for a cumulative dose of 1 mSv each. After TOF and sample-out background subtraction, the resultant spectra were analyzed to identify elements that showed statistically significant differences between the two samples. Concentration differences were found for Al, Br, Cl, Co, Fe, K, Rb, and Zn, of which Cl, Fe, K, and Zn were verified through NAA. The experiment demonstrates that NSECT can potentially be used to detect several cancer-marking elements in human tissue.

Table 15.2 gives a list of elements that showed statistically significant differences between the benign and malignant spectra. While several elements were identified with concentration differences in each of the two samples, NAA analysis of the samples determined that their concentration was of the order

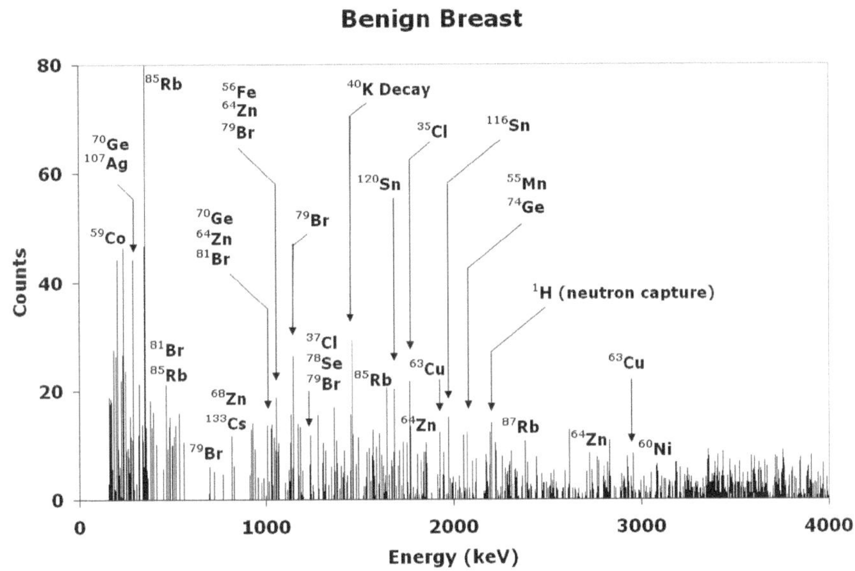

Fig. 15.7 NSECT spectrum of a benign breast sample showing elements identified through NSECT spectroscopy.
(Figure from Kapadia et al. [32], pp. 501–509. © 2008 IEEE)

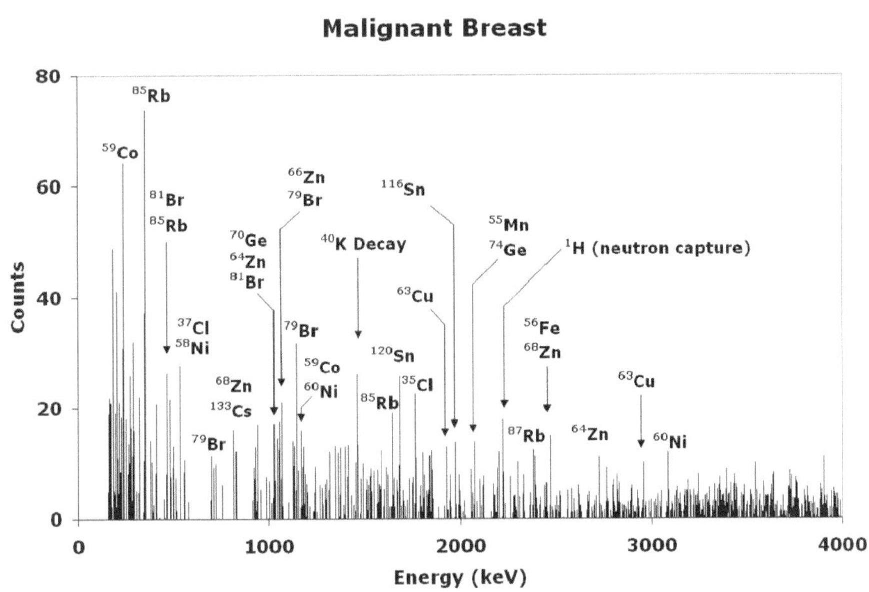

Fig. 15.8 NSECT spectrum of a malignant breast sample showing elements identified through NSECT spectroscopy.
(Figure from Kapadia et al. [32], pp. 501–509. © 2008 IEEE)

Table 15.2 List of elements showing statistically significant differences between benign and malignant spectra. Elements with negative differences showed a decrease in concentration in the malignant sample. Statistical significance was calculated using a z-score test for difference of means. (Table from Kapadia et al. [32], pp. 501–509. © 2008 IEEE)

Energy keV	Element match	Counts benign	Counts malignant	Diff	p-val
219	^{79}Br	6	19	13	0.01
397	^{59}Co, ^{79}Br	16	2	–14	0.01
1028	^{81}Br	13	29	16	0.05
1128	^{39}K, ^{68}Zn	0	13	13	0.001
1306	^{56}Fe	10	0	–10	0.01
2299	^{27}Al	0	13	13	0.001
2469	^{37}Cl, ^{56}Fe, ^{66}Zn	5	15	10	0.05
3635	^{35}Cl	3	14	11	0.01

of a few hundred micrograms, which is below the sensitivity of the current system, which is a few hundred milligrams [32]. Therefore, accurate quantification of these microgram concentrations will require a significant improvement in sensitivity.

Detailed analysis and a complete description of the experiment are available in [32].

15.4 Dose Analysis

Patient dose in NSECT is of significant concern due to the use of fast neutrons, which have a dose quality factor (Q-factor) of 10. For the technique to translate successfully to the clinical environment, NSECT dose must be comparable to the dose levels of other ionizing imaging modalities. As the technique is still in early stages of development, dose analysis in NSECT is currently performed using Monte-Carlo simulations. The process, described in [96, 100], can be summarized in the following three steps:

(a) a Monte-Carlo simulation is used to estimate two parameters for an incident neutron beam – the number of neutrons that interact in the volume of interest and the average energy deposited per interacting neutron,
(b) the resulting energy deposited in the volume is converted from MeV to J/kg using the known mass of the volume to give the absorbed energy in Gray (Gy), and
(c) the absorbed energy is multiplied by the quality factor for neutrons (10) and the weighting factor for the organ of interest to give the effective dose equivalent in Sieverts (Sv).

This technique has been used to calculate the patient dose for NSECT scans of the abdomen [96], liver [33, 96, 100], and breast [100, 101]. Table 15.3 summarizes the results of these dose-analysis simulations.

In comparison, an abdominal X-ray scan typically delivers 2 mSv, a mammogram delivers 0.7 mSv, while an abdominal CT exam delivers approximately 10 mSv [99]. The doses from NSECT scans appear comparable to, or even lower than, the other techniques. This is largely due to the modest number of neutrons

Table 15.3 NSECT dose delivered to organs in the body from NSECT scans

Organ	Spectroscopic scan	Tomography scan
Abdomen	1–2 mSv	1–5 mSv
Liver	0.02–1 mSv	0.5–3 mSv
Breast	0.02–0.5 mSv	0.5–1 mSv

required to create an NSECT image or extract sufficient information from the sample. Preliminary simulation studies indicate that as few as 10 million incident neutrons are required to obtain quantitative accuracy of 95% in a tomographic image [102]. A fluence of 10 million incident neutrons on the breast, for example, corresponds to a dose of less than 1 mSv. Dose levels in NSECT can potentially be reduced even further by increasing the number of gamma-ray detectors and using high-flux neutron sources to reduce scan time.

15.5 Summary

NSECT represents an exciting new imaging modality that has the potential for application in both medical and biological research. Several human disorders characterized by element changes can be diagnosed through noninvasive in vivo scanning using this technique. NSECT has the ability to obtain tomographic information about the spatial distribution of elements within a tissue or organ to make a quantitative and spatial diagnosis. A prototype of the NSECT acquisition system has been developed and built at Duke University using a Van-de-Graaff accelerator and HPGe detectors. As demonstrated through experiments with the prototype system, NSECT has the sensitivity to detect concentrations of iron that represent a clinically relevant liver iron overload condition. Sensitivity evaluation experiments indicate that concentrations as low as 3 mg/g may be quantifiable through NSECT. Although MRI is able to quantify moderate concentrations of iron overload, it suffers from a loss of accuracy for concentrations above 6 mg/g wet weight (20 mg/g dry weight) due to a reduction in the signal intensity caused by high concentrations of iron [89]. NSECT, on the other hand, shows an increase in the signal with increasing iron concentration. This facilitates iron overload detection in patients with severe overload, where MRI begins to lose accuracy.

The image resolution observed with the prototype system is approximately 1 cm, which is passable for imaging large organs such as the liver. However, as the resolution depends primarily on the width of the neutron beam, it is possible to improve the resolution significantly by using a narrow collimated beam (at the cost of additional scan time). Simulation experiments have demonstrated a resolution of 5 mm when scanning was performed with a 5 mm rectangular beam [96]. While the resolution of the NSECT system depends primarily on the size of the incident neutron beam, the best resolution achievable in a hydrogen-rich biological sample will also be limited by the noise component generated from neutron scatter by hydrogen. Neutrons that are meant to illuminate a certain voxel in the sample may scatter onto adjacent voxels containing an element of interest and induce inelastic gamma emission from that voxel. Detection of the adjacent-voxel gammas may lead to a spread in the tails of the resulting sinograms, which can lead to reduced resolution. The contribution of this effect to the limits of resolution is currently being investigated through Monte-Carlo simulations.

Patient dose for NSECT examinations of different organs have been found to vary between 0.02 mSv for a breast spectroscopic exam [101] to under 1 mSv [96] for a liver spectroscopic exam, evaluated through Monte-Carlo simulations. Tomographic doses are slightly higher depending on the number of projections used but are generally lower than the dose delivered from a corresponding X-ray CT exam [96].

15.6 Future

The current focus of NSECT has been on quantitative diagnosis of large-organ disorders (e.g., those of the liver), largely due to the limits of resolution and sensitivity observed in the prototype system. As these limiting effects are understood better, efforts to improve the sensitivity and resolution of the system are being made. Improved sensitivity will facilitate the diagnosis of a greater number of disorders, including several types of cancer. The use of portable neutron sources such as pulsed DD and DT tubes is being explored to develop a portable, high-flux scanning solution for use in a clinical environment. Such portable tomographic scanning systems with improved resolution and sensitivity will present an attractive method for human and small-animal imaging and for diagnosis and screening of several disorders in the human body.

Acknowledgment The author would like to acknowledge and thank all the members of the Duke Advanced Imaging Laboratories (DAILabs) and Triangle Universities Nuclear Laboratory (TUNL) who have been involved in the development of NSECT, especially Georgia Tourassi, Amy Sharma, and Janelle Bender for their analytical contribution and deep involvement in NSECT; Brian Harrawood for his unparalleled computing support; Calvin Howell, Alexander Crowell, Matthew Kiser, and Robert Macri for their help and guidance with NSECT acquisition experiments; and Anton Tonchev and Anthony Hutcheson for their help with gamma detector setup and management. Finally, the author would like to express his deep gratitude to Dr. Carey Floyd, the pioneer of NSECT, in whose memory and name this research continues.

References

1. J. Anderson, S.B. Osborn, R.W. Tomlinson, D. Newton, J. Rundo, L. Salmon, and J.W. Smith, Neutron-Activation Analysis in Man in Vivo. a New Technique in Medical Investigation, *Lancet* **2**, 1201–1205, (Dec 5 1964).
2. M.J. Chamberlain, J.H. Fremlin, D.K. Peters, and H. Philip, Total body sodium by whole body neutron activation in the living subject: further evidence for non-exchangeable sodium pool, *Br. Med. J.* **2**, 583–585 (Jun 8, 1968).

3. M.J. Chamberlain, J.H. Fremlin, D.K. Peters, and H. Philip, Total body calcium by whole body neutron activation: new technique for study of bone disease, *Br. Med. J.* **2**, 581–3, Jun 8 1968.
4. M.J. Chamberlain, J.H. Fremlin, D.K. Peters, and H. Philip, Measurement of whole body calcium and sodium by neutron activation analysis in the living subject, *Strahlentherapie [Sonderb]*, **67**, 178–85, 1968.
5. M.J. Chamberlain, Whole-body neutron activation analysis, *Proc. R. Soc. Med.* **62**, 370 3 (Apr 1969).
6. M.J. Chamberlain, J.H. Fremlin, I. Holloway, and D.K. Peters, Use of the cyclotron for whole body neutron activation analysis: theoretical and practical considerations, *Int. J. Appl. Radiat. Isot.* **21**, 725–34 (Dec 1970).
7. M.J. Chamberlain, J.H. Fremlin, D.K. Peters, and H. Philip, Applications of the whole-body counter in total-body neutron activation analysis, *Br. J. Radiol.* **43**, 287–8 (Apr 1970).
8. M.J. Chamberlain and J.H. Fremlin, Measurement of whole-body nitrogen by pulsed neutron activation analysis, *Strahlentherapie [Sonderb]* **72**, 88–93 (1972).
9. S.H. Cohn and C.S. Dombrowski, Measurement of total-body calcium, sodium, chlorine, nitrogen, and phosphorus in man by in vivo neutron activation analysis, *J. Nucl. Med.* **12**, 499–505 (Jul 1971).
10. D. Vartsky, K.J. Ellis, and S.H. Cohn, In vivo measurement of body nitrogen by analysis of prompt gammas from neutron capture, *J. Nucl. Med.* **20**, 1158–1155 (Nov 1979).
11. C.L. Hollas, L.E. Ussery, K.B. Butterfield, and R.E. Morgado, A method for in vivo determination of carbon and oxygen using prompt gamma radiations induced by 14.7-MeV neutrons, *Basic Life Sci.* **55**, 395–400 (1990).
12. H.C. Biggin, N.S. Chen, K.V. Ettinger, J.H. Fremlin, W.D. Morgan, R. Nowotny, M.J. Chamberlain, and T.C. Harvey, Cadmium by in vivo neutron activation analysis, in *J. Radioanal. Chem*, v. 19, no. 2, pp. 207–214; *International colloquium on activation analysis of very low amounts of elements; 2 Oct 1972; Saclay, France* United Kingdom (1974).
13. M.L. Arnold, F.E. McNeill, I.M. Stronach, A. Pejovic-Milic, D.R. Chettle, and A. Waker, An accelerator based system for in vivo neutron activation analysis measurements of manganese in human hand bones, *Med. Phys.* **29**, 2718–24 (Nov 2002).
14. A. Garg, V. Singh, et al., An elemental correlation study in cancerous and normal breast tissue with successive clinical stages by neutron activation analysis, *Biol. Trace Element Res.* **46**, 185–202 (1994).
15. A. Danielsen and E. Steinnes, A study of some selected trace elements in normal and cancerous tissue by neutron activation analysis, *J. Nucl. Med.* **11**, 260–4 (Jun 1970).
16. K.H. Ng, D.A. Bradley, and L.M. Looi, Elevated trace element concentrations in malignant breast tissues, *Br. J. Radiol.* **70**, 375–82 (Apr 1997).
17. K.H. Ng, D.A. Bradley, L.M. Looi, C.S. Mahmood, and A.K. Wood, Differentiation of elemental composition of normal and malignant breast tissue by instrumental neutron activation analysis, *Appl. Radiat. Isot.* **44**, 511–6 (Mar 1993).
18. K.J. Ellis, Human body composition: in vivo methods, *Physiol. Rev.* **80**, 649–80 (Apr 2000).
19. J.F. Sutcliffe, A review of in vivo experimental methods to determine the composition of the human body, *Phys. Med. Biol.* **41**, 791–833 (May 1996).
20. E. Witkowska, K. Szczepaniak, and M. Biziuk, Some applications of neutron activation analysis: A review, *J. Radioanal. Nucl. Chem.* **265**, 141–150 (2005).
21. S. Mattsson and B.J. Thomas, Development of methods for body composition studies, *Phys. Med. Biol.* **51**, R203–28 (Jul 7 2006).
22. K. Kyere, B. Oldroyd, C.B. Oxby, L. Burkinshaw, R.E. Ellis, and G.L. Hill, The feasibility of measuring total body carbon by counting neutron inelastic scatter gamma rays, *Phys. Med. Biol.* **27**, 805–17 (Jun 1982).

23. J.J. Kehayias, S.B. Heymsfield, A.F. LoMonte, J. Wang, and R.N. Pierson, Jr., In vivo determination of body fat by measuring total body carbon, *Am. J. Clin. Nutr.* **53**, 1339–44 (Jun 1991).
24. J.J. Kehayias, Aging and body composition: possibilities for future studies, *J. Nutr.* **123**, 454–8 (Feb 1993).
25. J.J. Kehayias, H. Zhuang, V. Hughes, and L. Dowling, Assessment of body fat and lean in the elderly by measuring body carbon and oxygen: validation against hydrodensitometry, *Appl. Radiat. Isot.* **49**, 723–5 (May-Jun 1998).
26. J.J. Kehayias, S. Valtuena, A.B. Waitekus, C.A. Sheahan, and M. O'Neill, In vivo elemental partition analysis using fast neutrons. A tool for testing the efficacy of new clinical interventions, *Ann. N Y. Acad. Sci.* **904**, 140–7 (May 2000).
27. L. Wielopolski, R.C. Ancona, R.T. Mossey, A.N. Vaswani, and S.H. Cohn, Nuclear resonance scattering measurement of human iron stores, *Med. Phys.* **12**, 401–4 (Jul-Aug 1985).
28. L. Wielopolski and E.C. Zaino, Noninvasive in-vivo measurement of hepatic and cardiac iron, *J. Nucl. Med.* **33**, 1278–82 (Jul 1992).
29. L. Wielopolski, Feasibility of measuring iron in vivo using fast 14 MeV neutrons, *Cooley's Anemia Foundation Inc.* BNL-7980–2005 (2005).
30. C.E. Floyd, C.R. Howell, B.P. Harrawood, A.S. Crowell, A.J. Kapadia, R. Macri, J.Q. Xia, R. Pedroni, J. Bowsher, M.R. Kiser, G.D. Tourassi, W. Tornow, and R. Walter, Neutron Stimulated Emission Computed Tomography of Stable Isotopes, in *SPIE Symposium on Medical Imaging*, San Diego, CA, pp. 248–254 (2004).
31. C.E. Floyd, J.E. Bender, A. Sharma, A. Kapadia, J. Xia, B. Harrawood, G.D. Tourassi, J. Lo, and C.R. Howell, Introduction to Neutron Stimulated Emission Computed Tomography. *Phys. Med. Biol.* **50** (14), 3375–90 (2005).
32. A.J. Kapadia, A.C. Sharma, J.E. Bender, G.D. Tourassi, C.R. Howell, A.S. Crowell, M.R. Kiser, B.P. Harrawood, and C.E. Floyd, Neutron Stimulated Emission Computed Tomography for Diagnosis of Breast Cancer, *IEEE Trans Nucl. Sci.* **55**, 501–509 (2008).
33. A.J. Kapadia, G.D. Tourassi, A.C. Sharma, A.S. Crowell, M.R. Kiser, and C.R. Howell, Experimental detection of iron overload in liver through neutron stimulated emission spectroscopy, *Phys. Med. Biol.* **53**, 2633–2649 (2008).
34. C.E. Floyd, J.E. Bender, A.C. Sharma, A.J. Kapadia, J.Q. Xia, B.P. Harrawood, G.D. Tourassi, J.Y. Lo, A.S. Crowell, and C.R. Howell, Introduction to neutron stimulated emission computed tomography, *Phys. Med. Biol.* **51**, 3375–3390 (2006).
35. M.N. Wernick and J.N. Aarsvold, *Emission Tomography: The Fundamentals of PET and SPECT*: Academic Press, San Diego (2004).
36. A. Kak and M. Slaney, *Principles of computerized tomographic imaging* vol. 33. Philadelphia: society for industrial and applied mathematics (2001).
37. J.E. Bender, C.E. Floyd, B.P. Harrawood, A.J. Kapadia, A.C. Sharma, and J.L. Jesneck, The effect of detector resolution for quantitative analysis of neutron stimulated emission computed tomography, in *SPIE Medical Imaging*, pp. 1597–1605 (2006).
38. L.W. Powell, Diagnosis of hemochromatosis, *Semin. Gastrointest Dis.* **13**, 80–8 (Apr 2002).
39. S. Joffe, Hemochromatosis, N. Lamki, B. Coombs, U. Schmiedl, R.M. Krasny, and J. Karani, Eds., *Medscape emedicine*, emedicine.medscape.com/article/369012-overview, Mar 11, 2005.
40. L. Powell, Hemochromatosis, in D. Kasper, Fawci, AS, Longo, DL, Braunwald, E, Hauser, SL, Jameson, JL, Ed.,*Harrison's Principles of Internal Medicine*, 16 ed. vol. 2, McGraw Hill, New York, NY, pp. 2298–2303 (2005).
41. G. Brewer, Wilson Disease, in D. Kasper, Fawci, AS, Longo, DL, Braunwald, E, Hauser, SL, Jameson, JL, Ed., *Harrison's Principals of Internal Medicine*, 16 ed. vol. 2, NY: McGraw Hill, pp. 2313–2315 (2005).
42. H. Mussalo-Rauhamaa, S. Piepponen, J. Lehto, R. Kauppila, and O. Auvinen, Cu, Zn, Se and Mg concentrations in breast fat of Finnish breast cancer patients and healthy controls, *Trace Elements Med.* **10**, 13–15 (1993).

43. K. Geraki, M.J. Farquharson, and D.A. Bradley, X-ray fluorescence and energy dispersive x-ray diffraction for the quantification of elemental concentrations in breast tissue, *Phys. Med. Biol.* **49**, 99–110 (Jan 7 2004).
44. U. Majewska, D. Banas, J. Braziewicz, S. Gozdz, A. Kubala-Kukus, and M. Kucharzewski, Trace element concentration distributions in breast, lung and colon tissues, *Phys. Med. Biol.* **52**, 3895–911 (Jul 7 2007).
45. K. Geraki, M.J. Farquharson, and D.A. Bradley, Concentrations of Fe, Cu and Zn in breast tissue: a synchrotron XRF study, *Phys. Med. Biol.* **47**, 2327–39 (Jul 7 2002).
46. S.L. Rizk and H.H. Sky-Peck, Comparison between concentrations of trace elements in normal and neoplastic human breast tissue, *Cancer. Res.* **44**, 5390–4 (Nov 1984).
47. S. Rizk and H. Sky-Peck, Comparison between Concetrations of Trace Elements in Normal and Neoplastic Human Breast Tissue, *Cancer. Research.* **44**, 5390–539 (1984).
48. P.M. Santoliquido, H.W. Southwick, and J.H. Olwin, Trace metal levels in cancer of the breast, *Surg. Gynecol. Obstet.* **142**, 65–70 (Jan 1976).
49. M. Brys, A.D. Nawrocka, E. Miekos, C. Zydek, M. Foksinski, A. Barecki, and W.M. Krajewska, Zinc and cadmium analysis in human prostate neoplasms, *Biol. Trace. Elem. Res.* **59**, 145–52 (Winter 1997).
50. A. Feustel, R. Wennrich, D. Steiniger, and P. Klauss, Zinc and cadmium concentration in prostatic carcinoma of different histological grading in comparison to normal prostate tissue and adenofibromyomatosis (BPH), *Urol. Res.* **10**, 301–3 (1982).
51. V. Zaichick, T.V. Sviridova, and S.V. Zaichick, Zinc in the human prostate gland: normal, hyperplastic and cancerous, *Int. Urol. Nephrol.* **29**, 565–74 (1997).
52. E. Andrasi, M. Suhajda, I. Saray, L. Bezur, L. Ernyei, and A. Reffy, Concentration of elements in human brain: glioblastoma multiforme, *Sci. Total. Environ.* **139/140**, 399–402 (Nov 1 1993).
53. M. Persigehl, H. Schicha, K. Kasperek, and H.J. Klein, Trace element concentration in human organs in dependence of age, *Beitr. Path.* **161**, 209–220 (1977).
54. E.J. Margalioth, J.G. Schenker, and M. Chevion, Copper and zinc levels in normal and malignant tissues, *Cancer* **52**, 868–72 (Sep 1 1983).
55. P. Ghadirian, P. Maisonneuve, C. Perret, G. Kennedy, P. Boyle, D. Krewski, and A. Lacroix, A case-control study of toenail selenium and cancer of the breast, colon, and prostate, *Cancer. Detect. Prev.* **24**, 305–13 (2000).
56. I. Kato, A.M. Dnistrian, M. Schwartz, P. Toniolo, K. Koenig, R.E. Shore, A. Zeleniuch-Jacquotte, A. Akhmedkhanov, and E. Riboli, Iron intake, body iron stores and colorectal cancer risk in women: a nested case-control study, *Int. J. Cancer.* **80**, 693–8 (Mar 1 1999).
57. R.L. Nelson, F.G. Davis, E. Sutter, L.H. Sobin, J.W. Kikendall, and P. Bowen, Body iron stores and risk of colonic neoplasia, *J. Natl. Cancer. Inst.* **86**, 455–60 (Mar 16 1994).
58. K. Witkowski, A. Kozlowski, M. Pardela, J. Piecuch, and P. Walichiewicz, [Level of copper in plasma and tissue of patients with esophageal and large bowel cancer], *Wiad. Lek.* **46**, 586–8 (Aug 1993).
59. K.Q. Xiao and W.J. Henderson, [Electron microscopy microanalysis and quantitative detection of trace elements in carcinoma of the colon], *Zhonghua Bing Li Xue Za Zhi* **21**, 142–5 (Jun 1992).
60. W.P. Banner, J.J. DeCosse, Q.H. Tan, and M.S. Zedeck, Selective distribution of selenium in colon parallels its antitumor activity, *Carcinogenesis* **5**, 1543–6 (Dec 1984).
61. G.C. Gregoriadis, N.S. Apostolidis, A.N. Romanos, and T.P. Paradellis, A comparative study of trace elements in normal and cancerous colorectal tissues, *Cancer* **52**, 508–19 (Aug 1 1983).
62. Z.M. Bataineh, I.H. Bani Hani, and J.R. Al-Alami, Zinc in normal and pathological human prostate gland, *Saudi. Med. J.* **23**, 218–20 (Feb 2002).
63. A. Feustel, R. Wennrich, and H. Dittrich, Investigations of trace elements in metastases and primary carcinoma of the prostate, *Urol. Res.* **17**, 107–9 (1989).

64. J.O. Ogunlewe and D.N. Osegbe, Zinc and cadmium concentrations in indigenous blacks with normal, hypertrophic, and malignant prostate, *Cancer* **63**, 1388–92 (Apr 1 1989).
65. I. Romics and L. Katchalova, Spectrographic determination of zinc in the tissues of adenoma and carcinoma of the prostate, *Int. Urol. Nephrol.* **15**, 171–6 (1983).
66. M. Yaman, D. Atici, S. Bakirdere, and I. Akdeniz, Comparison of trace metal concentrations in malign and benign human prostate, *J. Med. Chem.* **48**, 630–634 (2005).
67. F.K. Habib, G.L. Hammond, I.R. Lee, J.B. Dawson, M.K. Mason, P.H. Smith, S.R.Y.K. Stitch, A.G. Meade, E.P. Rack, and A.J. Blotcky, Metal-androgen interrelationships in carcinoma and hyperplasia of the human prostate, *J. Endocrinol.*, **61** (1), 133–41 (1976).
68. H. Kubo, S. Hashimoto, and A. Ishibashi, Simultaneous determinations of Fe, Cu, Zn, and Br concentrations in human tissue sections, *Med. Phys.* **3**, 204–9 (Jul-Aug 1976).
69. G.M. Brittenham and D.G. Badman, Noninvasive measurement of iron: report of an NIDDK workshop, *Blood* **101**, 15–9 (Jan 1 2003).
70. J.M. Alustiza, J. Artetxe, A. Castiella, C. Agirre, J.I. Emparanza, P. Otazua, M. Garcia-Bengoechea, J. Barrio, F. Mujica, and J.A. Recondo, MR quantification of hepatic iron concentration, *Radiology* **230**, 479–84 (Feb 2004).
71. W.F. Avrin and S. Kumar, Noninvasive liver-iron measurements with a room-temperature susceptometer, *Physiol. Meas.* **28**, 349–61 (Apr 2007).
72. H.L. Bonkovsky, R.B. Rubin, E.E. Cable, A. Davidoff, T.H. Rijcken, and D.D. Stark, Hepatic iron concentration: noninvasive estimation by means of MR imaging techniques, *Radiology* **212**, 227–34 (Jul 1999).
73. G.M. Brittenham, S. Sheth, C.J. Allen, and D.E. Farrell, Noninvasive methods for quantitative assessment of transfusional iron overload in sickle cell disease, *Semin. Hematol.* **38**, 37–56 (Jan 2001).
74. E. Cecchin, S. De Marchi, F. Querin, M.G. Marin, R. Fiorentino, and F. Tesio, Efficacy of hepatic computed tomography to detect iron overload in chronic hemodialysis, *Kidney. Int.* **37**, 943–50 (Mar 1990).
75. R.W. Chapman, G. Williams, G. Bydder, R. Dick, S. Sherlock, and L. Kreel, Computed tomography for determining liver iron content in primary haemochromatosis, *Br. Med. J.* **280**, 440–2 (Feb 16 1980).
76. J.L. Chezmar, R.C. Nelson, J.A. Malko, and M.E. Bernardino, Hepatic iron overload: diagnosis and quantification by noninvasive imaging, *Gastrointest. Radiol.* **15**, 27–31 (Winter 1990).
77. R.M. Dixon, P. Styles, F.N. al-Refaie, G.J. Kemp, S.M. Donohue, B. Wonke, A.V. Hoffbrand, G.K. Radda, and B. Rajagopalan, Assessment of hepatic iron overload in thalassemic patients by magnetic resonance spectroscopy, *Hepatology* **19**, 904–10 (Apr 1994).
78. Y. Gandon, D. Olivie, D. Guyader, C. Aube, F. Oberti, V. Sebille, and Y. Deugnier, Non-invasive assessment of hepatic iron stores by MRI., *Lancet* **363**, 357–362 (Jan 31, 2004 2004).
79. D. Guyader, Y. Gandon, J.Y. Robert, J.F. Heautot, H. Jouanolle, C. Jacquelinet, M. Messner, Y. Deugnier, and P. Brissot, Magnetic resonance imaging and assessment of liver iron content in genetic hemochromatosis, *J Hepatol* **15**, 304–8 (Jul 1992).
80. J.M. Howard, C.N. Ghent, L.S. Carey, P.R. Flanagan, and L.S. Valberg, Diagnostic efficacy of hepatic computed tomography in the detection of body iron overload, *Gastroenterology* **84**, 209–15 (Feb 1983).
81. P. Liu, M. Henkelman, J. Joshi, P. Hardy, J. Butany, M. Iwanochko, M. Clauberg, M. Dhar, D. Mai, S. Waien, and N. Olivieri, Quantification of cardiac and tissue iron by nuclear magnetic resonance relaxometry in a nvel murine thalassemia-cardiac iron overload model, *Can. J. Cardiol.* **12**, 155–64 (Feb 1996).
82. P. Nielsen, R. Engelhardt, M. Duerken, G.E. Janka, and R. Fischer, Using SQUID biomagnetic liver susceptometry in the treatment of thalassemia and other iron loading diseases, *Transfus. Sci.* **23**, 257–8 (Dec 2000).

83. P. Nielsen, R. Engelhardt, J. Dullmann, and R. Fischer, Non-invasive liver iron quantification by SQUID-biosusceptometry and serum ferritin iron as new diagnostic parameters in hereditary hemochromatosis, *Blood Cells Mol. Dis.* **29**, 451–8 (Nov-Dec 2002).
84. P. Nielsen, U. Kordes, R. Fischer, R. Engelhardt, and G.E. Janka, [SQUID-biosusceptometry in iron overloaded patients with hematologic diseases], *Klin Padiatr* **214**, 218–22 (Jul-Aug 2002).
85. H. Perrimond, C. Chagnon, I. Moulanier, G. Michel, H. Guidicelli, and P.J. Bernard, The value of nuclear magnetic resonance in the study of iron overload in thalassemia patients, *Ann. Pediatr. (Paris)* **38**, 175–84 (Mar 1991).
86. S. Sheth, SQUID biosusceptometry in the measurement of hepatic iron, *Pediatr. Radiol.* **33**, 373–7 (Jun 2003).
87. D. Vartsky, K.J. Ellis, D.M. Hull, and S.H. Cohn, Nuclear resonant scattering of gamma rays – a new technique for in vivo measurement of body iron stores, *Phys. Med. Biol.* **24**, 689–701 (Jul 1979).
88. D. Vartsky, L. Wielopolski, K.J. Ellis, and S.H. Cohn, The Use of Nuclear Resonant Scattering of Gamma-Rays for Invivo Measurement of Iron, *Nucl. Instrum. Meth. Phys. Res.* **193**, 359–364 (1982).
89. E. Angelucci, A. Giovagnoni, G. Valeri, E. Paci, M. Ripalti, P. Muretto, C. McLaren, G. M. Brittenham, and G. Lucarelli, Limitations of magnetic resonance imaging in measurement of hepatic iron, *Blood* **90**, 4736–42 (Dec 15 1997).
90. Z.J. Wang, J.C. Haselgrove, M.B. Martin, A.M. Hubbard, S. Li, K. Loomes, J.R. Moore, H. Zhao, and A.R. Cohen, Evaluation of iron overload by single voxel MRS measurement of liver T2, *J. Magn. Reson. Imaging* **15**, 395–400 (Apr 2002).
91. A.J. Kapadia, A.C. Sharma, G.D. Tourassi, J.E. Bender, A.S. Crowell, M.R. Kiser, C.R. Howell, and C.E. Floyd, Neutron Spectroscopy of Mouse Using Neutron Stimulated Emission Computed Tomography (NSECT), in *IEEE Nuclear Science Symposium, Medical Imaging Conference*, San Diego, CA, pp. 3546–3548 (2006).
92. A.C. Sharma, G.D. Tourassi, A.J. Kapadia, A.S. Crowell, M.R. Kiser, A. Hutcheson, B. P. Harrawood, C.R. Howell, and C.E. Floyd, Elemental Spectrum of a Mouse Obtained via Neutron Stimulation., in *2007 SPIE Symposium on Medical Imaging*, San Diego, CA, p. 65100 K (2007).
93. C.E. Floyd, A.C. Sharma, J.E. Bender, A.J. Kapadia, J.Q. Xia, B.P. Harrawood, G.D. Tourassi, J.Y. Lo, M.R. Kiser, A.S. Crowell, R.S. Pedroni, R.A. Macri, S. Tajima, and C. R. Howell, Neutron Stimulated Emission Computed Tomography: Background Corrections, *Nucl. Instrum. Meth. Phys. Res. Sect. B* **254**, 329–336 (2007).
94. C.E. Floyd, A.J. Kapadia, J.E. Bender, A.C. Sharma, J.Q. Xia, B.P. Harrawood, G.D. Tourassi, J.Y. Lo, A.S. Crowell, M.R. Kiser, and C.R. Howell, Neutron Stimulated Emission Computed Tomography of a Multi-Element Phantom, *Phys. Med. Biol.* **53**, 2313–2326 (2008).
95. K. Lange and R. Carson, EM reconstruction Algorithms for Emission and Transmission Tomography, *Journal of Computer Assisted Tomography* **8**, 306–316 (1984).
96. A.J. Kapadia, Accuracy and Patient Dose in Neutron Stimulated Emission Computed Tomography for Diagnosis of Liver Iron Overload: Simulations in GEANT4, in *Biomedical Engineering*. vol. PhD, Duke University, Durham, NC (2007).
97. W.A. Watson, III, J. Chen, G. Heyes, E. Jastrzembski, and D. Quarrie, CODA: a scalable, distributed data acquisition system, in *IEEE Transactions on Nuclear Science*, Vancouver, BC, Canada, pp. 61–68 (1994).
98. R. Fox, C. Bolen, K. Orji, and J. Venema, NSCLSpecTcl Meeting the Needs of Preliminary Nuclear Physics Data Analysis in *11'th Annual Tcl/Tk Conference* New Orleans, Louisiana (2004).
99. RSNA, Radiation Exposure in X-ray Examinations, in *American College of Radiology (ACR) and the Radiological Society of North America (RSNA)* (2007).

100. A.C. Sharma, B.P. Harrawood, J.E. Bender, G.D. Tourassi, and A.J. Kapadia, Neutron stimulated emission computed tomography: a Monte Carlo simulation approach, *Phys. Med. Biol.* **52**, 6117–31 (Oct 21 2007).
101. J.E. Bender, A.J. Kapadia, A.C. Sharma, G.D. Tourassi, B.P. Harrawood, and C. E. Floyd, Breast cancer detection using Neutron Stimulated Emission Computed Tomography: prominent elements and dose requirements, *Med. Phys.* **34**, 3866–3871 (2007).
102. A.J. Kapadia, C.E. Floyd, C.R. Howell, and B.P. Harrawood, Sampling Requirements for Neutron Stimulated Emission Computed Tomography, in *RSNA, Physics (Digital Imaging, PACS) session* Chicago, IL (2004).

Chapter 16
Visualizing Structures of Biological Macromolecules Through Indirect Imaging with Small-Angle Neutron Scattering and Modeling

W.T. Heller and G.A. Baker

Abstract The molecular machinery of living organisms is built of systems whose length scales range from nanometers to microns. While neutron imaging is a powerful tool for studying macroscopic systems, neutron optics is not capable of effectively resolving these macromolecular systems. Small-angle neutron scattering when combined with modeling and contrast variation methods enables indirect imaging for visualization of structure and function over these length scales. A brief overview of small-angle neutron scattering with contrast variation is presented with a description of available modeling methods. Applications of the methods are also presented to provide a survey of the wide variety of systems that are suitable for study.

Keywords Small-angle neutron scattering · Structural biology · Contrast variation · Modeling

16.1 Introduction

Neutron imaging has great potential in the biological sciences because it is a sensitive probe of hydrogen. Unfortunately, the lack of effective optics, such as those available for light and electron microscopes, limits the application of the technique to macroscopic systems, as opposed to the study of molecular systems where truly unique information can be obtained. While it is possible to focus neutrons, the focal lengths of neutron lenses are very long because neutrons do not interact strongly with matter. Therefore, it is difficult to refocus the neutrons to resolve the image of the object being studied as one does in a conventional microscope. Still, the unique properties of the neutron make it a very attractive probe of biological materials over a wide range of length scales.

W.T. Heller (✉)
Center for Structural Molecular Biology Oak Ridge National Laboratory
P.O. Box 2008, Oak Ridge, TN 37931, USA
e-mail: hellerwt@ornl.gov

I.S. Anderson et al. (eds.), *Neutron Imaging and Applications*,
Neutron Scattering Applications and Techniques, DOI 10.1007/978-0-387-78693-3_16,

Two closely related alternative approaches to extract structural information from macromolecular systems involve small-angle scattering (SAS) of either neutrons [small-angle neutron scattering (SANS)] or X-rays [small-angle X-ray scattering (SAXS)]. Small-angle scattering probes distances ranging from 1 to 100 nm, giving it a potential maximum resolution comparable to transmission electron microscopy. When working with dilute, homogeneous solutions, SAS becomes a powerful tool for indirectly imaging the structures and functions of proteins, nucleic acids, and their complexes, particularly when combined with computational methods for constructing models. Advances in modeling methods have generated considerable interest among researchers in the field of biological sciences who seek to understand complex biological systems, particularly those that have proven resistant to high-resolution structure determination by other means.

16.1.1 Theory

Small-angle scattering of dilute solutions of identical particles results from the interference of neutrons that interact with the nuclei of the atoms in the particles, much like a diffraction measurement. Unlike diffraction, however, the samples contain no crystalline ordering that produces well-defined intensity peaks at locations determined by the crystal lattice. Neutrons, which can be described as a propagating plane wave, are incident on the sample and scattered as shown in Fig. 16.1.

The scattered neutrons are then collected at an angle 2θ from the incident beam. The resulting pattern is due to the interference of neutrons, scattered from different locations, within the particle and therefore is a function of the particle shape. The scattering is described as a function of the momentum transfer, $\vec{Q}$, which has a magnitude $Q = 4\pi\sin(\theta)/\lambda$. The neutron wavelength, λ, is h/mv, where h is Plank's constant, m is the mass of the neutron and v is the neutron velocity. The neutrons commonly used for scattering experiments have wavelengths ranging from 2 to 20 Å, which corresponds to velocities

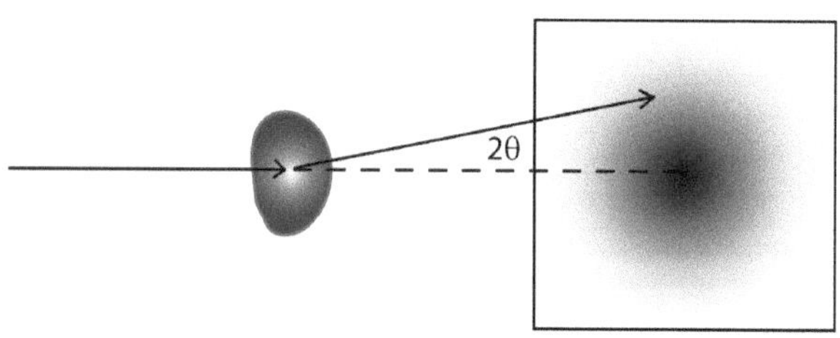

Fig. 16.1 Schematic of small-angle scattering. Neutrons as plane waves are incident on a sample particle and are scattered. The pattern observed results from the interference of neutrons, scattered from different positions, within the particle

ranging from a few hundred to a few thousand meters per second. For an isotropic sample, such as for a protein tumbling freely in solution, the pattern of scattered neutrons is symmetric about the beam and can be described according to Eq. (16.1) [1, 2].

$$I(Q) = n\left\langle \left| \int_V \left(\rho\left(\vec{R}\right) - \rho_s\right) e^{-i\vec{Q}\cdot\vec{R}} d^3R \right|^2 \right\rangle , \qquad (16.1)$$

where $I(Q)$ is the scattered intensity, n is the number of particles per unit volume, $\rho\left(\vec{R}\right)$ is the scattering length density of the particle at position $\vec{R}$, and ρ_s is the scattering length density of the solvent. The integral is taken over the volume of the particle V [1, 2]. The integral is averaged over time, all orientations, and the ensemble of structures present in the solution.

16.1.2 Contrast Variation

The fact that Eq. 16.1 describes the scattering signal relative to the background solvent is incredibly important because it enables the use of SANS with contrast variation [3]. The dramatic difference between the scattering lengths of hydrogen (–3.71 fm) and deuterium (6.67 fm) makes it possible to greatly alter the scattering length density of a biological macromolecule or the aqueous background by substituting for hydrogenated material ones that are isotopically labeled with deuterium. Contrast variation is a particularly powerful technique for studying multisubunit complexes in which one subunit is labeled with deuterium because it enables the separation of the scattering from the two components within the complex. In a contrast variation experiment, a complex containing a subunit with a different average scattering length density, such as a deuterium-labeled protein or nucleic acid, is measured in a series of solutions consisting of mixtures of H_2O and D_2O. Conceptually, the effect of changing the H_2O/D_2O mixture in the solution is illustrated in Fig. 16.2. The intensity profiles in the contrast variation series can be written in the following manner (Eq. 16.2):

$$I(Q, \Delta\rho_1, \Delta\rho_2) = \Delta\rho_1^2 I_1(Q) + \Delta\rho_2^2 I_2(Q) + \Delta\rho_1\Delta\rho_2 I_{12}(Q) \qquad (16.2)$$

Here, $\Delta\rho_1$ and $\Delta\rho_2$ are the differences between the hydrogenated and deuterated components of the complex relative to the solvent, respectively, and are often called the contrasts. The three functions, $I_1(Q)$, $I_2(Q)$, and $I_{12}(Q)$ are known as the basic scattering functions [3]. The basic scattering functions are the scattering signals from the components within the complex having two different scattering length densities and the cross term that provides information on the relative disposition of the components. The measured contrast variation series data define a set of linear equations that can be used to extract the basic scattering functions for further analysis.

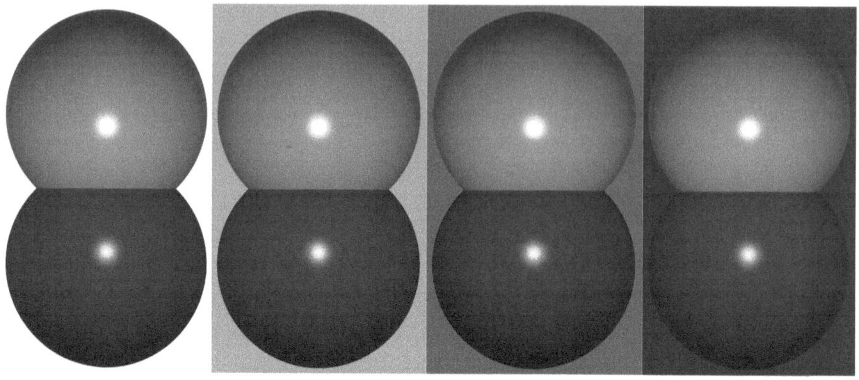

Fig. 16.2 Contrast variation. The scattering length density of the background solvent is varied relative to the two subunits having different scattering length densities, thereby differentially highlighting the structures of the subunits within the whole

16.1.3 Information Content and Basic Data Analysis

There are two common, basic methods of SAS data analysis. For a dilute, monodisperse solution of homogenous particles, the intensity I(Q) can be fit for the radius of gyration (R_g) according to Guinier [2]:

$$\ln[I(Q)] = \ln[I(0)] - \frac{(QR_g)^2}{3} \quad . \tag{16.3}$$

Here, $I(0)$ is the scattered intensity at $Q = 0$, which is a shape-independent function of the molecular weight of the particle and the average scattering length density relative to the background solution. The radius of gyration, R_g, is a shape-dependent function of the contrast-weighted distance from the center of a particle to scattering elements within the particle. While Guinier analysis may be used for any particle of unknown size or shape, there are instances when it is not valid. To use Guinier analysis, the following three conditions must be satisfied: (1) the system must be rotationally isotropic, (2) the particles must not interact with each other, and (3) $Q \times R_g$ must be less than 1.3, a condition resulting from the approximation used to derive Eq. (16.3) for globular particles. Condition (3) must be further restricted for other shapes, with $Q \times R_g$ less than 1.0 being commonly applied.

Additional analysis can be performed to obtain more information on the shape of the scattering particle. Equation (16.1) can be inverted to provide the distance distribution function $P(R)$ as a function of the experimental I(Q) using the Fourier transform in Eq. (16.4).

$$P(R) = \frac{1}{2\pi^2} \int_0^\infty I(Q) QR \sin(QR) \mathrm{d}Q \quad . \tag{16.4}$$

The actual approach uses inverse Fourier methods and includes the Glatter algorithm [4, 5], the Moore algorithm [6], and the method implemented in the software GNOM [7]. Fitting SAS data for $P(R)$ provides a measure of d_{max}, the maximum linear dimension of the particle.

An example SANS intensity profile calculated from the crystal structure of hemoglobin (PDB ID: 4HHB [8]) using the software ORNL_SAS [9] is shown in Fig. 16.3. As can be seen, the data have a relatively simple appearance that suggests they do not contain a great deal of information, in contrast to high-resolution crystallography data. Researchers have constructed arguments based on the Shannon Sampling Theorem [10] to estimate the appropriate number of degrees of freedom, n_f, suitable for fitting SAS data [6, 11–13]. The number is a function of the Q-range of the data collected and the size of the scattering particle. If data are collected for a particle of size d_{max} over a range that extends from Q_{min} to Q_{max}, then $n_f \sim d_{max}(Q_{max}\text{-}Q_{min})/\pi$.

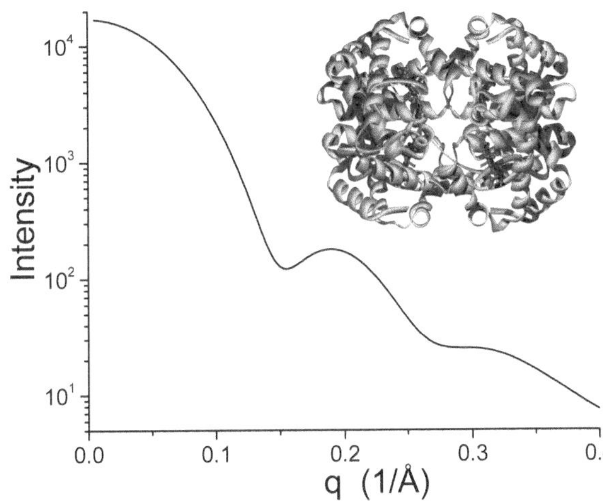

Fig. 16.3 Small-angle scattering profile calculated from the atomic-resolution structure of hemoglobin (PDB ID: 4HHB [8]) using the program ORNL_SAS [14]. The intensity is displayed in arbitrary units

16.2 Modeling for Visualization

Structural biology is a very visual science, but the basic analysis of SAS data does not provide a structure that can be used to understand the interplay between structure and function in complex macromolecular systems. Hence, methods have been developed for constructing models from SAS data. Much of the effort has focused on SAXS data, due to the greater availability of instruments. Still, tools for building models from SANS data have been developed that can be applied to contrast variation data of multisubunit complexes. The kind of modeling possible is often a function of the amount of existing structural information, particularly high-resolution structural information.

16.2.1 Shape Restoration

In cases where no structural information exists for the protein or complex being studied, several methods exist for building structural models that are referred to as shape restoration methods. These techniques encompass a broad range of approaches spanning simple shapes, spherical harmonics, and methods using aggregates of shapes. Each approach has its strengths and weaknesses. A large number of these methods are freely available to the scientific community, making them attractive choices for building models from SAS data.

Some of the most established modeling methods for SAS data on proteins and protein complexes use simple shapes. Model fitting using these approaches employs a limited number of degrees of freedom, providing a degree of confidence that the answer is uniquely determined by the data. Modeling using a simple shape has the further advantage that the associated intensity profile is quite often an analytical function [15], making it attractive for fitting data by hand. More importantly, many of these geometrical shapes are implemented in freely available software packages [16, 17]. For modeling using multiple shapes, the only analytical function available is the Debye formula for aggregates of spheres [18]. Other multishape structures lack analytical functions for the scattering curves, although methods have been developed for building models in this manner [19–27]. Such approaches are very well suited to modeling contrast variation series of SANS data from biological macromolecular complexes, where each subunit can be assigned to a single shape.

Shape reconstruction methods using spherical harmonics were among the first *ab initio* methods developed for extracting structural information from SAS data [11, 28–34]. In this approach, the structure is defined by a series superposition of spherical harmonics, an infinite set of orthogonal functions. It is trivial to limit the number of terms used in the reconstruction, which is typically assigned to n_f. The ability to readily restrict the number of free parameters in the fit provides a measure of confidence that the structure found is reasonable. Still, the resulting structures are restricted to those which can be represented by spherical harmonics, and significantly more complicated structures cannot be effectively modeled. The approach is readily applied to SAXS data, but it is also applicable to contrast variation experiments. These shape restoration methods can also be applied directly to basic scattering functions extracted from such data sets.

Currently, the most popular methods for shape restoration from SAS data are the aggregate *ab initio* techniques that use large sets of simple, relatively small shapes to define the volume of the scattering particles. Spheres are the most common base shape for aggregate techniques [35–38], but ellipsoids have also been used [14]. The programs DAMMIN [39] and GASBOR [40] use dummy atoms and dummy residues, respectively, for the shape reconstructions and are the most popular software packages currently available for this purpose. A related grid-based method has also been developed [41–43]. In general, any shape can be reproduced using these methods up to the limits of the resolution of the data and the base shape used to build the structure, in principle avoiding the limitations inherent in using spherical harmonics to define shapes. Unfortunately, the large number of degrees of freedom inherent in such modeling makes degenerate solutions possible that fit the SAS data equally well. For this reason, multiple independent structures are often generated and averaged into a consensus shape [35–37, 44]. An implementation of DAMMIN exists that is capable of directly modeling against contrast variation series of SANS data from complexes with selective deuterium labeling [39]. The other techniques can be applied to SAXS data, SAS data from structures that are not selectively deuterated, or to basic scattering functions extracted from contrast variation series data.

16.2.2 Modeling with High-Resolution Structures

The dramatic increase in the number of high-resolution structures deposited in the Protein Data Bank [45] has led to the development of methods that can leverage this data to build models from SAS data. The basic idea of comparing a measured SAS profile to the SAS intensity simulated from an atomic-resolution structure is not new, and several methods for doing so have been described throughout the literature [9, 28, 46–56]. Determining whether the solution structure of a protein is consistent with an available high-resolution structure is a form of modeling that allows one to visualize the structure. If the SAS data are inconsistent with the existing high-resolution structure, then additional modeling methods must be applied to understand the nature of these differences. For example, it is possible to understand the nature of the conformational changes by using the existing high-resolution structure as a starting point.

There are a variety of approaches that can use high-resolution structures as starting points. Rigid body modeling methods without advanced docking algorithms are the simplest, relying only on the SAS data to build models [21, 54, 57–60]. These relatively basic rigid body methods can also be applied to model building using low-resolution structures produced employing the *ab initio* methods described above. More recently, increased interest in the use of high-resolution structures for rigid body modeling of complexes has led to the development of methods that use more formal docking algorithms and potentials to produce more realistic models [61–64]. Mixed model building in which an *ab initio* density is added to an existing high-resolution structure to simulate missing mass is also a very useful tool [65].

It is important to point out that SAS modeling methods can be developed to address specific problems, and the amount of other structural information that is available truly dictates the best choice for the problem at hand. Time, creativity, and programming expertise are the only limiting factors.

16.3 Examples

Broader reviews of the applications of SAS and modeling to the study of protein structures have been published recently [66–67]. Rather than trying to provide a comprehensive listing of examples where SAS and modeling have been applied to visualize the structure of biological macromolecules in solution, some specific examples demonstrating the use of SANS with contrast variation are presented. Such studies are excellent showcases of the various ways in which models are built from scattering data and of the power of neutron scattering when combined with specific isotope labeling and contrast variation methods.

The structures of various ribosomes have been studied on multiple occasions using SANS with contrast variation, and these studies represent some of the earliest uses of modeling to visualize macromolecular complexes. For example, neutron scattering studies have been used to map out the relative distances between various proteins within the 30S ribosome complex from *E. coli* [68–70]. The culmination of these extensive studies was a low-resolution model showing the relative spatial relationships of the protein elements within the ribosome [71]. More advanced modeling methods were later applied to contrast variation experiments on larger ribosomal complexes. Small-angle neutron scattering with contrast variation experiments with the *E. coli* 50S ribosome, a larger complex than the 30S ribosome, resulted in a map of the distribution of the protein and RNA within the complex through the use of a two-phase model built using a spherical harmonic shape restoration approach [31, 32, 72]. A series of papers also investigated the even larger 70S ribosome from *E. coli* using the combined approach of SANS with contrast variation and selective deuteration. A total of 42 SANS data sets were collected and modeled with a four-phase spherical harmonic shape restoration approach [73, 74]. Later, this same data set was subjected to *ab initio* modeling using DAMMIN [75], providing a much more detailed structure of the complex macromolecular machine.

Other protein complexes have been subjected to extensive analysis using SANS with contrast variation and modeling. The activation of myosin light chain kinase by calmodulin was studied using selectively deuterated complexes [76, 77]. Simple shape modeling using ellipsoids was employed to visualize the structure and the conformational transitions that occur upon calcium binding to calmodulin. The interpretation of these results and subsequent X-ray scattering studies [78, 79] was further enhanced by docking high-resolution structures for the two proteins that showed the location of the calmodulin relative to the catalytic cleft of the myosin light chain kinase. Changes in the dimensions of the ellipsoids were used to infer that the cleft of the kinase opened in response to the binding of calmodulin.

The skeletal and cardiac variants of troponin complexes have also been studied by SANS with contrast variation. A variety of modeling approaches have been used to visualize the structure of the complex, including mixing high- and low-resolution structural information [80, 81], simple shapes [24], *ab initio* modeling of the basic scattering functions followed by rigid body modeling using the shapes [37, 58], as shown in Fig. 16.4, and investigating the function of the complex from an available high-resolution structure [82]. Protein kinase A is another dynamic multisubunit complex for which SANS plus contrast variation and modeling was used to obtain structural and function information [26, 83], as shown in Fig. 16.5. Protein–nucleic acid complexes benefit from the inherently different scattering length densities of hydrogenated protein and nucleic acids, which was leveraged in a study of the assembly of the methionine repressor protein, MetJ, on different lengths of DNA taken from regulatory sequences of different genes within the *met* regulon [84], which is shown in Fig. 16.6. By no means do these examples comprise an exhaustive listing of the multisubunit complexes that have been studied to great benefit by SANS with contrast variation, as other systems have also been studied recently [26, 85–90], a trend certainly to continue as interest in the method grows.

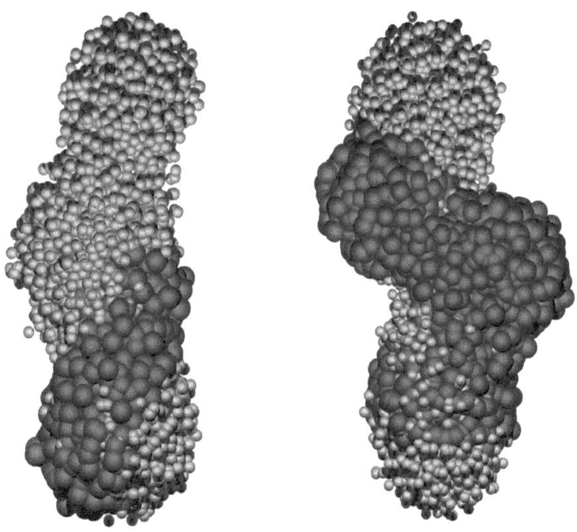

Fig. 16.4 Models of the ternary cardiac troponin complex (TnC–TnI–TnT(198–298) derived from SANS with contrast variation using two different isotopic labeling patterns [91]. In the left model, the TnC was deuterated (*large dark spheres*), while the TnI subunit was labeled (*small light spheres*) for the contrast variation series used to derive the model on the right. The density near the middle of the structure was inferred to be the TnT(198–298). The longest dimension of the structure is ~125 Å

Fig. 16.5 Model of the protein kinase A structure derived using SANS with contrast variation showing the catalytic domains (*gray*), cyclic AMP binding domains (*light gray*), the dimerization-docking domain (*dark gray*), and unstructured linker domains (*black*) [83]. The longest dimension of the structure is ~155 Å

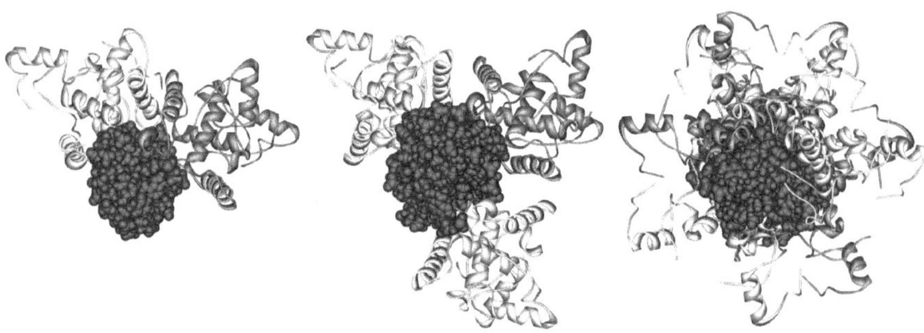

Fig. 16.6 Models of MetJ–DNA complexes using different lengths of DNA. MetJ dimers along the length of the DNA (*black spheres*) are colored light and dark gray in an alternating pattern to show differences in the length-dependent packing of the protein dimers on the DNA [84]. The radius from the axis of the DNA of the structure on the left is ~40 Å

16.4 Conclusions

Small-angle neutron scattering with contrast variation, when combined with modeling methods, is a powerful tool for visualizing the structure and internal organization of multisubunit biological complexes. While it is not a direct imaging method, it has the advantages of relatively simple sample preparation and very broad applicability to biological systems in near-native, solution environments. The methods presented here are readily applicable to protein–nucleic acid complexes, which benefit from an inherent difference in scattering length between the protein and nucleic acid constituents. Selective deuteration adds the powerful capability of introducing a protein subunit into a complex whose scattering can be separated from the remainder of the protein components. The continued development of modeling algorithms will further enhance the accessibility of SANS with contrast variation for nonexpert users and improve the value of the resulting models for further interpretation of the functions of the complexes studied.

Acknowledgments This work was supported by the Oak Ridge Center for Structural Molecular Biology (KP1102010) of the Office of Biological and Environmental Research of the U. S. Department of Energy under contract No. DE-AC05-00OR22725 with Oak Ridge National Laboratory, managed and operated by UT-Battelle, LLC. The submitted manuscript has been authored by a contractor of the U.S. government under Contract DE-AC05-00OR22725. Accordingly, the U.S. government retains a nonexclusive royalty-free license to publish or reproduce the published form of this contribution, or allow others to do so, for U.S. government purposes.

References

1. P. Debye and A. Bueche, Scattering by an Inhomogeneous Solid. *J. Appl. Phys.* **20**, 518–525 (1949).
2. A. Guinier and G. Fournet, *Small-angle Scattering of X-rays*. New York, Wiley (1955).
3. K. Ibel and H.B. Stuhrmann, Comparison of Neutron and X-Ray-Scattering of Dilute Myoglobin Solutions. *J. Mol. Biol.* **93**, 255–265 (1975).
4. O. Glatter, New Method for Evaluation of Small-Angle Scattering Data. *J. Appl. Crystallogr.* **10**, 415–421 (1977).
5. O. Glatter, Interpretation of Real-Space Information from Small-Angle Scattering Experiments. *J Appl Crystallogr* **12**: 166–175 (1979).
6. P.B. Moore, Small-Angle Scattering - Information-Content and Error Analysis. *J. Appl. Crystallogr.* **13**, 168–175 (1980).
7. D.I. Svergun, Determination of the Regularization Parameter in Indirect-Transform Methods Using Perceptual Criteria. *J. Appl. Crystallogr.* **25**, 495–503 (1992).
8. G. Fermi, M.F. Perutz, B. Shaanan, and R. Fourme, The Crystal-Structure of Human Deoxyhemoglobin at 1.74a Resolution. *J. Mol. Biol.* **175**, 159–174 (1984).
9. E. Tjioe and W.T. Heller, ORNL_SAS: software for calculation of small-angle scattering intensities of proteins and protein complexes. *J. Appl. Crystallogr.* **40**, 782–785 (2007).
10. C.E. Shannon, A Mathematical Theory of Communication. *Bell Sys. Tech. J.* **27**, 379–423, 623–656 (1948).
11. D.I. Svergun, V.V. Volkov, M.B. Kozin, and H.B. Stuhrmann, New developments in direct shape determination from small-angle scattering .2. Uniqueness. *Acta. Crystallogr. Sect. A* **52**, 419–426 (1996).
12. D. Taupin and V. Luzzati, Information-Content and Retrieval in Solution Scattering Studies .1. Degrees of Freedom and Data Reduction. *J. Appl. Crystallogr.* **15**, 289–300 (1982).
13. V.V. Volkov and D.I. Svergun, Uniqueness of *ab initio* shape determination in small-angle scattering. *J. Appl. Crystallogr.* **36**, 860–864 (2003).

14. D. Vigil, S.C. Gallagher, J. Trewhella, and A.E. Garcia, Functional dynamics of the hydrophobic cleft in the N-domain of calmodulin. *Biophys. J.* **80**, 2082–2092 (2001).
15. J.S. Pedersen, Analysis of small-angle scattering data from colloids and polymer solutions: modeling and least-squares fitting. *Adv. Colloid. Interface. Sci.* **70**, 171–210 (1997).
16. S.R. Kline, Reduction and analysis of SANS and USANS data using IGOR Pro. *J. Appl. Crystallogr.* **39**, 895–900 (2006).
17. M.V. Petoukhov, P.V. Konarev, A.G. Kikhney, and D.I. Svergun, ATSAS 2.1 – towards automated and web-supported small-angle scattering data analysis. *J. Appl. Crystallogr.* **40**, S223–S228 (2007).
18. P. Debye, Zerstreuung von Röntgenstrahlen. *Annalen der Physik* **351**, 809–823 (1915).
19. S. Fujiwara, F.J. Kull, E.P. Sablin, D.B. Stone, and R.A. Mendelson, The Shapes of the Motor Domains of 2 Oppositely Directed Microtubule Motors, Nod and Kinesin – a Neutron-Scattering Study. *Biophys. J.* **69**, 1563–1568 (1995).
20. S. Hansen Calculation of Small-Angle Scattering Profiles Using Monte-Carlo Simulation. J Appl Crystallogr **23**, 344–346 (1990).
21. S.P. Harris, W.T. Heller, M.L. Greaser, R.L. Moss, and J. Trewhella, Solution structure of heavy meromyosin by small-angle scattering. *J. Biol. Chem.* **278**, 6034–6040 (2003).
22. H. Hartmann, A. Bongers, and H. Decker, Small-angle neutron scattering reveals an oxygen-dependent conformational change of the immunogen keyhole limpet hemocyanin type 1 (KLH1). *Eur. Biophys. J. Biophys. Lett.* **30**, 471–475 (2001).
23. W.T. Heller, ELLSTAT: shape modeling for solution small-angle scattering of proteins and protein complexes with automated statistical characterization. *J. Appl. Crystallogr.* **39**, 671–675 (2006).
24. S.J. Henderson, Monte Carlo modeling of small-angle scattering data from non-interacting homogeneous and heterogeneous particles in solution. *Biophys. J.* **70**, 1618–1627 (1996).
25. D.B. Stone, P.A. Timmins, D.K. Schneider, I. Krylova, C.H.I. Ramos, F.C. Reinach, and R.A. Mendelson, The effect of regulatory Ca2+ on the in situ structures of troponin C and troponin I: A neutron scattering study. *J. Mol. Biol.* **281**, 689–704 (1998).
26. J.K. Zhao, E. Hoye, S. Boylan, D.A. Walsh, and J. Trewhella, Quaternary structures of a catalytic subunit-regulatory subunit dimeric complex and the holoenzyme of the cAMP-dependent protein kinase by neutron contrast variation. *J. Biol. Chem.* **273**, 30448–30459 (1998).
27. J. Zhou, A. Deyhim, S. Krueger, and S.K. Gregurick, LORES: Low resolution shape program for the calculation of small angle scattering profiles for biological macromolecules in solution. *Comput. Phys. Commun.* **170**, 186–204 (2005).
28. J.G. Grossmann, Z.H.L. Abraham, E.T. Adman, M. Neu, R.R. Eady, B.E. Smith, and S.S. Hasnain, X-Ray-Scattering Using Synchrotron-Radiation Shows Nitrite Reductase from Achromobacter-Xylosoxidans to Be a Trimer in Solution. *Biochemistry* **32**, 7360–7366 (1993).
29. F. Spinozzi, F. Carsughi, and P. Mariani, Particle shape reconstruction by small-angle scattering: Integration of group theory and maximum entropy to multipole expansion method. *J. Chem. Phys.* **109**, 10148–10158 (1998).
30. H.B. Stuhrman, Interpretation of Small-Angle Scattering Functions of Dilute Solutions and Gases – a Representation of Structures Related to a One-Particle-Scattering Function. *Acta Crystallogr. Sect. A-Crystal Phys. Diff. Theor. Gen. Crystallogr. A* **26**, 297–306 (1970).
31. D.I. Svergun, M.H.J. Koch, J.S. Pedersen, and I.N. Serdyuk, Structural Model of the 50-S Subunit of Escherichia-Coli Ribosomes from Solution Scattering .2. Neutron-Scattering Study. *J. Mol. Biol.* **240**, 78–86 (1994).
32. D.I. Svergun, M.H.J. Koch, and I.N. Serdyuk, Structural Model of the 50-S Subunit of Escherichia-Coli Ribosomes from Solution Scattering .1. X-Ray Synchrotron-Radiation Study. *J. Mol. Biol.* **240**, 66–77 (1994).

33. D.I. Svergun and H.B. Stuhrmann, New Developments in Direct Shape Determination from Small-Angle Scattering .1. Theory and Model-Calculations. *Acta. Crystallogr. Sect. A* **47**, 736–744 (1991).
34. D.I. Svergun, V.V. Volkov, M.B. Kozin, H.B. Stuhrmann, C. Barberato, and M.H.J. Koch, Shape determination from solution scattering of biopolymers. *J. Appl. Crystallogr.* **30**, 798–802 (1997).
35. P. Chacon, J.F. Diaz, F. Moran, and J.M. Andreu, Reconstruction of protein form with X-ray solution scattering and a genetic algorithm. *J. Mol. Biol.* **299**, 1289–1302 (2000).
36. P. Chacon, F. Moran, J.F. Diaz, E. Pantos, and J.M. Andreu, Low-resolution structures of proteins in solution retrieved from X-ray scattering with a genetic algorithm. *Biophys. J.* **74**: 2760–2775 (1998).
37. W.T. Heller, N.L. Finley, W.J. Dong, P. Timmins, H.C. Cheung, P.R. Rosevear, and J. Trewhella, Small-angle neutron scattering with contrast variation reveals spatial relationships between the three subunits in the ternary cardiac troponin complex and the effects of troponin I phosphorylation. *Biochemistry* **42**, 7790–7800 (2003).
38. D. Walther, F.E. Cohen, and S. Doniach, Reconstruction of low-resolution three-dimensional density maps from one-dimensional small-angle X-ray solution scattering data for biomolecules. *J. Appl. Crystallogr.* **33**, 350–363 (2000).
39. D.I. Svergun, Restoring low resolution structure of biological macromolecules from solution scattering using simulated annealing. *Biophys. J.* **76**, 2879–2886 (1999).
40. D.I. Svergun, M.V. Petoukhov, and M.H.J. Koch, Determination of domain structure of proteins from X-ray solution scattering. *Biophys. J.* **80**, 2946–2953 (2001).
41. H. Hartmann, A. Bongers, and H. Decker, Small-angle X-ray scattering-based three-dimensional reconstruction of the immunogen KLH1 reveals different oxygen-dependent conformations. *J. Biol. Chem.* **279**, 2841–2845 (2004).
42. H. Hartmann and H. Decker. Small-angle scattering techniques for analyzing conformational transitions in hemocyanins. *Energ. Biol. Macromol. Pt D.* **379**, 81–106 (2004).
43. H. Hartmann, T. Muller, and H. Decker, Modeling techniques for analysing conformational transitions in hemocyanins by small-angle scattering of X-rays and neutrons. *Micron* **35**, 11–13 (2004).
44. M.B. Kozin and D.I. Svergun, Automated matching of high- and low-resolution structural models. *J. Appl. Crystallogr.* **34**, 33–41 (2001).
45. H.M. Berman, J. Westbrook, Z. Feng, G. Gilliland, T.N. Bhat, H. Weissig, I.N. Shindyalov, and P.E. Bourne, The Protein Data Bank. *Nucl. Acids. Res.* **28,** 235–242 (2000).
46. B.A. Fedorov, O.B. Ptitsyn, and L.A. Voronin, X-Ray Diffuse Scattering of Globular Protein Solutions – Consideration of Solvent Influence. *FEBS Lett.* **28**, 188–190 (1972).
47. S.C. Gallagher, A.J. Callaghan, J.K. Zhao, H. Dalton, and J. Trewhella, Global conformational changes control the reactivity of methane monooxygenase. *Biochemistry* **38**, 6752–6760 (1999).
48. R. Langridge, D.A. Marvin, W.E. Seeds, H.R. Wilson, C.W. Hooper, M.H.F. Wilkins, and L.D. Hamilton, The Molecular Configuration of Deoxyriboucleic Acid: Molecular Models and their Fourier Transforms. *J. Mol. Biol.* **2**, 38–64 (1960).
49. F. Merzel and J.C. Smith, SASSIM: a method for calculating small-angle X-ray and neutron scattering and the associated molecular envelope from explicit-atom models of solvated proteins. *Acta. Crystallogr. Sect. D-Biol. Crystallogr.* **58**, 242–249 (2002).
50. J. Ninio, M. Yaniv, and V. Luzzati, Comparative Small-Angle X-Ray Scattering Studies on Unacylated, Acylated and Crosslinked Escherichia-Coli Transfer Rna Val(I). *J. Mol. Biol.* **71**, 217–229 (1972).
51. M.Y. Pavlov, M.A. Sinev, A.A. Timchenko, and O.B. Ptitsyn, A Study of Apo-Forms and Holo-Forms of Horse Liver Alcohol-Dehydrogenase in Solution by Diffuse-X-Ray Scattering. *Biopolymers* **25**, 1385–1397 (1986).
52. C.A. Pickover and D.M. Engelman, On the Interpretation and Prediction of X-Ray-Scattering Profiles of Biomolecules in Solution. *Biopolymers* **21**, 817–831 (1982).

53. D. Svergun, C. Barberato, and M.H.J. Koch, CRYSOL – A program to evaluate x-ray solution scattering of biological macromolecules from atomic coordinates. *J. Appl. Crystallogr.* **28**, 768–773 (1995).
54. D.I. Svergun, I. Aldag, T. Sieck, K. Altendorf, M.H.J. Koch, D.J. Kane, M.B. Kozin, and G. Gruber, A model of the quaternary structure of the Escherichia coli F-1 ATPase from X-ray solution scattering and evidence for structural changes in the delta subunit during ATP hydrolysis. *Biophys. J.* **75**, 2212–2219 (1998).
55. D.I. Svergun, Richard S, M.H.J. Koch, Z. Sayers, S. Kuprin, and G. Zaccai, Protein hydration in solution: Experimental observation by x-ray and neutron scattering. *Proc. Natl. Acad. Sci. USA* **95**, 2267–2272 (1998).
56. J. Zhou, S. Krueger, and S.K. Gregurick, A coarse graining approach to determine nucleic acid structures from small angle neutron scattering profiles in solution. *Nucl. Acids. Res.* **33**, 6361–6371 (2005).
57. A.W. Ashton, M.K. Boehm, J.R. Gallimore, M.B. Pepys, and S.J. Perkins, Pentameric and decameric structures in solution of serum amyloid P component by X-ray and neutron scattering and molecular modelling analyses. *J. Mol. Biol.* **272**, 408–422 (1997).
58. W.T. Heller, E. Abusamhadneh, N. Finley, P.R. Rosevear, and J. Trewhella, The solution structure of a cardiac troponin C-troponin I-troponin T complex shows a somewhat compact troponin c interacting with an extended troponin I-Troponin T component. *Biochemistry* **41**, 15654–15663 (2002).
59. P.V. Konarev, M.V. Petoukhov, and D.I. Svergun. MASSHA – a graphics system for rigid-body modelling of macromolecular complexes against solution scattering data. *J. Appl. Crystallogr.* **34**, 527–532 (2001).
60. M.B. Kozin and D.I. Svergun, A software system for rigid-body modelling of solution scattering data. *J. Appl. Crystallogr.* **33**, 775–777 (2000).
61. F. Gabel, B. Simon, and M. Sattler, A target function for quaternary structural refinement from small angle scattering and NMR orientational restraints. *Eur. Biophys. J. Biophys. Lett.* **35**, 313–327 (2006).
62. A. Grishaev, J. Wu, J. Trewhella, and A. Bax, Refinement of multidomain protein structures by combination of solution small-angle X-ray scattering and NMR data. *J. Am. Chem. Soc.* **127**, 16621–16628 (2005).
63. M. Nollmann, W.M. Stark, and O. Byron, A global multi-technique approach to study low-resolution solution structures. *J. Appl. Crystallogr.* **38**, 874–887 (2005).
64. M.V. Petoukhov and D.I. Svergun, Global rigid body modeling of macromolecular complexes against small-angle scattering data. *Biophys. J.* **89**, 1237–1250 (2005).
65. M.V. Petoukhov, N.A.J. Eady, K.A. Brown and D.I. Svergun, Addition of missing loops and domains to protein models by X-ray solution scattering. *Biophys. J.* **83**, 3113–3125 (2002).
66. M.H.J. Koch, P. Vachette, and D.I. Svergun, Small-angle scattering: a view on the properties, structures and structural changes of biological macromolecules in solution. *Q Rev. Biophys.* **36**, 147–227 (2003).
67. D.I. Svergun and M.H.J. Koch, Small-angle scattering studies of biological macromolecules in solution. *Rep. Prog. Phys.* **66**, 1735–1782 (2003).
68. J.A. Langer, D.M. Engelman, and P.B. Moore, Neutron-Scattering Studies of Ribosome of Escherichia-Coli – a Provisional Map of Locations of Proteins S3, S4, S5, S7, S8 and S9 in 30-S Subunit. *J. Mol. Biol.* **119**, 463–485 (1978).
69. V. Ramakrishnan, M. Capel, M. Kjeldgaard, D.M. Engelman, and P.B. Moore, Positions of Protein-S14, Protein-S18 and Protein-S20 in the 30-S Ribosomal-Subunit of Escherichia-Coli. *J. Mol. Biol.* **174**, 265–284 (1984).
70. V.R. Ramakrishnan, S. Yabuki, I.Y. Sillers, D.G. Schindler, D.M. Engelman, and P.B. Moore, Positions of Proteins S6, S11 and S15 in the 30-S Ribosomal-Subunit of Escherichia-Coli. *J. Mol. Biol.* **153**, 739–760 (1981).
71. M.S. Capel, D.M. Engelman, B.R. Freeborn, M. Kjeldgaard, J.A. Langer, V. Ramakrishnan, D.G. Schindler, D.K. Schneider, B.P. Schoenborn, I.Y. Sillers,

S. Yabuki, and P.B. Moore, A Complete Mapping of the Proteins in the Small Ribosomal-Subunit of Escherichia-Coli. *Science* **238**, 1403–1406 (1987).
72. D.I. Svergun, J.S. Pedersen, I.N. Serdyuk, and M.H.J. Koch, Solution Scattering from 50s Ribosomal-Subunit Resolves Inconsistency between Electron-Microscopic Models. *Proc. Natl. Acad. Sci. USA* **91**, 11826–11830 (1994).
73. D.I. Svergun, N. Burkhardt, J.S. Pedersen, M.H.J. Koch, V.V. Volkov, M.B. Kozin, W. Meerwink, H.B. Stuhrmann, G. Diedrich, and K.H. Nierhaus, Solution scattering structural analysis of the 70 S Escherichia coli ribosome by contrast variation .1. Invariants and validation of electron microscopy models. *J. Mol. Biol.* **271**, 588–601 (1997).
74. D.I. Svergun, N. Burkhardt, J.S. Pedersen, M.H.J. Koch, V.V. Volkov, M.B. Kozin, W. Meerwink, H.B. Stuhrmann, G. Diedrich, and K.H. Nierhaus, Solution scattering structural analysis of the 70 S Escherichia coli ribosome by contrast variation .2. A model of the ribosome and its RNA at 3.5 nm resolution. *J. Mol. Biol.* **271**, 602–618 (1997).
75. D.I. Svergun and K.H. Nierhaus, A map of protein-rRNA distribution in the 70 S Escherichia coli ribosome. *J. Biol. Chem.* **275**, 14432–14439 (2000).
76. J.K. Krueger, G.A. Olah, S.E. Rokop, G. Zhi, J.T. Stull, and J. Trewhella, Structures of calmodulin and a functional myosin light chain kinase in the activated complex: A neutron scattering study. *Biochemistry* **36**, 6017–6023 (1997).
77. J.K. Krueger, G. Zhi, J.T. Stull, and J. Trewhella, Neutron-scattering studies reveal further details of the Ca2+/calmodulin-dependent activation mechanism of myosin light chain kinase. *Biochemistry* **37**: 13997–14004 (1998).
78. J.K. Krueger, N.A. Bishop, D.K. Blumenthal, G. Zhi, K. Beckingham, J.T. Stull, and J. Trewhella, Calmodulin binding to myosin light chain kinase begins at substoichiometric Ca2+ concentrations: A small-angle scattering study of binding and conformational transitions. *Biochemistry* **37**, 17810–17817 (1998).
79. J.K. Krueger, S.C. Gallagher, G. Zhi, R. Geguchadze, A. Persechini, J.T. Stull, and J. Trewhella, Activation of myosin light chain kinase requires translocation of bound calmodulin. *J. Biol. Chem.* **276**, 4535–4538 (2001).
80. G.A. Olah, S.E. Rokop, C.L. WangA, S.L. Blechner, and J. Trewhella, Troponin-I Encompasses an Extended Troponin-C in the Ca2+-Bound Complex – a Small-Angle X-Ray and Neutron-Scattering Study. *Biochemistry* **33**, 8233–8239 (1994).
81. G.A. Olah and J. Trewhella, A Model Structure of the Muscle Protein Complex 4ca2+-Center-Dot-Troponin-C-Center-Dot-Troponin-I Derived from Small-Angle Scattering Data – Implications for Regulation. *Biochemistry* **33**, 12800–12806 (1994).
82. W.A. King, D.B. Stone, P.A. Timmins, T. Narayanan, von A.A.M. Brasch, R.A. Mendelson, and P.M.G. Curmi, Solution structure of the chicken skeletal muscle troponin complex via small-angle neutron and x-ray scattering. *J. Mol. Biol.* **345**, 797–815 (2005).
83. W.T. Heller, D. Vigil, S. Brown, D.K. Blumenthal, S.S. Taylor, and J. Trewhella, C Subunits binding to the protein kinase a RI alpha dimer induce a large conformational change. *J. Biol. Chem.* **279**, 19084–19090 (2004).
84. A.M. Augustus, P.N. Reardon, W.T. Heller, and L.D. Spicer, Structural basis for the differential regulation of DNA by the methionine repressor protein MetJ. *J. Biol. Chem.* **281**, 34269–34276 (2006).
85. P. Callow, A. Sukhodub, J.E. Taylor, and G.G. Kneale, Shape and subunit organisation of the DNA methyltransferase M.AhdI by small-angle neutron scattering. *J. Mol. Biol.* **369**, 177–185 (2007).
86. D. Comoletti, A. Grishaev, A.E. Whitten, I. Tsigelny, P. Taylor, and J. Trewhella, Synaptic arrangement of the neuroligin/beta-neurexin complex revealed by X-ray and neutron scattering. *Structure* **15**, 693–705 (2007).
87. D.L. Ho, W.M. Byrnes, W.P. Ma, Y. Shi, D.J.E. Callaway, and Z.M. Bu, Structure-specific DNA-induced conformational changes in Taq polymerase revealed by small angle neutron scattering. *J. Biol. Chem.* **279**, 39146–39154 (2004).

88. S. Krueger, S.K. Gregurick, J. Zondlo, and E. Eisenstein, Interaction of GroEL and GroEL/GroES complexes with a nonnative subtilisin variant: a small-angle neutron scattering study. *J. Struct. Biol.* **141**, 240–258 (2003).
89. A.E. Whitten, D.A. Jacques, B. Hammouda, T. Hanley, G.F. King, J.M. Guss, J. Trewhella, and D.B. Langley, The structure of the KinA-Sda complex suggests an allosteric mechanism of histidine kinase inhibition. *J. Mol. Biol.* **368**, 407–420 (2007).
90. J. Zimmer, D.A. Doyle, and J.G. Grossmann, Structural characterization and pH-induced conformational transition of full-length KcsA. *Biophys. J.* **90**, 1752–1766 (2006).
91. W.T. Heller, J.K. Krueger. and J. Trewhella, Further insights into calmodulin-myosin light chain kinase interaction from solution scattering and shape restoration. *Biochemistry* **42**, 10579–10588 (2003).

Chapter 17
Neutron Imaging Applied to Plant Physiology

T.M. Nakanishi

Abstract Until recently, the in situ study of roots and dynamic measurements of water uptake in plants or trees has relied on physical, and therefore invasive and often destructive, sampling or sensor installation in the surrounding soil. These methods can, at the very least, alter the equilibrium of the rhizosphere of the plant or require the removal of the tree or plant. The sensitivity of neutrons to hydrogen atoms provides a noninvasive, nondestructive experimental approach to understand real-time processes in plants (e.g., water uptake). This chapter attempts to illustrate the effectiveness of neutron imaging techniques in studying plant physiology, soils, and wood.

Keywords Botany · Neutrons · Cold Neutrons · Thermal Neutrons · 2D imaging · 3D imaging · Charge-coupled device · in situ · Soil · Water uptake · Wood · Flower · Carnation · Plant physiology · Roots · Growth · Sand

17.1 Introduction

As discussed in Chapter 1, neutrons are particularly sensitive to hydrogen atoms and can therefore provide information on their location with high spatial resolution ($\sim$50 μm). When a neutron beam passes through a sample, the attenuation coefficient of some light elements — including hydrogen as well as rare earth elements such as gadolinium, dysprosium, and samarium — is extremely high, that is, 100–1000 times higher than those of the other common elements [1, 2]. In the case of X-rays, the attenuation coefficient changes by factors rather than orders of magnitude, from a light element to a heavier one, making it difficult to distinguish the change in the contrast due to a

T.M. Nakanishi (✉)
Graduate School of Agricultural and Life Sciences, The University of Tokyo,
1-1-1 Yayoi, Bunkyo-ku, Tokyo 113-8657, Japan
e-mail: atomoko@mail.ecc.u-tokyo.ac.jp

I.S. Anderson et al. (eds.), *Neutron Imaging and Applications*,
Neutron Scattering Applications and Techniques, DOI 10.1007/978-0-387-78693-3_17,

specific element from other neighboring ones. Neutron imaging techniques are naturally suited for materials containing hydrogen atoms and other low atomic weight attenuating materials, permitting nondestructive in situ measurements of the spatial and temporal distribution of water (and other fluids rich in hydrogen) in plants and trees as well as in soils.

17.2 Nondestructive Water Observation in Plants and Soils Using Neutron Transmission Imaging Techniques

Nondestructive techniques for analyzing processes within tissues of plants have the potential to provide extremely important information for studying plant activity. Neutrons are efficient at probing plant roots, which have greater water content (between 70 and 95%) than their surrounding media (between 5 and 30%), as explained in [3]. Clearly the size of the root is limited to ensure a "reasonable" transmission thickness for neutrons, which may be up to several millimeters depending on the neutron wavelength (for an example of neutron transmission calculations, see Chapter 1). Water movement in particular, which plays an important role both in biochemical processes and in physical structure, has not been studied in detail. To acquire nondestructive in situ images of plant or soil water distribution, the use of X-ray computed tomography (CT) measurements has been reported [4]; however, the resolution obtained with this method was insufficiently high to acquire microscopic-level images. X-ray imaging has been able to resolve spatial patterns of root distribution in soil, although this requires destructive collection of soil slices [5]. The most promising approach to obtaining images of water, as well as images of other biological chemicals, is nuclear magnetic resonance spectroscopy. Although the resolution of these images can theoretically reach 10 μm [6], the sample size is limited and measurements are time-consuming.

Plant morphological development is reflected by water distribution within tissues, which can be monitored using neutron imaging techniques [7]. Neutron imaging of intact plant tissues can indeed yield exquisite microscopic images from a unique and novel perspective, and without causing much radiation damage.

Although neutron imaging is the most suitable method for the study of living plant behavior based on water uptake, historically, a few researchers in plant physiology have employed it [8–10]. Recently, however, with the development of neutron beam techniques such as cold neutron radiography [11–13], new technological advances such as thin neutron sensitive imaging scintillators [14], and advances in charge-coupled device cameras (Chapter 4), the use of neutrons for plant research has increased significantly. Dynamic imaging (i.e., the study of water movement) has especially increased in application but still remains limited compared to other techniques.

In thin plant tissues such as flower petals or leaves, changes in the water content are too small to image with thermal neutrons; therefore, cold neutrons can be applied to a thin tissue to enhance the contrast, as, for instance, in the case of a chrysanthemum leaf [13]. Although cold neutrons are limited in their use because of a low transmission, small veins in the leaf can be observed, and the contrast is approximately 70% higher than that of a thermal neutron image. In addition, the range of changes in water thickness observed in the leaf is smaller than 30 μm [15].

Neutron beam is perfectly suitable for analysis of the tissue level of the sample, such as water distribution within a cut flower [3, 16–20] or a wood sample [21–24]. However the resolution afforded by the technique, typically tens of microns, is insufficient to study the process taking place within a single cell because much higher resolution is needed for observation within a cell.

To obtain 2D neutron transmission images of living plants and their root system in soil, samples are often grown in thin aluminum containers (2–3 mm thick). The total cross section (absorption and scattering) of aluminum is approximately 1/50 that of hydrogen; therefore, aluminum is almost transparent to thermal neutrons, and images of samples are not much affected by it.

17.2.1 Plant Roots Imbedded in Soil

Understanding of root growth is very limited because of the difficulty in observing root systems without obvious damage of the surrounding media and/or the root system itself. Often, destructive physical sampling or invasive sensor or minirhizotron installation is necessary and ultimately perturbs the rhizosphere (i.e., immediate area around the root system). However, observation of growth rates and root systems can be successfully imaged with neutron [25] or X-ray [26] imaging.

Figure 17.1 shows a neutron radiograph of a soybean plant root, after (A) 8 and (B) 15 days of growth in an aluminum container (of size $10 \times 150 \times 2\ mm^3$) packed with standard sand (silt type, 197–203 μm grain size, in pore size), containing 18% (by weight) of water. Since the water content in the root (in white in Fig. 17.1) is much higher than that in the surrounding sand, the root image was clearly shown even in the presence of soil [27].

The water-deficient part of the plant (dark in the image) is clearly visible near the upper part of the main root. To distinguish the water content of the root more clearly, a part of the main root initiating the growth of lateral roots is magnified in Fig. 17.1C. The apparent 3D image in Fig. 17.1D is produced by converting the degree of whiteness, which is directly related to the water content and to the height. The lower area near the root indicates a large amount of water uptake by the root. The activity of the root is well seen in a 1-mm area from the root surface. In this region, not only nutrients or water uptake, but also a corporative interaction between root and microorganisms occurs. From the neutron radiograph, it can be seen that the amount of water was very low around some parts of the root, suggesting the possibility of water vapor absorption rather than water solution. Possible future studies include the measurement of heterogeneity in water distribution and uptake at the rhizosphere. The latter phenomenon is expected to play an important role in

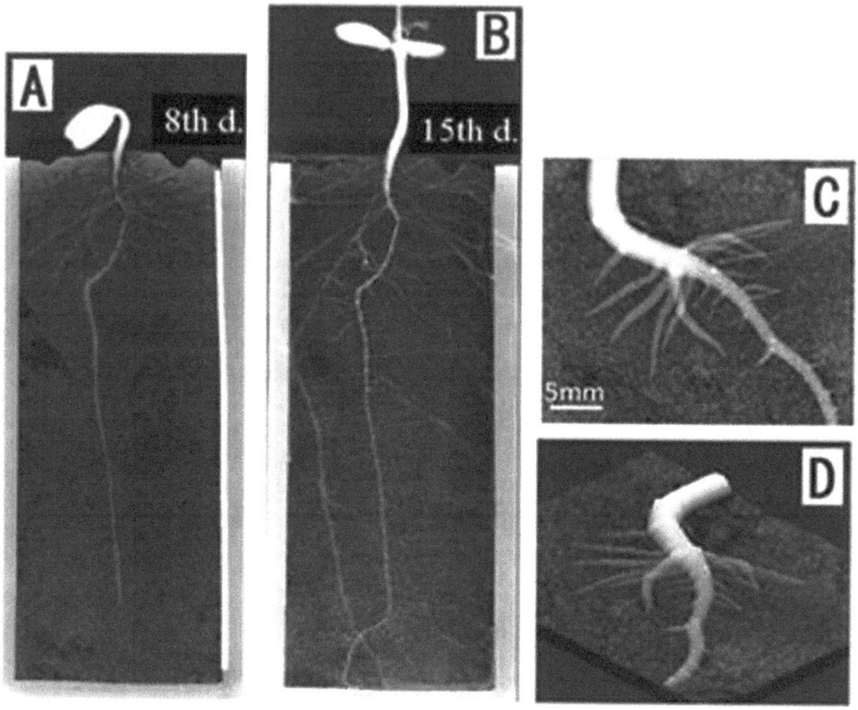

Fig. 17.1 A neutron radiograph of a soybean plant root embedded in soil after (**A**) 8 and (**B**) 15 days of growth; (**C**) magnification of the root; and (**D**) conversion of (**B**) to the 3D image

understanding the fundamental activities of the plant root system. Furthermore, neutron imaging may help explain the discrepancy between the amount of nutrients the plant actually absorbs from its environment and the amount "invasively" measured by soil sample extraction.

17.2.2 Observation of Root Development Under Different Soil Conditions

Noninvasive neutron imaging methods play an important role in the observation of growth rates and root development under different soil conditions, as discussed in [25]. In general, roots seek the most favorable place to grow, and adapt their growth orientation or pattern to the soil condition. For example, when water-rich soil is provided at one lateral side of the main root, the secondary roots preferably develop on that side. Similarly, Fig. 17.2A displays changes in root development due to the presence of vanadium in the soil. A soybean seedling is grown in an aluminum container where water-absorbing polymer material [28] containing 50 mM of vanadium is embedded in the soil. When vanadium is placed below the main root, the growth of the root ceases at

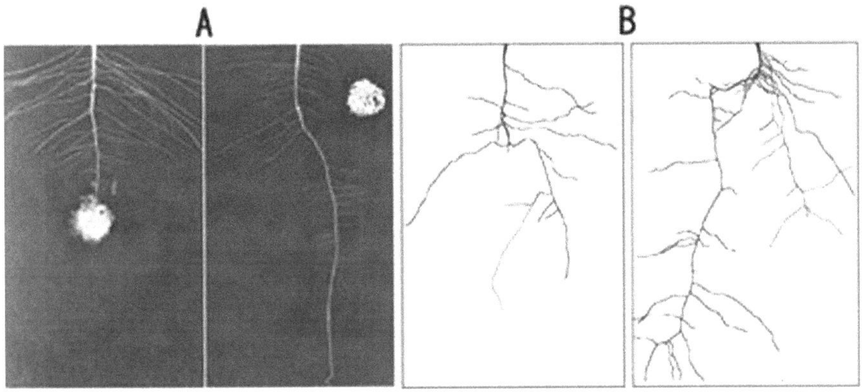

Fig. 17.2 Root development pattern. **A**: Soybean root development in the presence of 50 mM of vanadium (*white aggregate*). **B**: Radish root development. (*Left*) Low-pH soil containing 23 mM of aluminum; (*right*) Modified lignin mixed with the soil shown in the image on the left to reduce inhibition of growth by the aluminum

the polymer, but the side roots grows larger to compensate. On the other hand, when the polymer "doped" with vanadium is placed on one side of the root, the lateral root facing the polymer side does not grow, but the root system on the other side is longer. The position at which growth stops and the direction of root growth are dependent on the vanadium concentration inside the polymer [29]. It is also shown that the amount of water uptake at a certain height of the main root increases before side root growth begins.

Since root development is an indicator of the soil condition, neutron imaging can be applied to evaluate soil conditioning agents as well as the effect of a chemically modified fertilizer. To study lignin derivatives used as a soil conditioning agent, a radish plant is employed to observe the root profile in an acidic soil [30]. The root growth pattern captured in the neutron image in Fig. 17.2B clearly shows that the modified lignin is effective as a soil conditioning agent. Subsequently, the total root length is measured through image analysis, and the soil conditioning effect of different chemicals is analyzed. The root pattern shown in Fig. 17.2B is similar to patterns found in lightning or in the production of cracks in a pane of glass. There is no reproducibility in the pattern—a fundamental feature, especially for roots, for maintaining biological diversity. Because of this, experimental errors cannot be quantified properly when analysis of root growth is measured using invasive techniques such as the line intersection method, in which a lattice board is placed on the root and then removed from the soil, allowing the number of intersections between the root and the lattice to be counted. Neutron imaging, because it is nondestructive, can use the same sample to trace the root activity over time, which reduces the fundamental errors that arise from using different samples.

Recent laboratory experiments reproducing water infiltration in soil, plant, and the atmosphere have reported water redistribution processes as a function of different conditions such as, for example, root structure, soil structure, and moisture [31].

17.2.3 Neutron Tomography of Root Systems

Since water movement in a plant is slow (approximately several minutes to hours), computed tomography can be used to evaluate "static" spatial water distribution inside and around root systems. For example, Fig. 17.3 illustrates a soybean root grown in a 3-cm diameter aluminum container packed with standard sand [27].

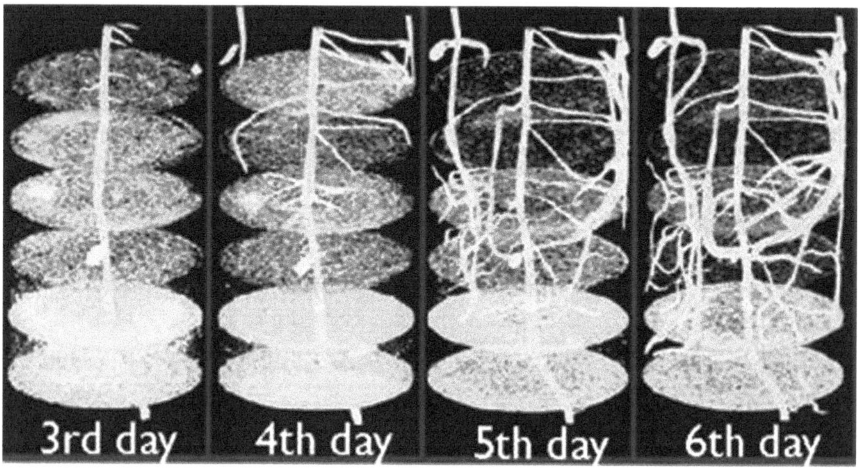

Fig. 17.3 In situ neutron tomography of the root system of a soybean during different days

Six CT reconstructions corresponding to different heights of the root system permit comparison of water content changes. Within a few days, the amount of water in the upper part of the container decreases, corresponding to an increased formation of roots. Using tomographic root images, positions and lengths of the side roots, as well as water movement around them, can be quantified (Fig. 17.4) [31].

There is an obvious increase in side root growth around the upper part of the container (up to 20 mm down from the air/soil interface) compared to side roots located farther down the container (Fig. 17.4, left). This correlates to a decrease in the water amount found in the soil after several days (6 days, for example) in the same region (Fig. 17.4, right).

Root surfaces and volumes can also be calculated using neutron imaging techniques. In the case of a soybean root, when 10 mM of $AlCl_3$ solution is applied to the soil, a decrease in both root surface area and root volume is observed [32]. The presence of aluminum ions is one of the main factors inhibiting plant growth

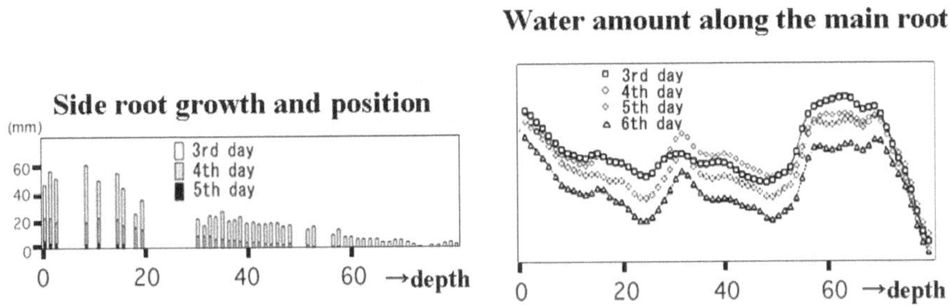

Fig. 17.4 Quantitative measurements of (*left*) side root growth according to their position along the main root and (*right*) water distribution around the main root days after watering the specimen. Depth is in mm

in an acidic soil. Moreover, root development in soil contaminated with heavy metals has also been analyzed via neutron imaging, as illustrated in [33].

17.2.4 The Aboveground Portion of the Plant

Neutron imaging is commonly used to analyze the aboveground portion of a plant. For example, neutron imaging of a cowpea plant can provide information on stem internode tissue, whose function is to store water [34]. Under water-deficient conditions, water is primarily moved from the internode to other tissues. A complementary study of water labeled with a positron emitter, ^{15}O, has measured real-time movement of water in a plant, as referenced in [35].

Extending the life of cut flowers is a key issue in the floral industry, and the amount of water in a flower plays a key role in maintaining its flowering stage. Neutron imaging of cut flowers shows detailed images of water content in flowers such as lilies, morning glories, and chrysanthemums. Figure 17.5 shows an example of carnation flowers imaged with thermal neutrons.

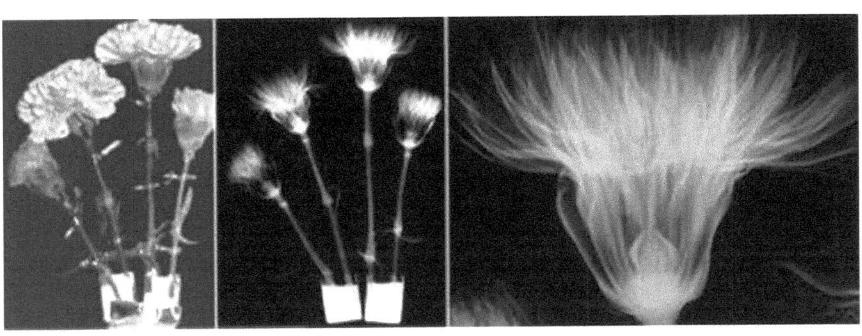

Fig. 17.5 Images of a carnation flower. A photograph is on the *left*, a neutron image in the *center*, and a magnification of the neutron image on the *right*. In the left and center images, the two flowers on the inside are controls and the two on the outside were dried before the neutron radiograph was taken [36]

Increasing the viscosity of water (i.e., xenon is dissolved under high pressure into water) can help prolong the life of a carnation flower after it has been harvested, as illustrated in Fig. 17.6A [36]. This helps control the metabolism of the flower, which slows down the deterioration process mediated by an enzymatic reaction. An example of a 3D neutron imaging study is shown in Fig. 17.6B. The flower respiration rate while conducting neutron imaging during the senescent (i.e., aging) stage has been evaluated [3, 18, 19]. Furthermore, the analysis of 3D images of a carnation flower demonstrates the importance of water inside the ovule to maintain the flowering stage longer. It is

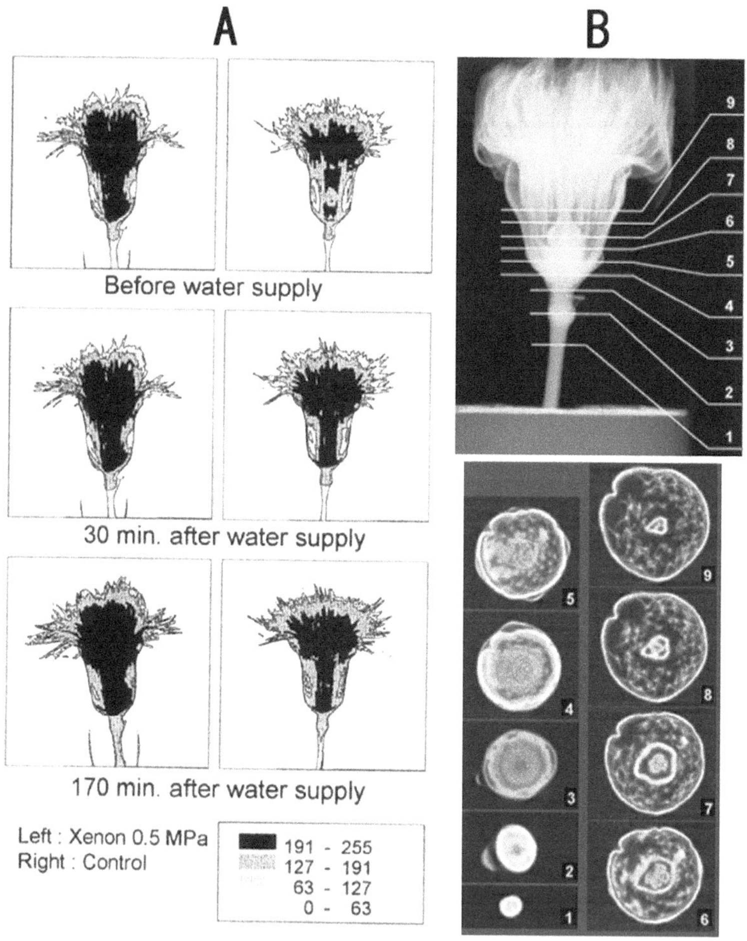

Fig. 17.6 Neutron image of a carnation flower. (**A**) Neutron radiographs of a carnation flower after being supplied with water containing xenon. (**B**) CT images at different heights along the flower head [36]

shown that the ovary and pistil areas continue to increase in size after the respiration rate stops increasing [36].

17.3 Investigation of Wood Samples

The study of physiological activity of a tree mainly relies on knowing the water distribution inside the wood. The example below illustrates the use of neutron imaging on wood disks extracted from a tree cut only a few hours before experiments, which therefore permits the observation of the green moisture image [23, 24]. A stem section, covered as a precaution to prevent water loss, is further cut to obtain 1 cm thick wood disks just prior to being exposed to neutrons. Figure 17.7 shows the neutron images of disks from a Japanese cypress (*Chamaecyparis obtusa*; Fig. 17.7A) and a Sugi (*Cryptomeria japonica*; Fig. 17.7B), whose ages are estimated to be about 19 and 24 years, respectively. The upper images in the figure are photographs of wood disks, and the

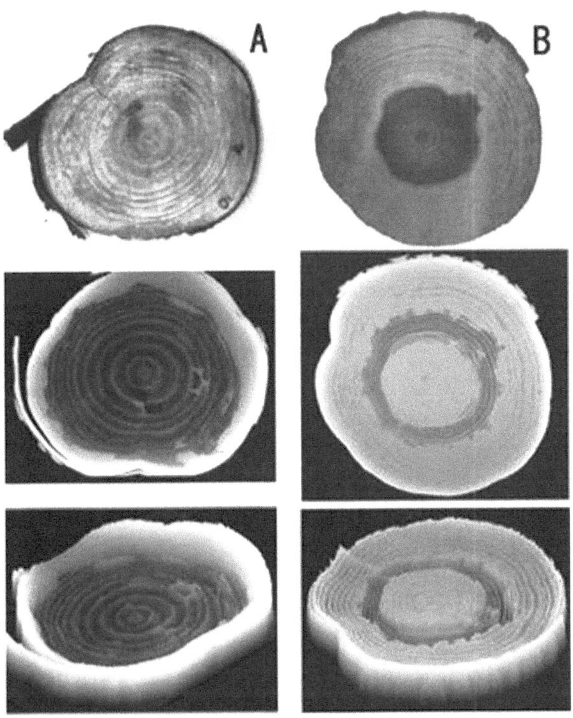

Fig. 17.7 Images of wood disks, each 1 cm thick. Photographs of disks are shown in the *top row*; neutron images in the *center row*; and reconstructions of the middle images, with high points corresponding to water-rich areas, in the *bottom row*

center images are the corresponding neutron images. The lower images are reconstructed ones based on the images in the center row with the degree of whiteness indicated as height.

In the Japanese cypress, the neutron image shows higher water content in the outer rings, adjacent to the bark. This zone includes the living phloem tissue and recently developed xylem tissue that transports water from the roots to the leaves. Both tissues have high water content. Water content declines radially from the bark into the older, less functional, air-filled xylem tissue. Water distribution in some areas shows a wavy pattern that might indicate tissue damage or buildup of specific chemicals within the rings. In the heartwood region, rings appear to have much lower water content, likely indicating buildup of decay-resistant extractive chemicals. Besides the large hydrogen-rich outer part of the disk, many rings corresponding to the annual rings of the wood are shown to contain a higher level of water than their immediate surroundings.

Sugi is a popular wood in Japan for building houses and furniture. Although there is only one species of Sugi, neighboring trees of the same cultivar show different water content at the heartwood. It is not known what causes the differences in moisture content at the heartwood. When lumber is processed, residual moisture causes serious problems. When the moisture content at the heartwood is high in green lumber, it is difficult to remove water completely during the drying process. However, moisture is gradually lost after the lumber is used in houses or furniture, causing the shape to warp over the years.

When neutron radiographs of wood discs are taken during the drying process, the mechanism by which moisture is lost can be directly analyzed [24]. In the case of Sugi, another feature in the disk is observed in the heartwood. As shown in Fig. 17.7B, there is always a water-deficient area between heartwood and sapwood. Because of the lighter color of this zone, it is called the white zone. As the tree grows, the area of heartwood increases. However, the water content throughout the heartwood is maintained at a high level as the tree grows. It is not known how the water in sapwood moves into heartwood across the white zone.

A rapid decrease in water content can be observed when the plant tissue is injured. For example, neutron imaging shows that when Sugi wood is inoculated with a canker fungus, a water deficiency occurs and spreads from the inoculated site [30]. The dynamic process of water absorption in a wood sample can be evaluated with neutron imaging techniques [22]. Imaging of solvent absorption and water-loss processes in small wood samples ($4 \times 4 \times 2.7$ cm) has been reported [21] when acryl resin solvents are applied to protect the surface of the wood, which is of importance for the development of wood-protective materials for preserving wood art with historical and cultural value.

Several other neutron imaging studies of wood can be found in [37–41].

17.4 Other Agricultural Applications

Besides the examples presented above, neutron imaging has been used in many other agricultural applications. The water uptake of seeds during germination provides useful information, for example, in investigating the method of storing seeds because in some seeds, physical damage or the storage condition may change the germination rate (Fig. 17.8) [42]. Even at an early stage of germination, the germination rate of seeds can be estimated using neutron images of water absorption.

Recently, a tomographic study of corn kernels has aimed at understanding the effect of *Aspergillus flavus* (*A-flavus* fungus) to resistant species. Anatomic changes are observed when resistant inoculated and uninoculated species are compared to their nonresistant inoculated counterparts [43].

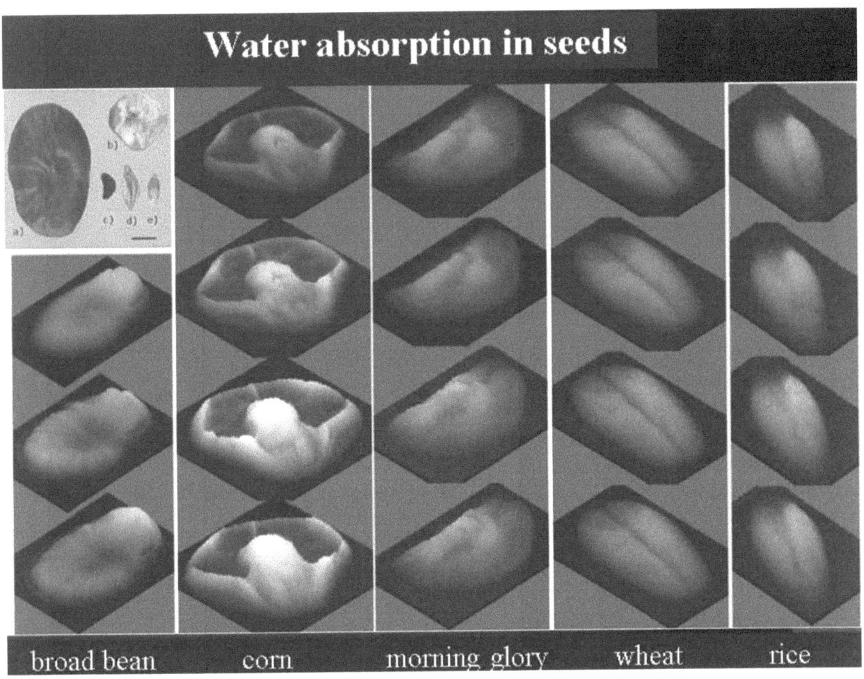

Fig. 17.8 Water absorption process in five kinds of seeds. Photographs of the seeds are shown in the *upper left corner* (*the black bar* is 5 mm long). Seeds are soaked in water and neutron images are taken after 0, 2, 4, and 6 h (*top* to *bottom*). For the broad bean, images taken after 0, 2, and 4 h are shown

17.5 Concluding Remarks

Nondestructive imaging methods for imaging water or H-rich fluids in living plants, soils, and wood using neutrons provide the highest resolution yet obtained for in situ water content analysis. Due to the high sensitivity to H atoms, this technique is capable of providing high-resolution images of water movement in seeds, in roots embedded in soil, in wood, and in meristems during development. The images presented in this chapter provide a taste of the intriguing new areas of investigation in the field of plant physiology that are being opened by new developments in neutron imaging methods.

Recently, the powerful methods derived from molecular genetics have resulted in a tendency to focus research on the molecular aspects of biology and to ignore important aspects of the intact plant. However, the intact plant itself has a high potential to integrate many functions and to respond to many diverse environmental conditions. Nondestructive techniques are essential to studying and understanding the activity of a living plant, its development, and its adaptation to environmental changes. The method reported here seems to be the most promising new tool for doing so.

References

1. K.A. Garrett and H. Berger, *Atom. Rev.* **15**, 125 (1977).
2. J.F.W. Markgraf and R. Matfield, Neutron beam, in J.C. Domanus, ed., *Practical Neutron Radiography*, Kluwar Academic Publishers, Dordrecht, Boston, London, pp. 26–50 (1992).
3. U. Matsushima, Y. Kawabata, C-M. Sim, K-Y. Nam, and T. Nishizawa, Proc. VIIIth IS Postharvest Phys. Ornamentals, Eds. N. Marissen et al., *Acta Hortic.* **669**, 111 ISHS (2005).
4. J.M. Hainsworth and L.A.G. Alymore, *Soil Sci. Soc. Am. J.* **50**, 841 (1986).
5. C.J. Moran, A. Pierret, and A.W. Stevenson, *Plant Soil* **223**, 101 (2000).
6. P.T. Callaghan, Diffusion-limited resolution, in P.T. Callaghan ed., *Principles of Nuclear Magnetic Resonance Microscopy*, Oxford Science, Oxford, pp. 203–206 (1991).
7. T.M. Nakanishi and M. Matsubayashi, *J. Plant Phys.* **151**, 442 (1997).
8. S.T. Willatt et al., *Agron. J.* **70**(4) 581–586 (1978).
9. S.T. Willatt et al., *Ann. Bot.* **43**, 415–422 (1979).
10. P. Couchat et al., *Agron. J.* **72**, 321–324 (1980).
11. A. Hilger, N. Kardjilov, M. Strobl, W. Treimer, and J. Banhart, *Physica B* **385**(86), 1213 (2006).
12. Y. Kawabata, M. Hino, T. Nakano, H. Sunohara, U. Matsushima, and P. Geltenbort, *Nucl. Instrum. Meth. Phys. Res. A* **542**, 61 (2005).
13. Y. Kawabata, U. Matsushima, T. Horie, T. Nakano, and R. Maruyama, *J. Radioanal. Nucl. Chem.* **264**, 319 (2005).
14. G. Frei and E. Lehmann, *Proceedings of 8th World Conference of Neutron Radiography*. Gaithersburg, USA, p. 21, October 16–19 (2006).
15. U. Matsushima, K. Kawabata, M. Hino, P. Geltenbort, and B. Nicolai, *Nucl. Instrum. Meth. Phys. Res. A* **542**, 76 (2005).
16. U. Matsushima, T. Ooshita, T.M. Nakanishi, M. Matsubayashi, Y. Seo, and Y. Kawagoe, *J. Jpn. Soc. Agr. Machin.* **62**, 70 (2000).

17. U. Matsushima, K. Kawabata, T. Nakano, M. Hino, P. Geltenbort, and B. Nicolai, *Proc. 5th Int. Postharvest Symp.* **682**, pp. 1411 ISHS (2005).
18. U. Matsushima, Y. Kawabata, and T. Horie, *J. Radioanal. Nucl. Chem.* **264**, 325 (2005).
19. U. Matsushima, E. Lehmann, P. Vontobel, G. Fre, B.M. Nicolai, T. Nishizawa, and Y. Kawamitsu, *Proceedings of the Asia-Pacific Symposium on Quality Management for Agri-Foods in Supply Chains*. Bangkok, Thailand, Aug. 7–10, p. 119 (2006).
20. T.M. Nakanishi, J. Furukawa, and M. Matsubayashi, *Nucl. Instrum. Meth. Phys. Res. A* **424**, 136 (1999).
21. E. Lehmann, S. Hartmann, and P. Wyer, *Nucl. Instrum. Meth. Phys. Res. A* **542**, 87 (2005).
22. D. Mannes, E. Lehmann, S. Oswald, and P. Niemz, *Proceedings of 5th Plant Biomechanics Conference*, Stockholm, August 28–September 1, pp. 393 (2006).
23. T.M. Nakanishi, I. Karakama, T. Sakura, and M. Matsubayashi, *Radioisotopes* **47**, 387 (1998).
24. T.M. Nakanishi, T. Okano, I. Karakama, T. Ishihara, and M. Matsubayashi, *Holzforschung* **52**, 673 (1998).
25. M. Menon, B. Robinson, S.E. Oswald, A. Kaestner, K.C. Abbaspour, E. Lehmann, and R. Schulin, *Euro. J. Soil Sci.* **58**, 802 (2007).
26. A. Pierret, M. Kirby, and C. Moran, *Plant Soil* **255**, 361–373 (2003).
27. T.M. Nakanishi, Y. Okuni, J. Furukawa, K. Tanoi, H. Yokota, N. Ikeue, M. Masubayashi, H. Uchida, and A. Tsuji, *J. Radioanal. Nucl. Chem.* **255**, 149 (2003).
28. T.M. Nakanishi, S. Matsumoto, and H. Kobayashi, *Radioisotopes* **42**, 26 (1993).
29. J. Furukawa, T.M. Nakanishi, and M. Matsubayashi, *Nucl. Instrum. Meth. Phys. Res. A* **424**, 116 (1999).
30. T. Yamada, Y. Aoki, M. Yamamoto, M. Komatsu, D. Kusumoto, K. Suzuki, and T.M. Nakanishi, *J. Radioanal. Nucl. Chem.* **264**, 329 (2005).
31. J. Furukawa, T.M. Nakanishi, and M. Matsubayashi, *Nondestr. Test Eval.* **16**, 335 (1999).
32. K. Saito and T.M. Nakanishi, *Mokuzai Gakkaishi* **43**, 669 (1997).
33. T.M. Nakanishi, S. Matsumoto, and H. Kobayashi, *Radioisotopes* **41**, 638 (1992).
34. T.M. Nakanishi, K. Don-Jin, T. Kitamura, R. Ishii, and M. Matsubayashi, 242, 353 (1999).
35. Y. Okuni, J. Furukawa, M. Matsubayashi, and T.M. Nakanishi, *Anal. Sci.* **17**(Supplement), i1499 (2001).
36. T.M. Nakanishi, J. Furukawa, and M. Matsubayashi, *Nucl. Instrum. Meth. Phys. Res. A* **424**, 136 (1999).
37. E.H. Lehmann, P. Vontobel, and P. Niemz, *Investigation of Moisture Distribution in Wooden Structures by Neutron Radiography*. PSI Annual Report Annex VI, p. 53 (1999).
38. E.H. Lehmann, P. Vontobel, P. Niemz, and P. Haller, The method of neutron radiography and its use for wood properties analysis. In *Proceedings of the International Conference on Wood and Wood Fibre Composites*, Stuttgart, Germany (2000).
39. E.H. Lehmann, P. Vontobel, P. Scherrer, and P. Niemz, Application of neutron radiography as method in the analysis of wood. *Holz als Rohund Werkstoff* **59**(6), 463–471 (2001).
40. P. Niemz, E.H. Lehmann, P. Vontobel, P. Haller, and S. Hanschke, Investigations using neutron radiography for evaluations of moisture ingress into corner connections of wood. *Holz als Roh-und Werkstoff* **60**(2), 118–126 (2002).
41. V. Bucur, *Nondestructive Characterization and Imaging of Wood*, Springer Series in Wood Science, (Hardcover), ISBN 3-540-43840–8 (2003).
42. T.M. Nakanishi and M. Matsubayashi, *Bioimages* **5**, 45 (1997).
43. T.E. Cleveland et al., *J. Cereal Sci.* **48**, 517 (2008).

Chapter 18
Homeland Security and Contraband Detection

R.C. Lanza

18.1 Introduction

18.1.1 Nature of Problem

Detection of contraband and illicit materials has become increasingly important, especially since the terrorist attacks in the United States on September 11, 2001. The nature of the detection problem embodies both physics issues and a set of operational constraints that limit the practical application of neutrons. The issue under consideration is detection of materials that are considered serious threats; these may include explosives, radioactive materials, fissile materials, and other materials associated with nuclear weapons, often referred to as special nuclear material (SNM). The overriding constraint is in the physics: systems must be based on clean physics; but unlike physics experiments, detection systems work under the limitation that materials must be identified nonintrusively, without interrupting the normal flow of commerce and with a high probability of detection and a low probability of false alarms. A great deal of work has been reported in the literature on neutron-based techniques for detecting explosives and drugs. The largest impetus by far for detecting explosives comes from aviation industry requirements for inspecting luggage and, to a lesser extent, cargo. The major alternative techniques are either X-ray–based or chemical trace detection methods that look for small traces of explosive residues. The limitations of the X-ray and trace methods in detecting explosives are well known, but currently (2008) it is safe to say that no neutron- or nuclear-based technique is being used routinely for security inspection, despite extensive development of these methods. Smuggling of nuclear materials has become a concern, and neutron techniques are particularly attractive for detecting them. Given

R.C. Lanza (✉)
Department of Nuclear Science and Engineering, MIT, NW13-221, Cambridge, MA 02139-4307, USA
e-mail: lanza@mit.edu

I.S. Anderson et al. (eds.), *Neutron Imaging and Applications*,
Neutron Scattering Applications and Techniques, DOI 10.1007/978-0-387-78693-3_18,

the limitations of X-ray techniques and the need for SNM detection, it is now useful to reexamine neutron methodologies, particularly imaging. A significant number of neutron-based techniques have been proposed and are under development for security applications, especially SNM detection, but describing how they work is beyond the scope of the chapter. Instead, one particular approach to neutron imaging, neutron resonance radiography (NRR), is discussed in detail as it illustrates many of the issues connected with imaging and detection.

18.1.2 Operational Issues

An important part of any proposed system design, beyond physics, is a description of how the actual measurement and detection will be done, how the system integrates into end-users' operations, and a means for testing systems under realistic conditions. The ability to reliably inspect containers without interrupting normal commerce is essential for any system to be deployed. It is necessary to have a reasonably detailed knowledge of various environments where a technology will operate to ensure that it will actually be usable.

Another important point is that the goal of these inspection systems is not to identify the contents of every container but rather to determine if the materials of interest are *not* present in the container. Consider the sheer volume of inspections compared with the number of actual threats that are present. There are estimates that, for example, the number of checked bags to be examined may be as high as 10^9 per year. The number of actual threats will be much smaller, perhaps ultimately only 10^3 or so. The difference, 10^6, means that system performance is dominated by false alarms; most "threat alarms" are in fact false alarms. Given this, we must devise means to clear containers rapidly; hence the idea is to determine that a threat is not present, rather than making a complete analysis of the container.

One approach is to simply consider a system with a single decision variable that is used to distinguish threats from nonthreats. In Fig. 18.1, the results of a series of measurements are shown. Clearly, to the extent that there are overlaps, it is not possible to have a decision variable with a high detection probability without generating false alarms. One way of expressing this tradeoff is to use the receiver–operator curve (ROC), Fig. 18.2, which shows the tradeoff in a clearer manner. This curve can be generated theoretically from assumptions about the distribution of the decision parameters in Fig. 18.1, but it is often necessary to generate it instead by experimental measurement. The line at 45° in Fig. 18.2 also illustrates the result of a purely random measurement. Unfortunately, the ratio of measurements (inspections) with no threat to measurements (inspections) with threats may be as high as 10^6 to 1, resulting in high false alarm rates. The goal, then, is to develop inspection methods that are specific to a given threat so as to make the separation as large as possible, as shown in Fig. 18.1. This is the driving force behind the use of neutrons in imaging.

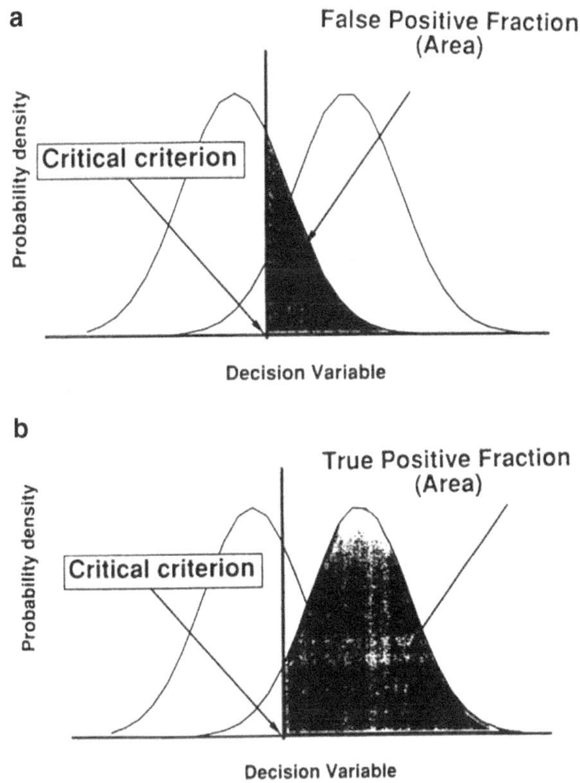

Fig. 18.1 Influence of decision threshold on false positive rate relative to true positive rate

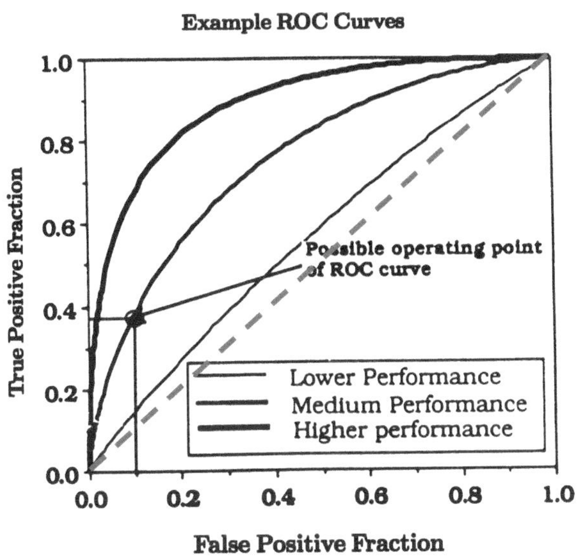

Fig. 18.2 Receiver–operator curve (ROC). The *dashed line* represents random performance

18.1.3 Practicality

The goal of all systems is their eventual use in a nonlaboratory environment, a goal that presents issues of safety, ease of operation, and reliability.

Generally, almost all systems share the requirement that they be relatively compact and operable by personnel relatively unskilled in nuclear instrumentation. This point is well known to designers of, for example, baggage X-ray scanners such as those used at airports and X-ray- and gamma-ray–based cargo container scanners. It is especially significant in techniques based on large accelerators and detector assemblies, which present a new level of complexity and complication. Most airports and ports are exceedingly crowded and either lack space for large instruments or would prefer to put the space for other uses that more directly impact the economic or operational aspects of the port.

All of these systems rely on the use of neutron sources, generally either a small accelerator [radio frequency quadrupole (RFQ) or electrostatic] or a sealed deuterium–tritium (DT) generator [1]. With the exception of the DT (or deuterium–deuterium) sealed-tube generators, accelerators remain a critical problem in deploying these techniques. Issues of size, weight, and general complexity must be solved before these systems can be placed in a typical inspection environment.

Practical systems also require sophisticated software for control and calibration and for automatic or semiautomatic threat detection algorithms. Most of these have been developed by vendors over years and form a substantial part of the cost of deploying scanners, even after the hardware has been built.

Performance specifications generally require specific detection probabilities, spatial resolution, scanning speed, and reliability. These are often couched in terms such as the ability to see a wire of a certain diameter behind a 25-cm steel plate or the ability to detect, for example, 100 cc of SNM behind 25 cm of steel.

Safety issues are especially important, since these systems are often large radiation producers that are used in semipublic areas, or at least areas in which it may be anticipated that there will be members of the general public and non radiation workers [2].

18.1.4 Special Nuclear Materials

The SNMs are the basic ingredients for nuclear weapons, and their control and interdiction are highly important. The current prevailing model for addressing threats associated with cargo-container-borne weapons of mass destruction, especially nuclear weapons, could be characterized as a customs-based approach. Under this scheme, new systems using either active or passive techniques to detect nuclear signatures, from hidden fissile material would be integrated into the existing customs infrastructure of ports. The customs-based approach is a natural extension of conventional strategies for finding and seizing incoming contraband. Nuclear weapons, however, are utterly unlike any other kind of contraband in that their power is so great, and their effects are so far-reaching, that even discovering them in port (or allowing them to enter port at all) cannot provide adequate protection from their destructive reach.

Despite decades of research and development and considerable expenditure on alternative approaches, the systems used to examine cargo are almost the same as those used in the 1970s with a few minor improvements. The most common nonintrusive inspection systems use X-ray transmission to form an image of the interior of a cargo container and search for anomalies. Some add X-ray backscattering to detect materials, such as drugs, hidden in the walls of containers. The major application of this technology is the verification of cargo manifests and the detection of large amounts of smuggled materials. There is, however, one overarching issue: no current X-ray system, whether transmission or transmission enhanced with backscattering, can distinguish fissile material from ordinary materials. Thus, X-ray systems are practically useless in intercepting nuclear weapons. The detection of fissile material requires a system designed for that purpose, and neutron imaging provides a particularly powerful tool.

18.1.5 Explosives

Explosive materials come in a large variety of physical and chemical forms, and their density and elemental composition vary widely [3]. Military-style high explosives have been used in the past; but in recent times the use of improvised and nonconventional explosives has increased, and the spectrum of potential threats has therefore expanded. Although some have argued that the primary common characteristic of these materials is that they explode, there remain characteristics of explosives that can be detected. Many commercial and military explosives have high concentrations of nitrogen; but recent experience has revealed a new set of nonnitrogenous explosives often fabricated by terrorists from readily available and fairly ordinary materials that can be purchased without arousing suspicion. It is essential that the designer of detection equipment recognize the vast variety of potential threats and thus clearly understand what the limitations of a particular technique may be. The implication of the widespread use of these improvised explosives is that detection of, for example, nitrogen alone will often be inadequate; indeed, the use of assumed elemental ratios for explosive composition may lead to difficulties in detection. Neutron imaging is an approach to detection that can be used to image elemental composition and hence has the potential to detect these newer explosives.

18.1.6 Others

Depending on circumstances, there may be other materials of interest for detection, such as drugs and other contraband or even smuggling of illicit goods to avoid customs taxes. Although these materials are of great concern, the current prevailing worry appears to have shifted from illicit drugs to materials likely to be associated with explosives and SNM.

18.1.7 Physics of Detection

Neutron techniques are based on detecting the nuclear properties of possible threats rather than their density or chemical properties. This has several advantages over X-ray techniques, such as the ability to determine elemental composition rather than just density, a high degree of penetration and potential for cargo inspection, the difficulties of shielding contraband against probing radiation, and the ability to detect SNM. However, in the interest of completeness, one should recognize that the presence of large absorbers in an object under inspection will also trigger a "shield alarm" in many X-ray imaging systems and thus is in itself an indication of a potential threat.

The capability to determine the nuclear or elemental properties of objects under inspection is the major advantage of nuclear techniques compared with conventional X-ray scanners, which generally measure the apparent density and shape of objects. This capability can be enhanced in the case of dual-energy X-ray systems. An approximation to the average atomic number (Z) of the material also can be obtained and thus used, along with spatial properties, to identify suspicious material. Neutron imaging techniques go beyond this capability to offer the possibility of determining the unique elemental composition of a suspicious object.

Imaging provides spatial and geometric information that can be used in recognizing and identifying suspect objects, and reduces the background due to extraneous objects in the container. Imaging often reduces the signal, but it reduces the "noise" by a much larger factor. Consequently, the signal-to-noise ratio improves; the ability to actually detect objects in the presence of background improves; and it is possible to note spatial relationships between SNM, explosives, and high-Z containers and shields.

18.2 Representative Technical Approaches

A number of approaches to detect explosives and SNM using neutron techniques exist, some of which can also be applied to imaging. Reviews of these applications for detecting explosives have been provided by Buffler [4] and Lanza [5]. Few of these methods have actually reached the prototype stage and some are beyond prototypes.

18.2.1 Pulsed Fast Neutron Analysis

One highly developed neutron approach that has been tested is pulsed fast neutron analysis (PFNA) [6–9]. In this system, shown in Figs. 18.3 and 18.4, fast neutrons are used to produce gamma rays from neutron reactions with elements of interest; for explosives, this means particularly C, N, and O, but other elements are also detectable. The approach in principle is also capable of detecting SNM by detection of delayed gammas and neutrons between pulses, but this has not been its primary design goal.

The operation is as follows: Neutrons in the range of ~8 MeV are generated by an electrostatic accelerator by the reaction D(d,n)He. These neutrons are collimated into a beam and the beam is mechanically scanned across a container. At the same time, the accelerator is pulsed to produce 1-ns pulses of neutrons with a repetition rate of ~1 MHz. Gamma rays from inelastic interactions are detected by an array of scintillation detectors. The time difference between the accelerator pulse and gamma detection is used to determine the position along the beam where the interaction took place. The pulsed approach also reduces background and clutter for another reason: the number of neutrons produced per pulse is low enough so that there is typically only one

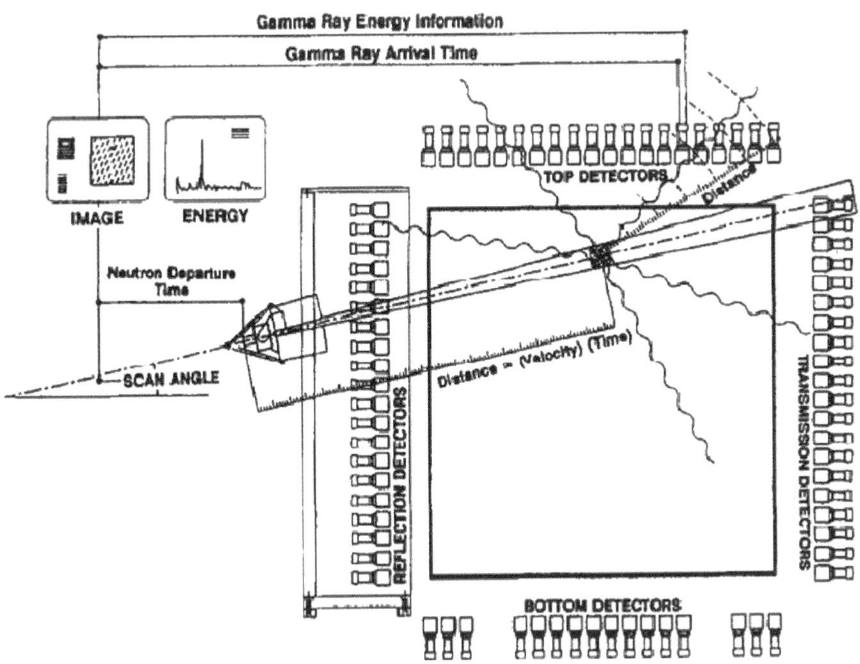

Fig. 18.3 Pulsed fast neutron analysis system; from Brown [7]

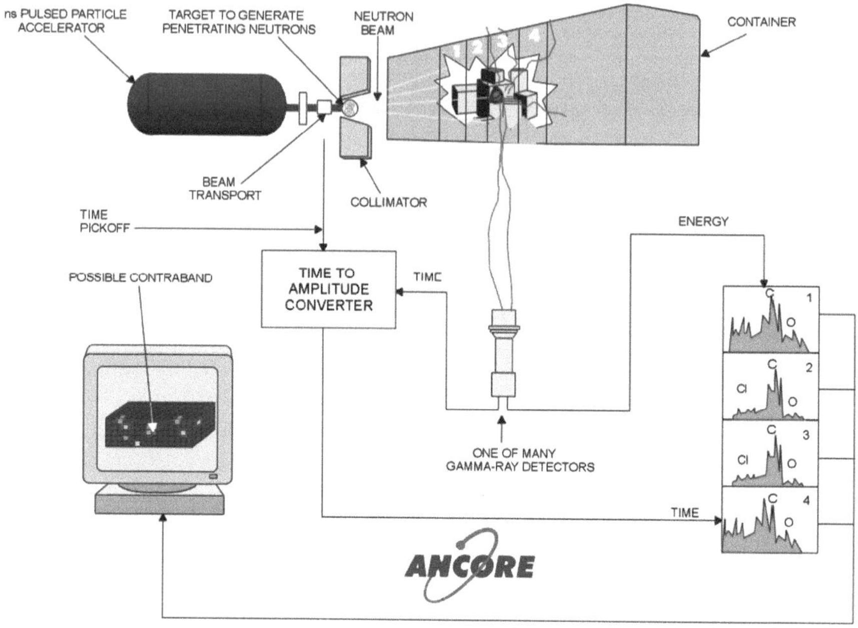

Fig. 18.4 Schematic of pulsed fast neutron analysis system

neutron interaction in the container during a given 1-ns pulse; the next potential event comes during the next beam pulse, 1 microsecond later.

The beam line is mechanically scanned so as to move the beam through the object, therefore the gammas can be correlated with the position of interaction in three dimensions, and a three-dimensional image is formed. The gamma-ray spectrum obtained for each voxel is used to determine relative elemental content in the object under inspection. For 8-MeV neutrons, a time-of-flight measurement of ~1 ns yields an accuracy of ~5 cm in space. The physics for this particular system has been refined over a number of years. Note that the choice of neutron energy, ~ 8 MeV, means that scattered neutrons will generally lose enough energy so that they fall below the threshold for gamma production and thus do not contribute to spatial clutter. Another point to observe is that the detectors, are conventional scintillators (NaI) rather than high-resolution detectors, such as HPGe. The choice of detectors was made after both experimental observation and cost analysis. NaI has considerably higher efficiency and count rate capability than does HPGe, but it has lower energy resolution. This disadvantage is overcome by the use of the entire spectrum rather than single peaks to determine elemental composition. Rather than fit the data to individual peaks, this approach fits the entire spectrum to a linear combination of whole spectra from each of the elements, a method long used in areas such as well-logging and online coal and mineral analysis.

18.2.2 The "Nuclear Car Wash"

Lawrence Livermore National Laboratory (LLNL) has developed a neutron system specifically for the detection of SNM in cargo containers, naming it after the system's resemblance to car-washing systems [10–12]. Cargo is inspected as it moves through a fan-beam of high-energy neutrons, and SNM is detected by the presence of fission neutrons and gammas. The neutron source is an RFQ that produces 7-MeV neutrons through the reaction $D(d,n)^3He$. The gammas are detected by means of large plastic scintillators that form walls on either side of a container as it moves through the inspection system (Fig. 18.5). Fission neutrons, either prompt or delayed, are often used as the signature of fission; however, in this system, the primary signature is the detection of gammas from decays of short-lived isotopes. These gammas are almost an order of magnitude more abundant than the delayed neutrons and are emitted on a time scale of $\sim$1 min. The distribution of energies and times produces a unique signature of fission, and further, high-energy gammas are more likely to escape from the container to be detected. It is estimated that the intensity of such delayed gammas may be orders of magnitude greater than the intensity of delayed neutrons.

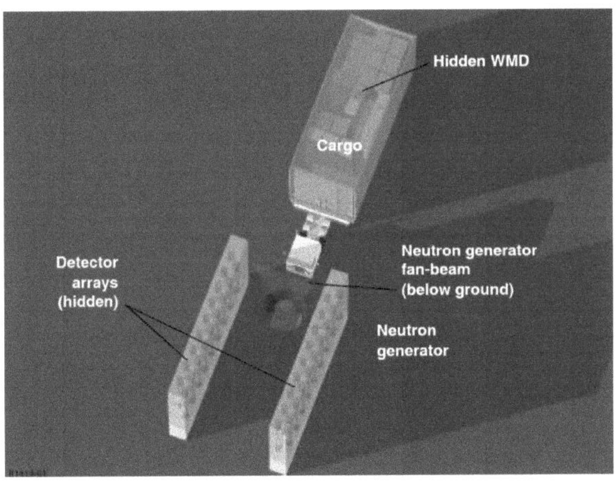

Fig. 18.5 The Lawrence Livermore National Laboratory (LLNL) nuclear "car wash"

Extensive modeling and preliminary experiments have confirmed the performance potential of this approach as a detector for SNM in cargo containers. Figure 18.6 shows the high-energy gamma-ray spectrum with a U target absent (lower curve) and present (upper curve).

18.2.3 Combined Neutron and Gamma Radiography

Combined neutron and gamma radiography has been extensively developed by the Commonwealth Scientific and Industrial Organization and is currently in

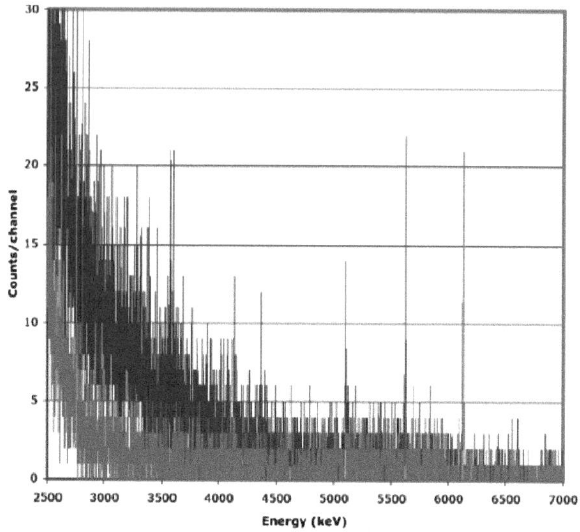

Fig. 18.6 Delayed gamma spectrum with special nuclear material (SNM) (*black*) and without SNM (gray)

production and deployment as an air cargo inspection system [13, 14]. Its basic approach is to image a container using a combination of 14-MeV neutrons generated from a sealed DT neutron generator and gamma rays from a ^{60}Co source (1.17, 1.33 MeV). The ratio, *R*, of the neutron and gamma attenuation coefficients can be determined on a pixel by pixel basis. This ratio shows a significant correlation with material type and can thus be used to discriminate among various materials. Figure 18.7 shows this ratio for various materials of interest in cargo. The ratio is color-coded and displayed as an image for visual inspection by the operator (Fig. 18.8).

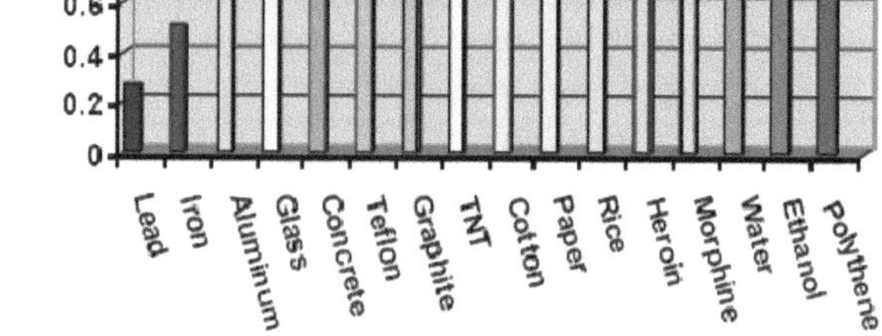

Fig. 18.7 Ratio (R) of neutron to gamma attenuation coefficients for various materials

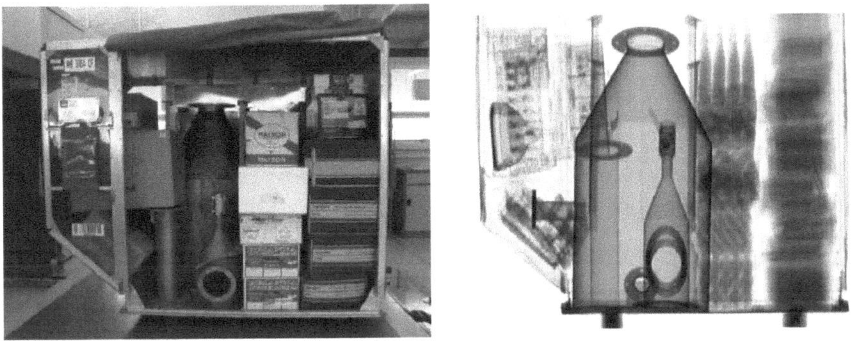

Fig. 18.8 Optical picture (*left*) and R-ratio image of aircraft container (*right*)

18.2.4 Neutron Resonance Radiography

Fast neutron resonance radiography (NRR) is a neutron imaging technique proposed as a method for elemental imaging [15–19]. NRR exploits the fact that the total neutron cross section exhibits unique resonance structure for many elements (Fig. 18.9). In a two-dimensional projection image, the

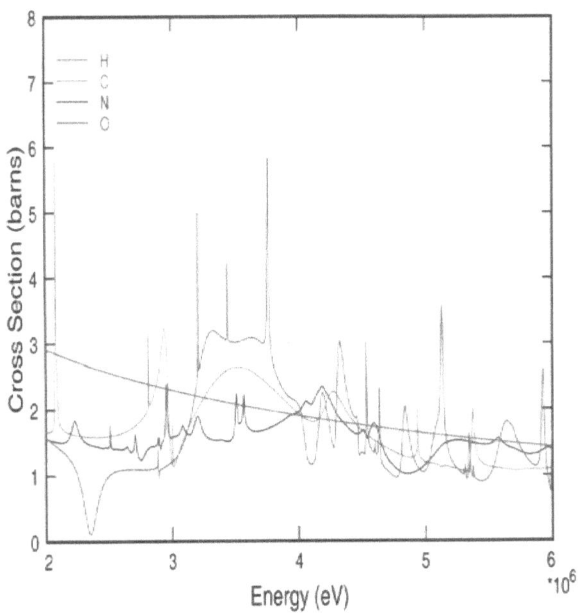

Fig. 18.9 Total neutron cross section for H, C, N, and O in the MeV energy region. With the exception of hydrogen, clear and unique resonance structures exist for each element

attenuation image is the sum of elemental contents. Attenuation images taken at different neutron energies form a set of linear equations that can be solved to map individual elemental contents; in principle, this makes detection possible based on the presence of specific elements or combinations of elements.

The attenuation of a beam of neutrons traversing a slab of element a with thickness in the beam direction x_a is governed by

$$-\ln\left(\frac{I_x}{I_{\text{air}}}\right) = N_a \sigma_a(E) x_a \;,$$

where $-\ln(I_x/I_{\text{air}})$ represents the ratio of the neutron beam attenuated by the object to the unattenuated beam, N_a represents the atom density of element a, and $\sigma_a(E)$ represents the total neutron cross section of material a at energy E. All equations in this section apply to a single pixel. Aside from the intrinsic energy variation in the neutron cross section, factors related to the neutron production mechanism will introduce a smearing in the cross section as a function of energy. The notation σ_a^1 is used to represent the effective neutron cross section for material a at energy setting *1*. Notationally, superscripts will represent the energy setting while subscripts represent the material. The measurement $-\ln(I_x/I_{\text{air}})$at energy setting 1 is defined as b^1. Finally, $N_a \cdot x_a$ is combined into χ_a. In a two-dimensional projection image, the attenuation image is the sum of elemental contents. For a material comprising several different elements *(a, b, c, d)*, the observed attenuation becomes

$$\sigma_a^1 \chi_a + \sigma_b^1 \chi_b + \sigma_c^1 \chi_c + \sigma_d^1 \chi_d = b^1 \;.$$

Several attenuation images taken at different neutron energies form a set of linear equations that can be solved to map individual elemental contents. After performing measurements at, for example, six energies, the set of equations is

$$\begin{aligned}
\sigma_a^1 \chi_a + \sigma_b^1 \chi_b + \sigma_c^1 \chi_c + \sigma_d^1 \chi_d &= b^1 \\
\sigma_a^2 \chi_a + \sigma_b^2 \chi_b + \sigma_c^2 \chi_c + \sigma_d^2 \chi_d &= b^2 \\
\sigma_a^3 \chi_a + \sigma_b^3 \chi_b + \sigma_c^3 \chi_c + \sigma_d^3 \chi_d &= b^3 \\
\sigma_a^4 \chi_a + \sigma_b^4 \chi_b + \sigma_c^4 \chi_c + \sigma_d^4 \chi_d &= b^4 \\
\sigma_a^5 \chi_a + \sigma_b^5 \chi_b + \sigma_c^5 \chi_c + \sigma_d^5 \chi_d &= b^5 \\
\sigma_a^6 \chi_a + \sigma_b^6 \chi_b + \sigma_c^{61} \chi_c + \sigma_d^6 \chi_d &= b^6
\end{aligned}$$

This can be cast into a matrix form $A\chi = b$ as

$$\begin{pmatrix} \sigma_a^1 & \sigma_b^1 & \sigma_c^1 & \sigma_d^1 \\ \sigma_a^2 & \sigma_b^2 & \sigma_c^2 & \sigma_d^2 \\ \sigma_a^3 & \sigma_b^3 & \sigma_c^3 & \sigma_d^3 \\ \sigma_a^4 & \sigma_b^4 & \sigma_c^4 & \sigma_d^4 \\ \sigma_a^5 & \sigma_b^5 & \sigma_c^5 & \sigma_d^5 \\ \sigma_a^6 & \sigma_b^6 & \sigma_c^6 & \sigma_d^{61} \end{pmatrix} \cdot \begin{pmatrix} \chi_a \\ \chi_b \\ \chi_c \\ \chi_d \end{pmatrix} = \begin{pmatrix} b^1 \\ b^2 \\ b^3 \\ b^4 \\ b^5 \\ b^6 \end{pmatrix}.$$

As long as the number of equations is greater than the number of elements, these equations can be solved by a least squares method. Measurement of the matrix A and the set of observed attenuations b will allow determination of χ. This in principle makes detection possible based on the presence of specific elements and enables the formation of a two-dimensional image with elemental separation.

In addition to its multielement detection capability, NRR offers the distinct advantage that attenuation images are governed by a $1/R^2$ falloff in the neutron flux, whereas other methods ("neutron in, photon out") are governed by $1/R^2 * 1/R^2 = 1/R^4$. Using the kinematics of the reaction $D(d,n)^3He$ [20], a single energy deuteron beam can be used to generate 2- to 6-MeV neutrons, as shown in Fig. 18.9. An alternative approach to obtain energy-resolved neutron spectra uses a pulsed "white" neutron source with a broad range of neutron energies and separates them out by time of flight [21].

Despite its conceptual simplicity, NRR is technically difficult in practice. The challenges include producing a variable-energy neutron source with sufficient flux and minimal energy spread at each energy, detecting fast neutrons in the presence of photons, and producing basic measurements of pure elements of sufficient quality so that reliable fits can be obtained in a complex image environment typical of actual cargo. This work was the result of a collaboration between L-3 Communications, the Massachusetts Institute of Technology (MIT), and LLNL.

18.2.5 Apparatus

The primary system components are shown in Fig. 18.10. A fixed-energy deuteron beam is provided by a commercial 3-MeV RFQ accelerator. The beam is focused and transported to the deuterium gas target by the high-energy beam transport system. An adjustable source collimator comprising borated-polyethylene wedges surrounds the deuterium gas target. The entire neutron source sits on a platform that rotates from 0 to 120° to allow selection of the neutron energy. Following the source collimator are a set of fixed front and rear object collimators. The front object collimator

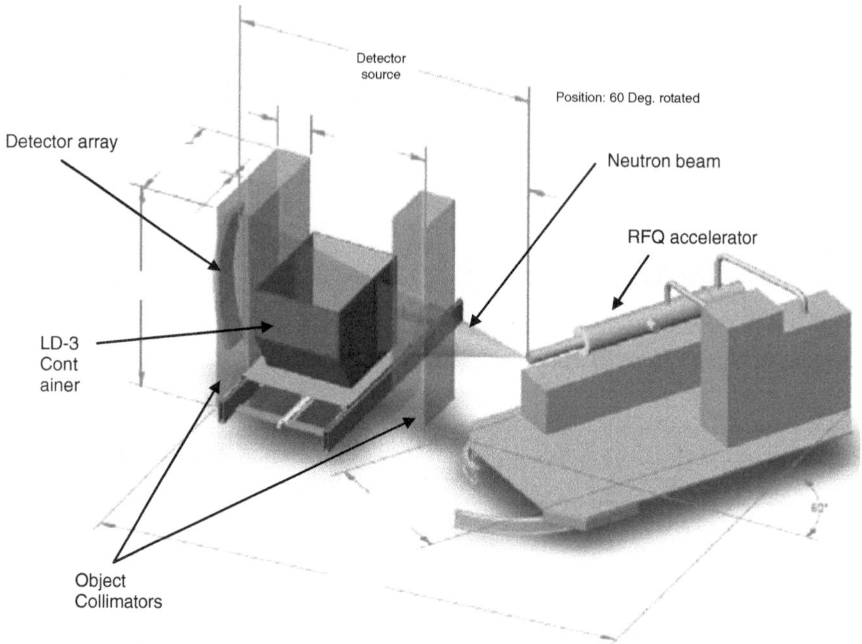

Fig. 18.10 Neutron resonance radiography imaging system. The deuteron accelerator on the rotating platform (in gray) rotated to the 60° position. Front and rear object collimators are shown with the object being imaged (typically an LD-3 air cargo container) between the collimators, resting on the object transport system. The detector array is behind the rear object collimator

prevents stray radiation from the source from reaching the object, and the rear object collimator prevents radiation scattered from the object from reaching the detectors. The neutron beam is defined by aligning the adjustable source collimator with the fixed object collimators and the detector array. The object to be imaged (typically a standard aircraft LD-3 container) is transported through the beam on a motorized trolley. The system is housed in a warehouse with minimal climate control (heat only) and no running water. The prototype occupies a radiation safety exclusion zone of about 25 by 15 m.

The detector system is based on a combination of plastic scintillator and photomultiplier tube in which the fast neutron is detected by proton recoil in the plastic. Typically, gamma background signals are larger than those produced by neutrons, and the electronics are based on a window discriminator for gamma rejection.

The success of the NRR technique is largely determined by the accuracy of the elemental cross sections. Initially, Monte Carlo methods were used to simulate neutron scattering on elements. The cross sections extracted from the simulations were then used to study the possibility of creating elemental maps. However, it was quickly realized that the ideal detector in the type of background-free environment assumed in the Monte Carlo studies does not reflect actual experimental conditions. Hence, a precise experimental extraction of the elemental cross sections was necessary as photon background can significantly change the extracted neutron cross section. Overwhelming photon background washes out the resonant features in the cross sections and makes NRR difficult; reducing the photon-to-neutron ratio is essential to the success of the method.

Figure 18.11 shows experimentally determined carbon and nitrogen cross sections. All elemental cross sections were lower than those calculated using Monte Carlo simulations, largely as a result of gamma contamination. However, the resonant structure of these elemental cross sections is still observed, allowing for a good elemental separation in compounds.

Once the elemental calibration had been completed, a set of angles was selected for performing radiographic analysis. The best set of angles is the one that allows the most precise determination of elemental composition in compounds. Analytically, these angles would correspond to the most pronounced features on elemental cross sections. For example, the oxygen cross section exhibits the most unique behavior at high angles (100 through 115°), whereas the nitrogen cross section has a distinct structure at lower angles (around 35°). The carbon cross section is most distinctive at around 90°. Hydrogen and, to a lesser extent, silicon cross sections exhibit no distinct resonance structure. A nontrivial relationship among these five cross sections in determining elemental composition is best studied quantitatively using a computer optimization code.

To find the best set of angles, compounds composed solely of the basis materials (such as melamine, $H_6C_3N_6$) were scanned at 5° increments in accelerator angle. To resolve a set of five basis elements, scans of six or more angles were required. Since a larger number of angles linearly increase the scan time, it was found that seven angles represented a good balance

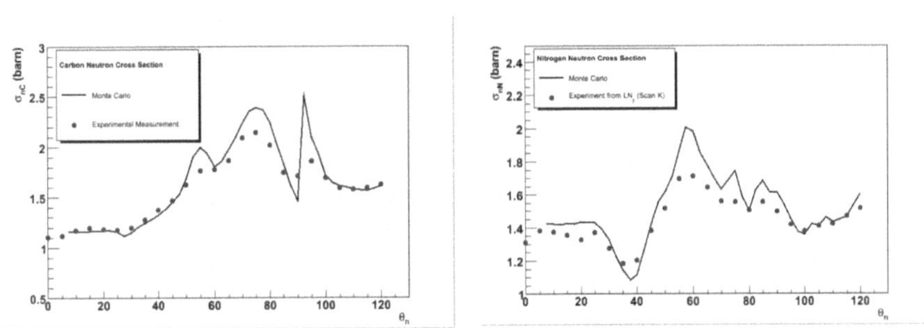

Fig. 18.11 Measurement of total neutron cross section for carbon (*left*) and for nitrogen (*right*) as a function of accelerator angle (degrees)

between quality and scan time. For each unique set of 7 angles picked from a set of 24, the square of the difference between radiographically determined and known elemental composition in each compound was calculated. The sum of these differences was then minimized to produce the best set of angles. Figure 18.12 (left) shows the results of the angle optimization studies, comparing the theoretical atom fraction for melamine with the one determined from the set of optimal angles. The resulting calibration was also tested with various compounds not included in the optimization. A scan of one such compound, ammonium nitrate (NH_4NO_3), in air is shown in Fig. 18.12 (right). The relative amounts of hydrogen and nitrogen are in good agreement with expected values. The amount of oxygen is somewhat lower than the expectation, but still pronounced in comparison to carbon and "other" represented by silicon.

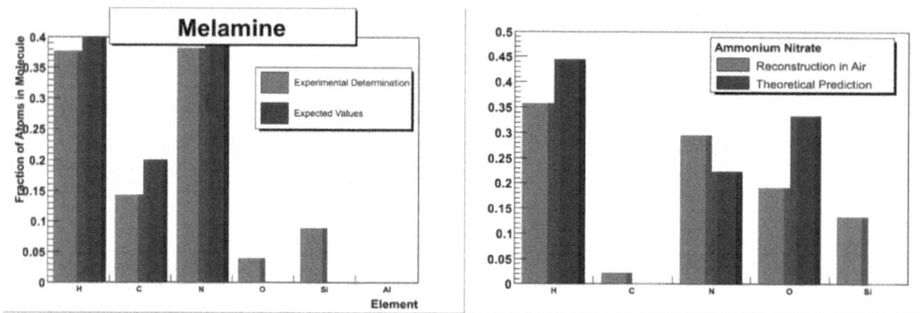

Fig. 18.12 *Left*: Atom fraction of melamine sample as determined from the calibration procedure. The basis elements are H, C, N, O, and Si. The *black* (*right*) *bars* are the expected values and the *gray* (*left*) bars represent the best values obtained during the optimal angle selection. The correctness of the melamine atom fraction was one of the constraints of the angle selection. *Right*: Atom fraction of ammonium nitrate sample. The basis elements are H, C, N, O, and Si. The *black* (*right*) *bars* are the expected values and the *gray* (*left*) bars represent the best values obtained after the optimal angle selection. The correctness of the ammonium nitrate atom fraction was not constrained by the angle selection

18.2.6 Elemental Maps

The results in Fig. 18.12 are derived from blocks of pixels on isolated objects rather than from a more realistic situation with multiple elements. Figures 18.13 and 18.14 show images developed by the system together with their elemental maps. Images presented are rotated by 90° from their actual scan direction: top to bottom in the image is left to right in the scan; left to right in the image is bottom to top in the scan. The leftmost part of Fig. 18.13 represents the attenuation image at one of the 7 energies scanned

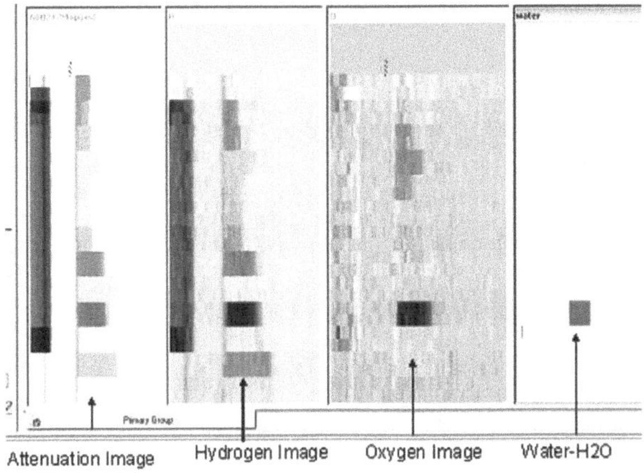

Fig. 18.13 Identification of water sample from 11 other targets. Shown in the panes (*left* to *right*) are a single energy attenuation image, a hydrogen elemental map, an oxygen elemental map, and identification of water target through the coincident presence of hydrogen and oxygen

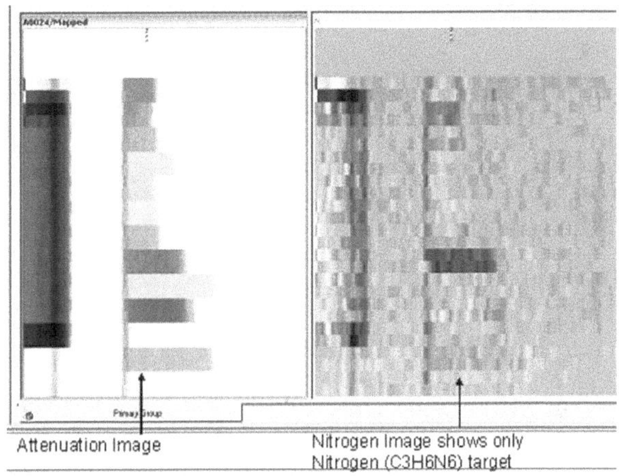

Fig. 18.14 Identification of melamine sample from 11 other targets. Shown in the panes (*left* to *right*) are a single energy attenuation image and the nitrogen elemental map. The melamine sample is clearly distinguished from the other samples in the nitrogen elemental map

with the presence of 12 different targets. Also shown are the hydrogen and oxygen elemental maps and finally, on the right, the result of a simple algorithm that correctly identifies the water sample by identifying the coincident presence of hydrogen and oxygen. Similarly, Fig. 18.14 shows the identification of the nitrogen-rich melamine.

In the next set of images, Figs. 18.15 and 18.16, a series of targets are superimposed into typical air cargo contents. As an example of algorithm development in this project, a preliminary algorithm was developed to detect graphite. The carbon elemental image appears to be very stable and have low noise. The carbon block is detected with high efficiency in an environment of typical air cargo contents is seeded with other potential targets.

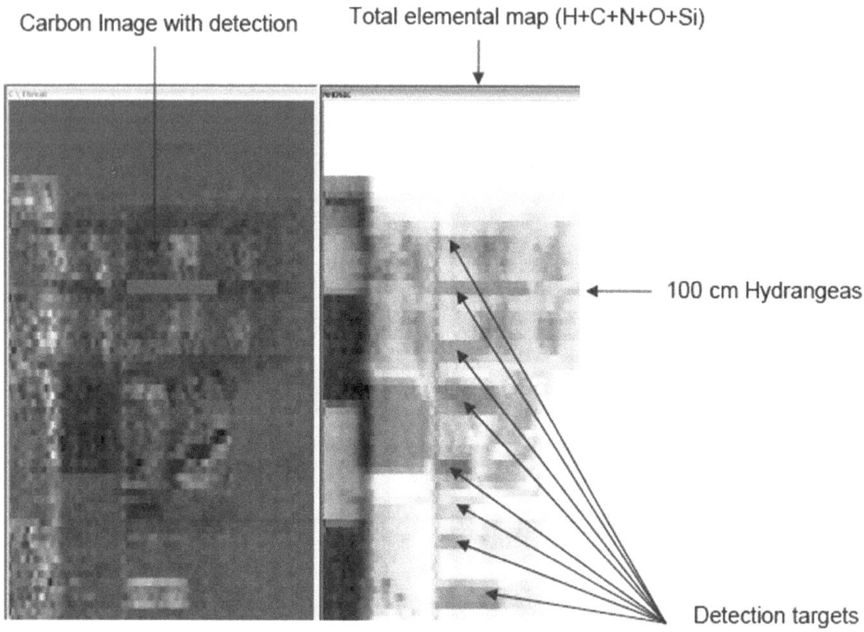

Fig. 18.15 Automated detection of carbon object in an air cargo of cut flowers. *The left pane* is the carbon image and the right pane is the sum of the elemental map for all five basis elements. The carbon block alone is correctly identified by the algorithm among the clutter (100 cm hydrangeas) and other seeded targets

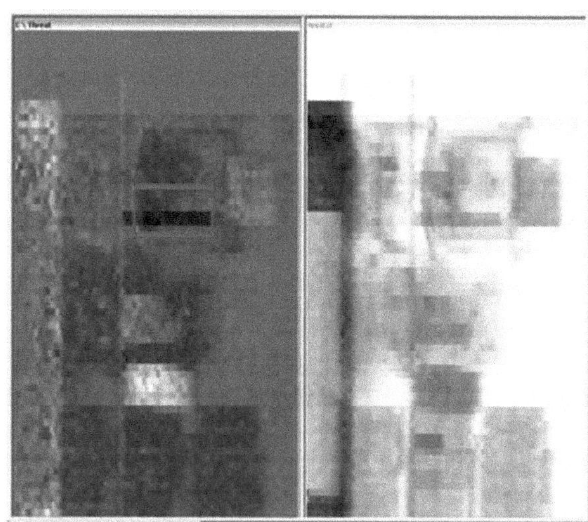

Fig. 18.16 Automated detection of carbon object in air cargo. The left pane is the carbon image and the right pane is the sum of the elemental map for all five basis elements. The carbon block alone is correctly identified by the algorithm amongst the clutter (electronics) and other seeded targets

18.3 Summary and Future Outlook

All of the methods described in this chapter have demonstrated the potential, but final deployment will require a significant amount of engineering to make them faster, more reliable, and capable of automatic, or at least semi-automatic, operation with a minimal number of skilled technical level operators. This work has concentrated only on one aspect of this problem—those applications that involve both neutrons and imaging. Until recently, much of the work was concentrated in the area of detection of contraband and explosives; however, more recently, there has been increased interest in the detection of SNM. The NRR system is now being modified to enable it to detect SNM as well as explosives; this work will be reported in the future. Active interrogation techniques generally have relied on relatively large accelerators, and that has limited their practical deployment. However, if more compact and less expensive neutron sources were developed, there would be a greater impetus toward deployment, especially for SNM detection. Solving the source problem will not in itself lead to deployment, as the issues of cost, complexity, and capability to rapidly detect threats in complex backgrounds still must be overcome.

Acknowledgments The work on NRR was a collaboration between the MIT Department of Nuclear Science and Engineering, the MIT Laboratory for Nuclear Science, L3-Communications, and LLNL. The author especially thanks David Perticone (L-3), who was the principle investigator on the NRR project and was responsible for much of the progress made on it, as well as Vitalyi Ziskin (L-3 and MIT), Gongyin Chen (L-3 and Varian), Whitney Raas (MIT), and Gordon Kohse (MIT) for their contributions throughout the project. Jim Hall (LLNL), Brian Rusnak (LLNL), and John Watterson (University of the Witwatersrand) made invaluable contribution in the early parts of the project.

Others include the staff of the MIT-Bates Laboratory for their excellent contributions to the construction and operation of the NRR prototype, especially Bob Fisk, Peter Binns, Jim Kelsey, Peter Goodwin, Christoph Tschalaer, and Jan Van Der Laan; and the following engineering staffs of L-3 Communications for the design of system components: Tim Hart, Ken Prather, Jim Karon, Tony Antoniou, John Price, Keith McClelland, Brain Pearson, Steve McDevitt, Jeff Stillson, and Sal Gargiulo.

References

1. D.L. Chichester and J.D. Simpson, Compact accelerator neutron generators, *The Industrial Physicist*, Dec-Jan 2003/2004 **9**(6), 22–25 (2004).
2. ANSI 43.14, *Radiation Safety Guidelines for Active Interrogation Systems for Security Screening of Cargo*. American National Standards Institute (2008).
3. J. Zukas and W. Walters, eds., *Explosive Effects and Applications*, Springer-Verlag, New York, NY (1997).
4. A. Buffler, Contraband detection with fast neutrons, *Rad. Phys. Chem.* **71**(3–4), 853–861 (2004).
5 R. Lanza, Neutron techniques for detection of explosives in Y. Yinon, ed., *Counterterrorist Detection Techniques of Explosives*, Elsevier, Amsterdam, pp. 131–155 (2007).

6. T. Gozani, Novel applications of fast neutron interrogation methods, *Nuc. Instrum. Meth. A* **353**(1–3), 635–640 (1994).
7. D.R. Brown and T. Gozani , Cargo inspection system based on pulsed fast neutron analysis, *Nucl. Instrum. Meth. B* **99**(1–4), 753–756 (1995).
8. R. Loveman, J. Bendahan, T. Gozani and J. Stevenson ,Time of flight fast neutron radiography, *Nucl. Instrum. Meth. B* **99**(1–4), 765–768 (1995).
9. D.R. Brown, T. Gozani, R. Loveman, J. Bendahan, P. Ryge, J. Stevenson, F. Liu and M. Sivakumar, Application of pulsed fast neutrons analysis to cargo inspection, *Nucl. Instrum. Meth. A* **353**(1–3), 684–688 (1994).
10. D. Slaughter, M. Accatino, A. Bernstein, J. Candy, A. Dougan, J. Hall, A. Loshak, D. Manatt, A. Meyer, B. Pohl, S. Prussin., R. Walling,and D.Weirup, *Detection of special nuclear material in cargo containers using neutron interrogation*, UCRL-ID-155315 (2003).
11. D.R. Slaughter, M.R. Accatino, A.Bernstein, A.D. Dougan, J. M. Hall, A. Loshak,D.R. Manatt, B.A. Pohl, R.S. Walling, D.L. Weirup, and S.G. Prussin, The "Nuclear Car Wash": A Scanner to Detect Illicit Special Nuclear Material in Cargo Containers, *IEEE Sens. J.* **5**(4), 560–564 (2005).
12. J.M. Hall, S. Asztalos, P. Biltoft, J. Church, M.-A. Descalle, T. Luu, D. Manatt, G. Mauger, E. Norman, D. Petersen, J. Pruet, S. Prussinb, and D. Slaughter, The Nuclear Car Wash: Neutron interrogation of cargo containers to detect hidden SNM, *Nucl. Instrum. Meth. B* **261,** 337–340 (2007).
13. J.E. Eberhardt, Y. Liu, S. Rainey, G.J. Roach, R.J. Stevens, B.D. Sowerby and J.R. Tickner, Fast Neutron and Gamma-Ray Interrogation of Air Cargo Containers in *Proceedings of International Workshop on Fast Neutron Detectors and Applications (FNDA2006)* in Proceedings of Science http://pos.sissa.it/
14. J.E. Eberhardt, Y. Liu, S. Rainey, G.J. Roach, R.J. Stevens, B.D. Sowerby and J.R. Tickner, Fast neutron radiography scanner for the detection of contraband in air cargo containers, *Appl. Rad. Isot.* **63**, 179 (2005).
15. G. Chen, *Fast Neutron Resonance Radiography for Elemental Imaging: Theory and Applications*, Ph.D. Thesis, MIT, April (2001).
16 G. Chen, R.C. Lanza, Fast neutron resonance radiography for elemental imaging: Theory and applications, *IEEE Trans. Nucl. Sci.* **49**(4), 1919–1924 (2002).
17. W. L. Raas, B.W. Blackburn, E.M. Boyd, G.E. Kohse, and R.C. Lanza, Neutron interrogation for threat interdiction: technical challenges, *IEEE Nuclear Science Symposium Conference Record* (2005).
18. B.W Blackburn, G.Chen, W.A Franklin, E. E Ihloff, G. E Kohse, R.C. Lanza, B. McAllister, D. Perticone, V. Ziskin, Fast Neutron Resonance Radiography for Homeland Security, *IEEE Nuclear Science Symposium Conference Record* (2007).
19. B.W. Blackburn, G. Chen, W.A. Franklin, E.E Ihloff, G.E Kohse, R.C. Lanza, B. McAllister, D. Perticone, and V. Ziskin, Fast Neutron Resonance Radiography for Elemental Imaging: First Results, *IEEE Trans. Nucl. Sci.* (in press) (2008).
20. H. Lisikien and A. Paulson, Neutron Production Cross Sections and Energies for the Reactions T(p,n)^{3}He, D(d,n)^{3}He, and T(d,n)^{4}He, *Nucl. Data Tables* **11**, 569–619 (1973).
21. P.K. Van Staagen, T.G. Miller, B.C.Gibson, R.A. Krauss, High speed data acquisition for contraband identification using neutron transmission, *AIP Conference Proceedings*, n 392, pt.2, pp. 853–856 (1997).

Index

SPRINGER NATURE

GPSR Compliance

The European Union's (EU) General Product Safety Regulation (GPSR) is a set of rules that requires consumer products to be safe and our obligations to ensure this.

If you have any concerns about our products, you can contact us on ProductSafety@springernature.com

In case Publisher is established outside the EU, the EU authorized representative is:

Springer Nature Customer Service Center GmbH
Europaplatz 3
69115 Heidelberg, Germany

Kolibri 360 GmbH
Ferdinand-Jühlke-Straße 7
99095 Erfurt, Deutschland
produktsicherheit@kolibri360.de